Lecture Notes
in Control and Informat

Editors: M. Thoma, F. Allgöwer, M. Morari

Daniel Alberer, Håkan Hjalmarsson,
and Luigi del Re (Eds.)

Identification for Automotive Systems

 Springer

Editors

Dr. Daniel Alberer
Institute for Design and Control of
Mechatronical Systems,
Johannes Kepler University Linz
Altenbergerstr. 69
4040 Linz
Austria
E-mail: daniel.alberer@jku.at

Prof. Luigi del Re
Institute for Design and Control of
Mechatronical Systems,
Johannes Kepler University Linz
Altenbergerstr. 69
4040 Linz
Austria
E-mail: luigi.delre@jku.at

Prof. Håkan Hjalmarsson
School of Electrical Engineering
Automatic Control
KTH Royal Institute of Technology
Osquldas v. 10
SE-100 44 Stockholm
E-mail: hjalmars@ee.kth.se

ISBN 978-1-4471-2220-3   e-ISBN 978-1-4471-2221-0

DOI 10.1007/978-1-4471-2221-0

Lecture Notes in Control and Information Sciences   ISSN 0170-8643

Library of Congress Control Number: 2011938110

© 2012 Springer-Verlag London Limited

Typeset & Cover Design: Scientific Publishing Services Pvt. Ltd., Chennai, India.

Printed on acid-free paper

9 8 7 6 5 4 3 2 1

springer.com

Preface

The increasing complexity of automotive systems and the high number of variants make the exploitation of available degrees of freedom more difficult. Model based control is generally advocated as a solution or at least as a support in this direction, but modeling is not a simple issue, at least for some problems in automotive applications, e.g. in emissions based control. Classical first principle methods are extremely helpful because they provide a good insight into the operation of the systems, but frequently require too much effort and/or do not achieve the required precision and/or are not suitable for online use. There is much know-how available in the identification community to improve this, either by purely parameter estimation based approaches or by mixed models, but this know-how is frequently not tailored for the needs of the automotive community, and the cooperation is actually quite limited.

Against this background, a workshop organized by the Austrian Center of Competence in Mechatronics (ACCM) took place at the Johannes Kepler University Linz (Austria) from July 15-16, 2010, which brought together users, in particular from the industry, identification experts, in particular from the academy, and experts in both worlds. The contents of this book are peer reviewed versions of the workshop contributions and are structured into four parts, starting with an assessment of the need for (nonlinear) identification methods, followed by a presentation of suitable methods, then a discussion on the importance of data is addressed and finally the description of several applications of identification methods for automotive systems is presented.

Neither the workshop nor this collection of contributions would have been possible without the support of several people (in particular of Daniela Hummer and Michaela Beneder). Thanks are due also to the reviewers of the single chapters who have done an important and essential work.

Organization

Steering Organization

Austrian Center of Competence in Mechatronics, Linz, Austria

Hosting Organization

Johannes Kepler University Linz, Austria

Program Committee

Daniel Alberer	Johannes Kepler University Linz, Austria
Rolf Johansson	Lund University, Sweden
Ilya Kolmanovsky	University of Michigan, USA
Sergio Savaresi	Politecnico di Milano, Italy
Greg Stewart	Honeywell, Canada

Organizing Committee

Daniel Alberer	Johannes Kepler University Linz, Austria
Daniela Hummer	Johannes Kepler University Linz, Austria

Referees

J.C. Agüero	X.J.A. Bombois	M. Deistler
R. Backman	A. Brown	P. Dickinson
O. Bänfer	D. Cieslar	M. Enqvist
W. Baumann	M. Corno	L. Eriksson
C. Benatzky	A. Darlington	D. Filev

Contents

**8   Parameter Identification in Dynamic Systems Using
     the Homotopy Optimization Approach** 129

Chandrika P. Vyasarayani, Thomas Uchida, Ashwin Carvalho,
and John McPhee

Part III: The Importance of Data

**9   A Tutorial on Applications-Oriented Optimal
     Experiment Design** 149

Cristian R. Rojas, Jonas Mårtensson, and Håkan Hjalmarsson

Chapter 1
System Identification for Automotive Systems: Opportunities and Challenges

Daniel Alberer, Håkan Hjalmarsson, and Luigi del Re

Abstract. Without control many essential targets of the automotive industry could not be achieved. As control relies directly or indirectly on models and model quality directly influences the control performance, especially in feedforward structures as widely used in the automotive world, good models are needed. Good first principle models would be the first choice, and their determination is frequently difficult or even impossible. Against this background methods and tools developed by the system identification community could be used to obtain fast and reliably models, but a large gap seems to exist: neither these methods are sufficiently well known in the automotive community, nor enough attention is paid by the system identification community to the needs of the automotive industry. This introduction summarizes the state of the art and highlights possible critical issues for a future cooperation as they arose from an *ACCM Workshop on Identification for Automotive Systems* recently held in Linz, Austria.

1.1 Introduction

It is well known that every control includes in some way a model of the plant to be controlled - even tuning a PID control according to the Ziegler-Nichols rules includes a phase of "identification" of a very simple model, though not every technician is aware of it.

Daniel Alberer · Luigi del Re
Institute for Design and Control of Mechatronical Systems,
Johannes Kepler University Linz, Altenbergerstr. 69, 4040 Linz, Austria
e-mail: daniel.alberer@jku.at

Håkan Hjalmarsson
ACCESS Linnaeus Center, School of Electrical Engineering,
KTH – Royal, Institute of Technology, S-100 44 Stockholm, Sweden
e-mail: hakan.hjalmarsson@ee.kth.se

D. Alberer et al. (Eds.): Identification for Automotive Systems, LNCIS 418, pp. 1–10.
springerlink.com   © Springer-Verlag London Limited 2012

Automotive industry, in particular engine control, is characterized by multiple, conflicting and extremely tight requirements together with a very high cost pressure. As a consequence, control needs to achieve extremely high performance with little sensor information and this leads to feedforward control structures calibrated on test benches with measurement possibilities not available in production (see Figure 1.1). At the end, the real problem – a constrained optimization of mostly unmeasurable quantities – is replaced by a much simpler tracking problem on measurable quantities. Of course, all possible conditions and deviations must be taken into account – traditionally, this has been done by measuring a huge number of cases and saving the control parameters, essentially the inverse system model, in an enormous number of maps.

The economy of scale, especially of passenger cars, has made this method – essentially the experimental calibration of countless maps in the feedforward structure – affordable, at least so far.

Nevertheless, there is a wide consensus that new approaches are needed, and they are coming. In a simplified form, they can be classified in three categories:

- Automation of calibration by (static) Design Of Experiments (DOE)
- Automation of the tuning of first principle models and
- Direct modeling by system identification (or universal approximators)

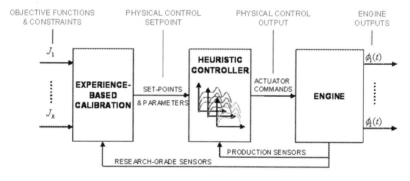

Fig. 1.1 Structure of actual control systems in development and production

The first category, as presented, *e.g.*, in chapter 10, has already received much attention from the industrial community as it does not really represent a change of the classical calibration procedure, but uses systematic experiment design to improve its speed and possibly the quality of the resulting controller.

First principle models are usually the first choice for control system engineers, as they yield both physical insight and a good control design basis. Unfortunately first principle models are approximations, and tuning their parameters in order to have the model behavior to correctly reproduce measurements usually leads to aggregate parameters, whose numerical values take into account as far as possible the unmodeled dynamics and distributed nature of the real system. This approach, however, is quite useful for many applications, a wide presentation being given, *e.g.*, by chapter 15.

For a control engineer with less need of a physical understanding of the underlying reality the third option, direct modeling, seems to be the most natural one. Indeed, critical target quantities, like emissions, frequently cannot be modeled satisfactorily with physical models, *e.g.*, particulate emissions need a large set of chemical equations for their description, typically some hundreds, which can be solved only if the chemical and physical conditions are known - at every place in the combustion chamber and at every instant.

Building dynamical systems from experimental data is known as system identification. There are two principal ways to identify a model from data: Black-box modeling, where a model is selected from a set of general purpose models through some model fitting procedure, and gray-box modeling, where physical insight is used when constructing the model set and where model parameters often have physical meaning.

Generally speaking, gray-box modeling is often considered superior as such models often have better generalization abilities than black-box models, *i.e.*, they often perform better when confronted with new experimental conditions. The reason being that, unlike black-box models, they are based on the underlying physics of the system.

As indicated in Figure 1.2, the process of system identification is iterative in nature. Having designed an experiment and collected data, a model set has to be selected together with a criterion of fit and a model (or a set of possible models) has to be calculated. The model then has to be validated and if the model does not pass the validation test, previous choices have to be revised. All steps are quite involved and in addition closely linked, but well established and efficient methods exist for linear models. Using linear methods for nonlinear systems, possibly in a multimodel setup, *e.g.*, in the form of gain scheduling, is a wide-spread practice, and works in many practical cases, also because feedback improves the linearity of the controlled system [8]. Unfortunately, this does not help much in the case of most automotive applications, due to their essentially feedforward structure. In other words: the nonlinearity must be explicitly accounted for.

While nonlinear system identification has not yet reached the same maturity as its linear counterpart, there do exist many approaches, which can be roughly divided into uniform approximators and gray-box models.

Uniform approximators, like neural networks, but also, *e.g.*, bilinear systems or their counterpart in discrete time, the state affine systems [1], are essentially those functions which fulfill the conditions of the Stone-Weierstrass theorem [17], *i.e.*, are able to approximate to any degrees of precision another function at least over a limited set. The usual price to be paid for is the increase of parameters, with the consequences well known from

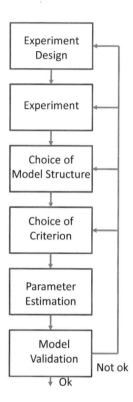

Fig. 1.2 System identification – an iterative procedure.

overparametrization in the linear framework, in particular the exploding number of parameters, but additionally with the non-convexity problems usually arising from the nonlinear nature of the system. Nonlinearity has also another critical effect on the identification procedure [6]: while in the linear case all signals with a few common features, essentially zero mean value and sufficient bandwidth, lead to the same parameter estimation after a sufficiently long time, the same is not true for nonlinear systems: different signals tend to lead to different parameters and thus to different models.

Against this background, the choice of the right data becomes possibly the most critical issue in nonlinear identification, especially if coupled with limited measuring time, as it is always the case in real life. Of course, different numerical approaches, like the use of regularization, can improve its performance, but at the end the data is the limiting factor. Not surprisingly, the commercially available automatic tuning methods do rely almost completely on DOE and very little on the specific model approach.

The other way to cope with nonlinearity is the already mentioned use of gray-box models. If it is possible to export known nonlinearities [11, 4], it becomes possible to split the problem into two parts, on one side a (quasi) linear estimation of an unknown dynamics derived from a more complex setup.

In the rest of this chapter we shall shortly analyze the state of the art in data driven system identification and locate the specific contributions presented in this book.

1.2   Methods for System Identification of Automotive Systems

While system identification is a well explored field described in several textbooks (see, *e.g.*, [10]), the same is not yet the case for nonlinear identification. While more than 15 years old, the reference [15] still provides a good overview of techniques for nonlinear black-box modeling. In recent years so-called kernel methods, a class of methods very popular in statistical learning theory, have been adapted to identification of nonlinear dynamical models, see, *e.g.*, [3].

There are a number of model characteristics that significantly affect which identification methods are suitable, and the associated computational complexity. First, a model may be either in continuous time or discrete time, or even a hybrid between these two. As data typically come in sampled form, black-box models are often in discrete-time, while physically motivated gray-box models generically are in continuous time. Second, unmeasured external system stimuli can be assumed to be either deterministic or stochastic in nature. In the latter case, these "disturbances" are endowed with some assumptions about their statistical distributions.

Generally, continuous time models require numerical solution of differential equations. Stochastic models require optimal (non-linear) filtering, or smoothing as an intermediate in the parameter estimation. By far the most computationally expensive models are stochastic continuous time models. Chapter 5 provides a tutorial on numerical methods for state and parameter estimation for such models. The chapter considers the approach where the disturbance sequences are explicitly estimated, thus bypassing the nonlinear optimal filtering problem.

Often it is difficult to find a suitable model structure. The number of possible structures grows combinatorially with the number of regressor candidates to include in the model. Many different selection schemes have been developed and recently also sparse estimation methods have appeared as interesting alternatives. The latter methods employ a special regularization which ensures that parameters with little influence automatically are set to zero. Another way of handling this problem is presented in chapter 6. Here Genetic Programming is used to automatically generate and to prune the set of regressors. Thus what is produced is a code with rules for how to generate the regressors. The authors report several successful applications, covering engine modeling, quality assessment in steel industry and medical data analysis.

Much effort has been devoted to develop methods that provide good initial parameter estimates for optimization based methods (such as the prediction error method). For ready-made linear black-box models, such as Output Error and Box-Jenkins, instrumental variable methods [16] and, more recently, subspace methods (see below) provide adequate solutions. There exist also methods for certain nonlinear model sets, *e.g.*, Hammerstein models [14]. However, this is still very much an open issue for gray-box models despite some attempts [12]. An interesting alternative and complementary approach is presented in chapter 8. A homotopy is constructed which at one end of the morphing parameter corresponds to the desired problem and where the other end corresponds to a problem with known solution. The construct is tailored to mechanical systems. The morphing parameter affects an observer such that the model output becomes close to the true output at one end point of the morphing parameter. Several interesting examples are used to illustrate the method.

Subspace identification is a method which avoids the problem of local minima. This methodology has its roots in (stochastic) realization theory and instead of relying on numerical optimization of a cost function, it explores the geometrical structure of state-space models leading to numerically robust algorithms based on singular value-decomposition. The methodology was first developed for linear time-invariant systems operating in open loop [19, 18], and subsequently extended to system operating in closed loop, linear time-varying systems and certain classes of nonlinear systems. Recently, algorithms able to handle also linear parameter varying (LPV) systems have appeared. Chapter 4 presents recent results on subspace identification of LPV systems. The method is illustrated by identifying the lateral dynamics of a vehicle.

From a system identification perspective, one interesting feature of automotive systems is that it is relatively easy to generate and collect data. For example, today's engines are well equipped with sensors and actuators and are also relatively inexpensive. Also, the time-constants are short allowing for limited test durations. This is in sharp contrast with, for example, paper mills and refineries where processes tend to be either very slow and/or very expensive (or even dangerous) when perturbed from their normal operating points. However, with increasing performance requirements enforcing the development of more accurate models of, in turn, very complex automotive systems requires experimental data from a wide range of operating conditions. Optimizing experiments can help reduce experiment time and cost. An interesting framework for this is least-costly identification [8]. It has been recognized for long that optimal experiment design should take the intended model used into account, *e.g.*, [2], and recently a quite general framework for applications oriented experiment design has been developed [7, 5]. This framework allows quite general criteria to be relaxed to convex optimization problems. A key contribution from this line of research has been the insight that, not only does optimal experiment design reduce experimental cost, but it also simplifies the modeling problem. The reason is that an experiment that is optimal for a particular application, ensures that system properties that are important for that application are visible in the data, but also that unimportant properties are not excited more than necessary. The implication of this from a modeling perspective is that it is sufficient that the model captures system properties that are of importance for the application.

Applications oriented experiment design is now being introduced in a variety of contexts such as training (or pilot) signal design in communication systems and model predictive control [9]. Experiment design is certainly used and recognized as important in the area of modeling of automotive systems. Approaches here are often limited to standard design criteria, such as D-optimal design, which only involves the accuracy of the parameters, without explicit reference to the application [13]. In chapter 9 the basics of applications-oriented experiment design are outlined. Chapter 18 considers design of experiments for determining transient trajectories for engine parameter estimation. This chapter also points to the link from experiment design to adaptive control.

1.3 Applications

The subject *automotive systems* covers a huge variety of applications, ranging from detailed descriptions of the combustion process in reciprocating engines over vehicle body control tasks to questions of vehicle guidance and navigation, or driver modeling. Obviously, this book cannot present examples for all possible tasks in this field, but contains a collection of several state of the art applications motivated by questions from academia as well as from industry.

As it is often the case, modeling turns out to be the crucial issue in the design phase of a model based controller for engine control. Moreover, due to tightening emission limits and high demands towards fuel efficiency, engines have become complex systems with lots of actuators. In other words, models need to be accurate and frequently the effects of numerous inputs have to be captured by a single model, resulting in high dimensional multi-input multi-output representations. Chapter 15 presents an industrially motivated approach for gray box mean value modeling of turbocharged Diesel engines with a particular focus on stability of models both in simulation and for identification. Furthermore, low complexity black box models suitable for model predictive control of a heavy duty engine are shown in chapter 13 and fuel system modeling of an offroad heavy duty Diesel engine is illustrated in chapter 12.

The high complexity of engines has also increased the calibration effort considerably (in order to fulfill the legislation requirements). Therefore automatic tuning methods have found their way into practice. Chapter 10 presents an example of such an automated engine calibration using nonlinear dynamic programming, with the goal of reducing calibration time and consequently costs during engine and vehicle development. The importance of tailoring the method to the modeling task is emphasized in chapter 2, e.g. to systematically integrate first principles models and empirical models.

Mean value models are a valuable choice for many engine control applications, *e.g.*, engine airpath control, however, since they are associated with simplifications of the geometry, the time scale and the physical phenomena involved, one has to be aware of their limitations. To this end, chapter 11 analyzes through several real-life examples the effects of the most important simplifications done in mean value engine modeling.

As mentioned above, mainly due to the cost pressure, typically only little sensor information is available for engines. Nevertheless, several physical quantities need to be estimated on-line, see chapter 17, or the feed-forward maps of the controller need to be adapted to take into account the impact of aging and wear on the control system – chapter 14 presents an estimator based approach for an off- and on-line identification of these maps.

Besides the desirable high efficiency, the heterogeneous combustion taking place in Diesel engines unfortunately results in high levels of nitric oxide and soot emissions. Over the last two decades so called alternative combustion modes, like homogeneous charge compression ignition (HCCI), have been developed, leading to promising results concerning efficiency and emissions. However, HCCI is highly sensitive to operating conditions and lacks direct actuation, which is a challenge for closed-loop control. Therefore, chapter 16 shows the design of a control model and, furthermore depicts its satisfactory use of the model during model predictive HCCI control of a heavy duty Diesel engine.

In a similar manner as for engine control, modeling turns out to be the critical phase for vehicle chassis control applications. Chapter 3 provides a survey on several identification issues in vehicle control and discusses the advantages and disadvantages of the identification approaches by means of four examples. The performance of Markov chains is analyzed in chapter 7, in particular their application to vehicle speed and road grade estimation.

1.4   Conclusions and Outlook

System identification could have a paramount importance in automotive control as it can be the solution for one of the most practical problems – how to get a model with little effort.

While the number of applications is still limited, and frequently more related to the academic than to the industrial world, it is growing and a clear indicator of the potential relevance of the method.

On the way to more generalized use there are mainly two obstacles.

On one side, automotive engineering has been evolving from a classical mechanical engineering branch to a mechatronical one only in the last two decades, and the main background of most people in that community, but in particular of those in responsibility positions, is clearly mechanical and thus mainly first-principle oriented. Data driven methods are less easy to understand and frequently abused as a kind of universal problem solver – which they are not – and therefore seen with distrust. To cope with this problem, we need applications which clearly show the advantages, but even more, as discussed in chapter 2), the integration of first principles and data based models must receive more attention.

An effort in this direction is proposed, *e.g.*, by [11] where emission models are developed combining complex thermodynamics and data based approaches according to the available information. However, much more is needed to allow a breakthrough from the industrial side.

On the other side, we need more support from the academic community. Inside this community little interest is given to the specific setups that arise at automotive systems (closed loop nonlinear systems identification for virtual sensing but also inverse control under data richness or even excess). Data richness is more of a problem than can be thought of – data is not equivalent to information and too much data tend to hide more than offer useful information. This calls both for DOE – static and dynamic – to produce the right data, but also for statistical methods to extract information from data. Both of course in the nonlinear framework.

References

[1] Fliess, M., Normand-Cyrot, D.: On the approximation of nonlinear systems by some simple state-space models. In: Bekey, G., Saridis, G. (eds.) Proceedings of the Sixth IFAC Symposium on Identification and System Parameter Estimation 1982. IFAC, vol. 1, pp. 511–514. Pergamon, Oxford (1983); Proceedings of the Sixth IFAC Symposium on Identification and System Parameter Estimation 1982, Washington, DC, USA, June 7-11 (1982)

[2] Gevers, M., Ljung, L.: Optimal experiment designs with respect to the intended model application. Automatica 22(5), 543–554 (1986)

[3] Goethals, I., Pelckmans, K., Suykens, J.A.K., Moor, B.D.: Identification of MIMO Hammerstein models using least squares support vector machines. Automatica 41(7), 1263–1272 (2005)

[4] Hirsch, M., Alberer, D., del Re, L.: Grey-box control oriented emissions models. In: Proc. 17th IFAC World Congress, Seoul, South Korea (2008)

[5] Hjalmarsson, H.: System identification of complex and structured systems. European Journal of Control 15(4), 275–310 (2009); Plenaryaddress. European Control Conference

[6] Hjalmarsson, H., Mårtensson, J.: Optimal input design for identification of non-linear systems: Learning from the linear case. In: American Control Conference, New York City, USA (2007)

[7] Jansson, H., Hjalmarsson, H.: Input design via LMIs admitting frequency - wise model specifications in confidence regions. IEEE Transactions on Automatic Control 50(10), 1534–1549 (2005)

[8] Krener, A., Isidori, A., Respondek, W.: Partial and robust linearization by feedback. In: Proceedings of the 22nd IEEE Conference on Decision and Control, New York, NY, USA, vol. 1, pp. 126–130 (1983); Proceedings of the 22nd IEEE Conference on Decision and Control, San Antonio, TX, USA, December 14-16 (1983)

[9] Larsson, C., Rojas, C., Hjalmarsson, H.: MPC oriented experiment design. In: 18th IFAC World Congress, Milano, Italy (to appear, 2011)

[10] Ljung, L.: System Identification: Theory for the User. Prentice-Hall, Upper Saddle River (1998)

[11] Oppenauer, K.S., del Re, L.: Hybrid 2-zone Diesel combustion model for NO formation. In: SAE ICE 2009 – 9th International Conference on Engines and Vehicles, SAE 2009-24-0135. Capri, Italy (2009)

[12] Parrilo, P., Ljung, L.: Initialization of physical parameter estimates. In: van der Hof, S.W.P., Wahlberg, B. (eds.) Proc. 13th IFAC Symposium on System Identification, pp. 1524–1529. Rotterdam, The Netherlands (2003)

[13] Röpke, K., von Essen, C.: DoE in engine development. Quality and Reliability Engineering International 24(6), 643–651

[14] Schoukens, J., Widanage, W., Godfrey, K., Pintelon, R.: Initial estimates for the dynamics of a hammerstein system. Automatica 43(7), 1296–1301 (2007)

[15] Sjöberg, J., Ljung, Q.Z.L., Benveniste, A., Deylon, B., Glorennec, P.Y., Hjalmarsson, H., Juditsky, A.: Nonlinear black-box models in system identification: a unified overview. Automatica 31, 1691–1724 (1995)

[16] Söderström, T., Stoica, P.: System Identification. Prentice-Hall International, Hemel Hempstead (1989)

[17] Stone, M.H.: The generalized weierstrass approximation theorem. Mathematics Magazine 4(21), 167–184; 21(5), 237–254 (1948)

[18] Van Overschee, P., Moor, B.D.: Subspace Identification for Linear Systems: Theory-Implementation-Applications. Springer, New York (1996)

[19] Verhaegen, M.: Identification of the deterministic part of mimo statespace models given in innovations form from input-output data the deterministic part of mimo state-space models given in innovations form from input-output data. Automatica 30(1), 61–74 (1994)

Part I
Needs and Chances of Nonlinear Identification for Automotive Systems

Chapter 2
A Desired Modeling Environment for Automotive Powertrain Controls

Akira Ohata

Abstract. A desired modeling environment for automotive powertrain control development is proposed in this paper to timely develop the required models for control system developments. Our aim is to systematically derive a grey box model consisting of physical and empirical models with the minimum order and number of the parameters. The environment consists of physical and empirical modeling, model simplification, system modeling, model/data managements, optimization methodologies and physical law libraries. In this paper, physical model is defined as the one satisfying the considered conservation laws and empirical model is defined as the one having adjusted parameters. In this paper, two approaches from a physical model to the target and a pure empirical model to the target are introduced. Physical modeling based on the constraints and the considered conservation laws is also introduced.

2.1 Introduction

The automotive industry has encountered the complexity issue of control system development. Model-Based Development (MBD) has been highly expected to resolve the issue [1, 2]. MBD is a development where state of the art simulation technologies are efficiently used. Therefore, modeling controlled objects is one of the important elements of MBD. However, multi physics modeling and systematic grey-box modeling [3] haven't been well established. We have to deal with mechanical, chemical, fluid, thermo dynamics simultaneously at least in powertrain control susyetem developments but there are few engineers who have sufficient knowledge of various physical fields although sophisticated modeling methods have been well developed

Akira Ohata
Toyota Motor Corporation
e-mail: ohata@control.tec.toyota.co.jp

D. Alberer et al. (Eds.): Identification for Automotive Systems, LNCIS 418, pp. 13–34.

in each specific physical field. The automotive industry has developed many engine and component models based on first principle. But, those models require combining experimental data and that has been done heuristically. Therefore, reusing and maintaining the models are usually difficult. From the background, a desired modeling environment is introduced associated with MBD in order to promote the development of efficient modeling environment shared among developers of pwertrain control systems.

First, the author tries to show how to use models in MBD. The concept of MBD is introduced in Figure 2.1. A control system in the actual world consists of the controlled hardware and the Electronic Control Unit (ECU). After the required validation tests, the developed control system is put into the production. The structure combining the controlled hardware and the ECU in the actual world is modeled in the virtual world. Therefore, there are the controlled hardware and the ECU models in virtual world. The closed loop system is validated through the required simulation tests. This is called Software In the Loop Simulation (SILS) [4, 5]. An interesting thing is that there are two links between the actual and virtual worlds. One is Rapid Prototyping ECU that the actual hardware is controlled with the virtual ECU that means the control logic works on a general purpose PC with the sufficient execution speed and the memory sizes. The other link is Hardware In the Loop Simulation (HILS) that a real time hardware model is controlled by the actual ECU [6]. That can make debugging the developed ECU remarkably efficient because of easy recreation of observed bugs. Moreover, some actuators can be inserted into the closed loop when the models don't exist and the model accuracies are hardly guaranteed.

The key of MBD is to use the hardware and ECU models as the functional specifications. In the ECU development, embedded codes can be generated from the ECU model because the specification is sufficiently accurate. In the hardware development, it must be corrected when some differences are detected between simulations and experiments although that may be unacceptable for many hardware developers. However, they would produce trifle errors causing tough and urgent iterated works in the latter part of the development. The early detection of produced defects can considerably increase the productivity of the development and also increase the quality of the control system. The comparison between the product and the specification is the basic method to detect the error and the model can make it very accurate and efficient.

Figure 2.2 shows the development process of MBD. It consists of three V-cycles [2]. The process starts from the system requirements and constraints analysis. The system design is performed according to the result. Control and hardware designs are done also according to the requirements and constraints transferred from the system design. The outputs of each design are the models as the functional specifications of the components. The following processes are prototyping hardware parts and embedded codes developments. The behaviors of the outputs are compared with the models in the verification processes. Thus, defects are detected immediately. These components are combined and the hardware system and ECU are constructed. The closed loop system validations are performed with Rapid Prototyping ECU and HILS. Here, validation means to guarantee the correctness of the developed products. It can be expected that the quality of the actual closed loop system is sufficiently high because defects can be detected and fixed immediately. The development is completed after the final validation test. The purpose of this MBD process is to continuously improve the process by the evaluation of the process quality and removing iterated works as much as possible.

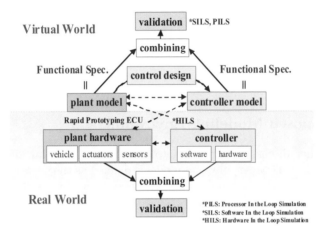

Fig. 2.1 Concept of model-based development

We want to use plant models from the beginning of development process. However, it is not easy to develop the required plant models timely. Sometimes, they are completed after the development. Models are categorized into white/black/grey-box models or physical/empirical models [7, 8, 9, 10]. Almost all may agree that grey-box models are practically effective and constructed with the combination of physical and empirical modeling methods. However, the combination has been carried out heuristically. Thus, the model quality including the description style highly depends on the developer. That causes the difficulty to apply models to other developments. This paper is structured as follows. The proposed modeling environment is introduced connected with the definitions of physical and empirical models in the section 2.2. Multi physics modeling environment is described according to the definition of physical model in the section 2.2. Empirical modeling environment is briefly summarized and the issue of parameter redundancy is shown in the section 2.3. In the following section, the importance of integrating physical and empirical models is discussed and the some integration methods are also introduced. Finally, the summary of this paper is placed.

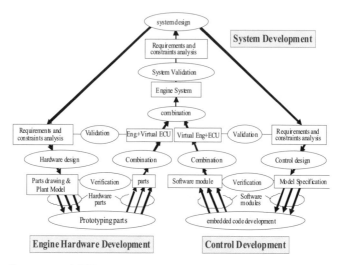

Fig. 2.2 Concurrent MBD process

2.2 Proposed Modeling Environment

Before further discussions, we have to define what physical and empirical models are. In this paper, physical model is defined as the one satisfying considered conservation laws and empirical model is defined as the one having adjusted parameters. According to the definitions, we can rigorously distinguish the physical models from the ones that are not. Figure 2.3 shows the definitions. There is the overlap of physical and empirical models in the figure. A model with adjusted parameters is a physical model if it satisfies the considered conservation laws. However, conservation laws are too strict constraints which should be relaxed practically. Thus, approximated physical model is placed around physical one. Our target is in the set of approximated physical model. There are two approaches to reach the target,

Approach A: from physical to the target → how to combine empirical models
Approach B: from empirical model to the target → how to insert physical structure. Both approaches should be systematic.

Figure 2.4 shows the proposed plant modeling environment in this paper. The solid lines show information transfers including models and the dotted lines indicate the model/data managements. Physical modeling environment is connected with optimization and physical law libraries. 3D simulations can be simplified with model simplification technologies. HLMT in this figure stands for "High Level Modeling Tool" described in the section 2.3. System identification technologies are used in the empirical modeling environment associated with optimization methods. Equations of empirical model may be derived from physical models with a model simplification method. Physical and empirical models are integrated to construct component models. In the system modeling environment, component models are assembled. However, the accuracy may be insufficient because highly accurate component modeling is not easy and the compensation with empirical models may be necessary. Model/data managements are also essential for the automotive industry.

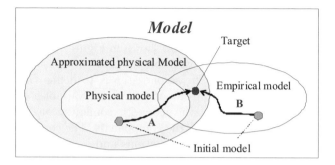

Fig. 2.3 Definition of physical and empirical models

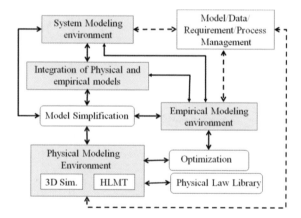

Fig. 2.4 Desired plant

2.3 Physical Modeling Environment

Mechanical and electric circuit modeling have been well developed in both physical domains. We have used these methods combining with other physical domain modeling methods. Internal combustion engine modeling requires fully combining mechanics, electric circuits, chemical reaction, thermal conduction and fluid dynamics. Bond Graph has been popular method as physical modeling among researchers and engineers [11]. However, it is based on the energy conservation laws and deal with two types of variables of which multiplication results in energy flow. That is the reason why it isn't easy for Bond Graph to deal with fluid dynamics because three variables, for instance velocity, pressure and temperature, are treated at least. In this section, a multi physical modeling based on the constraints and the considered conservation laws, for instance mass, energy, momentum and molecules conservation laws, is proposed [12].

According to the definition of physical model, a physical modeling method is proposed in this section. The process shown in Figure 2.5 is taken in physical modeling. The first step is "partitioning" that the considered system is divided into the components. The second step is to define the constraints which determine the relationships among the state variables of the model. They determine, for an example, the geometric configuration. The next step is to define the interactions among the components. The constraints and the interactions are described with the equations and a physical model means the set of the equations.

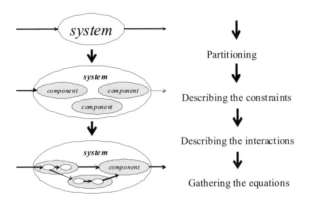

Fig. 2.5 Modeling process

Figure 2.6 shows examples of constraint described above. Example 1 shows the rotational joint mechanism and it has the position constraints. Example 2 shows a combustion model dividing the gas in the chamber into the burnt gas and the mixture gas portions. It is supposed that the pressures of the burnt and mixture gas portions are equal to each other and the summation of two volumes is equal to the chamber one. The top figure shows the general description of constraint. Two components in these cases are corresponding to the circles and the constraints are corresponding to the square block. The equations of the constraints are put into the block.

The idea to describe the interactions in this paper is to use the considered conservation laws as shown in Figure 2.7. It is considered that the interactions are caused by the exchanges of the conservation quantities such as mass, momentum, energy, molecules and so on. As mentioned above, three

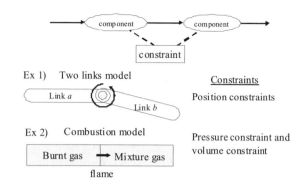

Fig. 2.6 Description of constraints

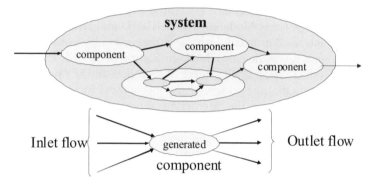

Fig. 2.7 Description of interactions

conservation laws of mass, momentum and energy are required to describe fluid systems because it can include three independent variables, the velocity, the pressure and the temperature. The conservation flows shows the arrows in Figure 2.7 and the direction of arrow means the sign convention of conservation flow. The interactions are described with

$$\frac{dE_i}{dt} = \sum_{j=1}^{N} e_{ji} - e_{gi} \tag{2.1}$$

where $E_i \in R^{n_i}$: Conserved quantities of the component i, $e_{ji} \in R^{n_i}$: conserved quantity flows from the component $j \neq i$ to the i, $e_{gi} \in R^{n_i}$: generated conserved quantity rate in the component i, n_i: the number of the considered conserved quantities of the component i, N: the number of the components. The conserved quantities aren't used to show the behavior of the model generally and they must be transferred to other popular variables, such as the positions, the velocities, the pressures and the temperatures. Such variables are indicated with $X_i \in R^{n_{xi}}$. The flows are defined with the functions of $E = \left[E_1^T, E_2^T, \cdots, E_N^T\right]^T$ and $X = \left[X_1^T, X_2^T, \cdots, X_N^T\right]^T$ of both components connecting with each other. The constraints are described with the algebraic equation (2.2).

$$h\left(E, X\right) = 0 \tag{2.2}$$

Where h is $R^{n_1+n_2+\cdots+n_N} \times R^{n_{X1}+n_{X2}+\cdots+n_{XN}} \rightarrow R^p$, $\sum\limits_{i=1}^{N} n_i + \sum\limits_{j=1}^{N} n_{Xi} > R^p$

and p is the number of the constraints. Therefore, the model is described with the high index differential algebraic equations not solved numerically in general. Thus, it is required to reduce the index [13]. The model description based on the constraints and the considered conservation laws are called HLMD (High Level Model Description) and the tool supporting GUI, the simplification of the differential equations generated from HLMD and simulation is called HLMT (High Level Modeling Tool). HLMD and HLMT guarantee the developed models are consistent with the conservation laws but don't guarantee the accuracies are sufficient. HLMD in Figure 2.8 reveals the substantial problem of physical model fidelity level. We suppose in the upper figure that the heat generation at the spring and the mechanical friction are neglected. Thus, we can only consider the interactions among the mass, the spring and the ground and the HLMD is the left figure. In the bottom figure, let's consider the lubrication oil between the mass and the ground. The motion of the mass causes the heat generation at the oil and the heat conducts to the mass and the ground. The heat also conducts from the mass and the ground to the atmosphere. The heat is also generated at the spring by the damping effect, raises the temperature and conducts to the atmosphere. In this case, the HLMD becomes the left bottom figure. To calculate the temperature, we can't neglect the mass of the spring. We can consider a higher fidelity model than any models. We may not neglect the resistance force from the air. Like this, a physical model is always perturbed by a higher fidelity model than itself. This means that identified model parameters loose the physical meanings and tend toward the different values from the expected ones.

Here, a question has risen if the ranges of the parameters are restricted by the physical meanings. For example, masses and spring coefficients may be negative as the results of the parameter optimization by using experimental data. We can set the assumption that a physical model describes experimental data to a certain extent. Thus, it is considered that our target model is around the model. That is reason why the approximated physical model is defined in Figure 2.3.

One of the advantages of HLMD is the readability of the diagram that is highly formalized and allows the tool to generate the model equations. The first prototype of HLMT was already developed [14]. Figure 2.9 shows what HLMT looks like and Fig. 2.10 shows the physical model of three masses and two springs connected in series. The damping effect generates heat at the springs that conducts to the atmosphere and raises the temperature of the

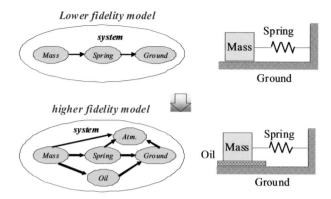

Fig. 2.8 Unlimited hierarchies of model fidelity level

springs. Fig. 2.11 shows an example of the model execution results. The left figure is the velocity of the left mass and the right figure is the temperature of the left spring.

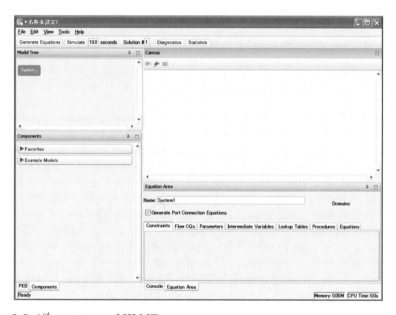

Fig. 2.9 1st prototype of HLMT

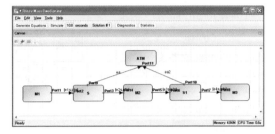

Fig. 2.10 Physical model with three masses and two springs on HLMT

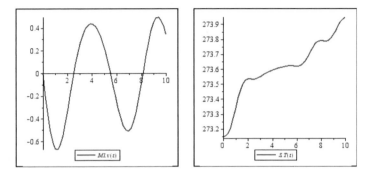

Fig. 2.11 Example of Simulation Result

2.4 Empirical Modeling Environment

Empirical model is defined as the one having adjusted parameters as shown in Fig. 2.3. Usually, we consider that pure empirical model is based on function approximation, for instance Taylor series and linear combination of radial basis functions. We want the proof that the error becomes sufficiently small when we follow a systematic procedure. Taylor series and the linear combination of basis functions have this feature theoretically. DoE (Design of Experiments) has become popular in the calibration process of powertrain control [15]. It has been introduced from steady state calibration, called base map calibration, and reduced the experiments of base map calibration by 50%. According to the success, it has started to expand the usage to transient calibrations. Now, parametric Volterra series and Wiener-Hammerstein model [16] are popular in the powertrain calibration area.

Fig. 2.12 shows an example of parametric Volterra series applied to HC emission of the FTP cold start mode. It shows five test cases with the different throttle, the fuel injection and the spark advance controls. The solid lines are experimental data and the dotted lines are simulation results of identified models. The model equation is as shown below:

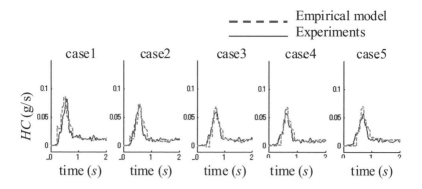

Fig. 2.12 HC emission of FTP cold start

$$
\begin{aligned}
HC\left(k\right) = & \; \alpha_1 \, s_a\left(k\right) + \alpha_2 \, f_{it}\left(k\right) + \alpha_3 \, f_{im}\left(k\right) + \alpha_4 \, \theta\left(k\right) + \alpha_5 \, s_a\left(k\right)^2 + \\
& \alpha_6 \, s_a\left(k\right) \, f_{it}\left(k\right) + \alpha_7 \, s_a\left(k\right) \, f_{im}\left(k\right) + \alpha_8 \, s_a\left(k\right) \, \theta\left(k\right) + \\
& \alpha_9 \, f_{it}\left(k\right)^2 + \alpha_{10} \, f_{it}\left(k\right) \, f_{im}\left(k\right) + \alpha_{11} \, f_{it}\left(k\right) \, \theta\left(k\right) + \\
& \alpha_{12} \, f_{it}\left(k\right)^2 + \alpha_{13} \, f_{im}\left(k\right) \theta\left(k\right) + \alpha_{14} \, \theta\left(k\right)^2 + \alpha_{15} \, s_a\left(k - 10\right) + \\
& \alpha_{16} \, f_{it}\left(k - 10\right) + \alpha_{17} \, f_{im}\left(k - 10\right) + \alpha_{18} \, \theta\left(k - 10\right) + \\
& \alpha_{19} \, s_a\left(k - 10\right)^2 + \alpha_{20} \, s_a\left(k - 10\right) \, f_{it}\left(k - 10\right) + \\
& \alpha_{21} \, s_a\left(k - 10\right) \, f_{im}\left(k - 10\right) + \alpha_{22} \, s_a\left(k - 10\right) \, \theta\left(k - 10\right) + \\
& \alpha_{23} \, f_{it}\left(k - 10\right)^2 + \alpha_{24} \, f_{it}\left(k - 10\right) \, f_{im}\left(k - 10\right) + \\
& \alpha_{25} \, f_{it}\left(k - 10\right) \, \theta\left(k - 10\right) + \alpha_{26} \, f_{im}\left(k - 10\right)^2 + \\
& \alpha_{27} \, f_{im}\left(k - 10\right) \, \theta\left(k - 10\right)^2 + \alpha_{28} \, \theta\left(k - 10\right)^2 + \alpha_{29}
\end{aligned}
\tag{2.3}
$$

Where, s_a is the spark advance, f_{it} is the fuel injection timing, f_{im} is the amount of fuel injection and θ is the throttle opening angle.

Generally, physical model of HC emission isn't easy to construct because many physical phenomena are involved. On the other hand, empirical model can easily recreate experimental data even for very complex phenomena. Like this example, many parameters appear in nonlinear identification generally. We have often encountered the exponentially increase of model parameter in dynamical empirical models although there are methods to mitigate the issue such as kernel approach [17]. The model is applied to limited engine operation conditions close to the measurement condition. It seems that inserting physical structure is effective.

2.5   Integration of Physical and Empirical Models

Grey-box modeling is widely used in many applications. It is a kind of the integration of physical and empirical models. However, it can be said that grey box models are derived heuristically. In this paper, we discuss a systematic integration. Now, let's suppose the equation,

$$\frac{dx}{dt} = f\left(x, u, \theta\right), \quad y = g\left(x\right) \tag{2.4}$$

is a physical model. Where $x \in R^n$ is the state vector, $u \in R^m$ is the input vector, $y \in R^p$ and $\theta \in R^l$ is the constant vector corresponding to physical and geometric constants. In this section, two approaches A and B shown in Fig. 2.3 are introduced.

2.5.1 Approach A

Approach A means to combine empirical models to a physical model. We categorize the methods into the following three types.

2.5.1.1 Type 1

$$\frac{dx_r}{dt} = f_r\left(x_r, u, \theta_r\right) \approx \sum_{i=1}^{N_r} \theta_{ri} f_{r1}\left(x_r, u\right) \quad y = g_r\left(x_r\right) \tag{2.5}$$

Where $x_r \in R^{n_r} \left(n_r < n\right)$ is the reduced state vector and $f_{r1}\left(x_r, u\right)$, $(i = 1, 2, \cdots, N_r)$ are the basis functions. Type 1 includes approximated description of

$$F_r\left(\frac{d^{n_r}y}{dt^{n_r}}, \frac{d^{n_r-1}y}{dt^{n_r-1}}, \cdots, y, \frac{d^{m_r}u}{dt^{m_r}}, \frac{d^{m_r-1}u}{dt^{m_r-1}}, \cdots, u, \theta_r\right) = 0. \tag{2.6}$$

This method can be applied to the discrete time systems. It can be said that Type 1 includes model order reduction and function approximation. Type 1 is highly connected with empirical model and approximated models can be empirical models.

An empirical model from the throttle to the intake pressure is constracted as an example of the type 1 model in this section. Fig. 2.13 shows the 46th order V6 engine model [18] that is not a mean value model but an instantaneous model. We tried to identify the nonlinear empirical model from the throttle to the intake pressure by using simulation data [19]. By the analysis of Hankel matrix consisting of the input and the output sequences, we decided to adopt the 1st order affine model. Thus, each local model around the intake pressure is described by

$$p\left(k + 1\right) = \alpha_1\left(p_0\right) p\left(k\right) + \alpha_2\left(p_0\right) u\left(k\right) + \alpha_3\left(p_0\right) \tag{2.7}$$

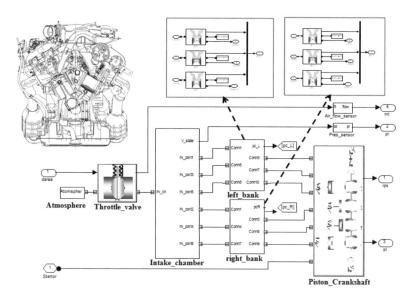

Fig. 2.13 V6 engine model

Figure 2.14 shows the identification result for the simulation data shown in Figure 2.15. The throttle angle fed to the engine model was the summation of the random and the gradually increasing signals. The green line is the curves of α_1, α_2 and α_3 approximated with the polynomial functions with the intake pressure. Thus, the global model is described with

$$p(k+1) = \left\{ \beta_8\, p(k)^8 + \beta_7\, p(k)^7 + \cdots + \beta_0 \right\} p(k) + \\ \left\{ \gamma_8\, p(k)^8 + \gamma_7\, p(k)^7 + \cdots + \gamma_0 \right\} u(k) + \\ \left\{ \delta_8\, p(k)^8 + \delta_7\, p(k)^7 + \cdots + \delta_0 \right\} \qquad (2.8)$$

Figure 2.14 shows the equation (8) is well fitted to the local models and Figure 2.16 shows a good correlation between the globalized and the original models. The model was successfully applied to an actual engine data and the re-identified model by using experimental data shows the good correlation as shown in Figure 2.16.

However, there are 27 parameters in the model that easily change in identifications with small perturbations although each correlation between experimental and simulation data is good. We loose the physical structure during the global model construction and that causes the parameter redundancy described in the following section. It can be said generally that models with many parameters are flexibly fitted to many data but it is difficult to determine them uniquely. According to function approximation theory, what we have to do is to increase terms and coefficients when the model isn't

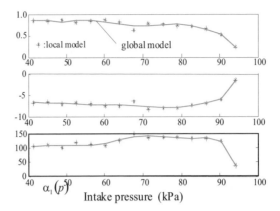

Fig. 2.14 Coefficients of local affine model

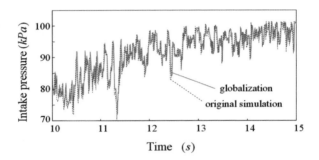

Fig. 2.15 Correlation between globalized model and original simulation

sufficiently accurate. Thus, empirical models don't require the effort to develop physical models. However, this doesn't mean that an obtained empirical model can be applied to other data measured on different operating conditions. The required parameters and the terms would exponentially increase when we try to fit the model to many data.

2.5.1.2 Type 2

$$\frac{dx}{dt} = f\left(x, u, \theta\left(x, u\right)\right)$$
$$y = g\left(x\right) \tag{2.9}$$

The argument θ in the equation (4) is a theoretical constant vector. But, it is possible to deal with θ as the function with x and u, practically. For example, we can deal with the engine rotational inertia as the function with the engine speed although it is a constant or a function with crank angle physically. That would be reflected from the simplified assumption that we neglected the coolant and lubrication oil flows. We can also deal with the plenum chamber volume is a function with the intake pressure although it is a constant theoretically. However, the method is a simple and effective way to improve the accuracy.

2.5.1.3   Type 3

$$y_{experiment} = y_{model} + e_{add}$$
$$e_{add}(x, u) \approx f_e(x_r, u, \theta_r) \tag{2.10}$$

$f_e(x_r, u)$ is called error function. Error function includes

$$y_{experiment} = e_{mul}\, y_{model}$$
$$e_{mul} \approx f_e(x_r, u, \theta_r) \tag{2.11}$$

where, the function $f_e(x_r, u, \theta_r)$ is an approximated model error from the experimental data. The combination of the equation (2.10) and (2.11) is possible.

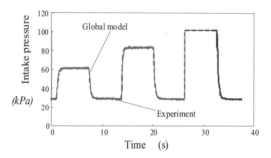

Fig. 2.16 Comparison of simulation with global model and measured data

Figure 2.17 shows an example of error function applying to the air charge estimation. The function of the air charge estimation. The function of the air charge is expressed by

$$f_{expreiment}(X) = error(X)\, f_{model}(X), \tag{2.12}$$

where $X = [\, rpm, intake\ pressure, intake\ valve\ lift, intake\ valve\ phase\ angle\,]^T$ and the air charge is indicated by $f_*(X)$. The right figure shows that the integrated model can improve the accuracy considerably.

For types 1 to 3, θ_r is optimized such that the error between the experiments and the approximated model is sufficiently small. These are applied to both component models and system models. For system modeling, it is important to notice the possibility of the parameter redundancy. An empirical model of the heat transfer from the gas in the cylinder to the cylinder wall affects the air charge and the air flow model of the intake valve also affects it.

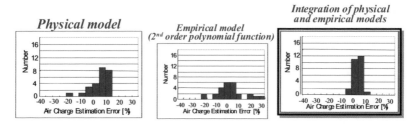

Fig. 2.17 Example of error function application

There are other component models affecting the air charge. Thus, the parameter redundancy easily happens. It is very difficult to adjust the parameters correctly or uniquely. Moreover, the heuristic integration tends to make the error complex. We have a risk that the compensation with empirical models tends to make the model error complex. We need a lot of experimental data to compensate the complex one. Thus, the important point of physical modeling is to make the model error simple.

2.5.2 Approach B

However, nonlinear identification has been still tough area in control engineering although various methods have been proposed. Figure 2.18 shows one of the difficulties of empirical model. It often happens that the distance between two models in the parameter space may be long even if the behaviors of the models are almost same as shown in Figure 2.17. The right figure shows the parameter distance in the plane of p_1 and p_2. This image is extended to the models with more parameters than three. The parameter distance L_p in Figure 2.17 is, for an example, defined by

$$L_p = \sum_{i=1}^{N_p} (p_{Ai} - p_{Bi})^2. \qquad (2.13)$$

In the equation (2.13), N_p indicates the number of the parameters.

That can be caused by the redundancy of model parameter that the other parameters can adjust the behavior to match the required data even if the parameter p_1 of a model takes an arbitrary value. The problem is that the model parameters can't be determined uniquely. This causes the difficulty when a global model is constructed from local models.

Let's suppose that the target system described by model (4) is transferred to the SISO discrete-time description,

$$F\left(y\left(k\right), y\left(k-1\right), \cdots, u\left(k\right), u\left(k-1\right), \cdots, p\right) = 0. \qquad (2.14)$$

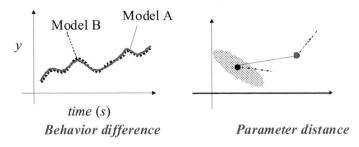

Fig. 2.18 Inconsistency of model parameter distance and behavior similarity

Where, $y \in R$ is the output, $u \in R$ is the input and $p \in R$ is the model parameter. From the Jacobean of F, we obtain

$$y(k) + \alpha_1 y(k-1) + \cdots + \beta_0 u(k) + \beta_1 u(k-1) + \gamma = 0. \tag{2.15}$$

Many parameters appear in the equation (2.15) even if p is a scalar. The coefficients of the equation (2.15) are the functions with the $y(k) = y(k-1) = \cdots = y_0$, $u(k) = u(k-1) = \cdots = u_0$ and p because the linearization is done around the condition. This means that the coefficients depend upon y_0, u_0 and p. Thus, there are the relationships between the coefficients

$$\Phi_i(\theta) = 0 \quad i = 1, 2, \cdots, \tag{2.16}$$

where $\theta = [\alpha_1, \alpha_2, \cdots, \alpha_n, \beta_0, \beta_1, \cdots, \beta_m, \gamma]^t$. They want to make some coefficients zero especially in the machine learning area [20]. Practically, such conditions are derived from physical considerations associated with AIC [21]. In such cases, $\alpha_j = 0$ and $\beta_l = 0$ are the special case of the equation (2.16) [21]. Therefore, we can define the relationship among empirical model parameters as physical structure and we define the relationship expressed by (2.16) is SPS (Strong Physical Structure). Because, the model adopting SPS is almost same as the original model and the flexibility of empirical model may be lost. Thus, the relaxation of SPS is considered. For the purpose, the constraints with inequality is used as

$$|\Phi_i(\theta)| < \delta_i \quad i = 1, 2, \cdots. \tag{2.17}$$

δ_i is determined according to the confidence level. The constraints expressed by (2.17) are called WSP (Week Physical Structure). This concept can be applied to nonlinear empirical models. Another way to introduce physical constraints is to insert the constraint (2.16) into the criterion as shown in

$$J = \|y_{\text{exp}} - y_{model}\| + \sum_i^{N_c} \lambda_i \|\Phi_i(\theta)\|, \tag{2.18}$$

where each $\|\bullet\|$ means a norm including L_2, L_1 and L_∞. This method follows the concept of WPS because $\Phi(\theta) = 0$ isn't strictly required. We can make some of $\Phi_i(\theta)$ equal to zero when we take the norm L_1 for the second term. This is an application of LASSO [20].

To show this concept clearly, the following toy example,

$$\begin{bmatrix} I & \varepsilon_1 \\ \varepsilon_2 & \varepsilon_3 \end{bmatrix} \frac{d}{dt} \begin{bmatrix} x_1 \\ x_2 \end{bmatrix} = \begin{bmatrix} f_1(x_1, u) \\ f_2(x_1, x_2) \end{bmatrix}$$
$$y = g(x_1) + w \tag{2.19}$$

is used. It becomes

$$\frac{dx_1}{dt} = f_1(x_1, u)$$
$$y = g(x_1) + w \tag{2.20}$$

when $\varepsilon_1, \varepsilon_2, \varepsilon_3 \to 0$. Thus, (2.19) means x_1 is perturbed by x_2. The purpose here is to investigate the effect of the perturbation and the inserted physical structure to system identification.

The nominal model investigated here is the transfer function

$$\left(s^3 + \theta_4 s^2 + \theta_5 s^1 + \theta_6\right) Y(s) = \left(\theta_1 s^2 + \theta_2 s^1 + \theta_3\right) U(s) \tag{2.21}$$

Where $\Theta = [\theta_1, \theta_2, \theta_3, \theta_4, \theta_5, \theta_6]^t = [2p, 0.5p, 1, 1, p - p]$ and $p = 1$. The equation (2.21) is transferred to the state equation,

$$\frac{dx_1}{dt} = \begin{bmatrix} -2 & -0.5 & -1 \\ 1 & 0 & 0 \\ 0 & 1 & 0 \end{bmatrix} x_1 + \begin{bmatrix} 1 \\ 0 \\ 0 \end{bmatrix} u$$
$$y_0 = \begin{bmatrix} 1 & 1 & -1 \end{bmatrix} x_1 \tag{2.22}$$

and the perturbed output is

$$\frac{dx_2}{dt} = -2 x_2 + 2 y_0$$
$$y = x_2 + w \tag{2.23}$$

where $E(w) = \lambda \delta$ ($\lambda > 0$ *and* δ is Dirac delta function) is a white noise signal.

Figure 2.19 shows the comparisons of identification results when $\varepsilon_1 = 0.01$, $\varepsilon_2 = 0.01$, $\varepsilon_3 = 0.1$ and $\lambda = 0.005$. The top figure is the comparison of the nominal output y_0 and the identification result of ARX. The correlation is very good because noise isn't involved. The second figure is for the identification of y. In this case, singular perturbation and noise are involved. The identification is crushed down. The third figure shows the identification minimizing the error between the y with no constraint and the simulation of the indentified model. We obtained a good correlation. The bottom figure shows the identification minimizing the error with the constraint

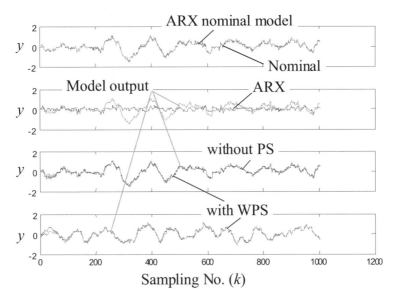

Fig. 2.19 Comparisons of identification results

$$
-\begin{bmatrix} 0.1 \\ 0.2 \\ 0.1 \\ 0.1 \\ 0.1 \\ 0.1 \end{bmatrix} < \Theta\left(\theta\right) = \begin{bmatrix} 1 & 0 & 0\ 0\ 0\ 0 \\ 0 & 1 & 0\ 0\ 0\ 0 \\ 0 & 1 & 1\ 0\ 0\ 0 \\ 0 & -2 & 0\ 1\ 0\ 0 \\ 0 & -0.5 & 0\ 0\ 1\ 0 \\ 0 & 0 & 0\ 0\ 0\ 1 \end{bmatrix} \begin{bmatrix} \theta_1 \\ \theta_2 \\ \theta_3 \\ \theta_4 \\ \theta_5 \\ \theta_6 \end{bmatrix} - \begin{bmatrix} 1 \\ p \\ 0 \\ 0 \\ 0 \\ 1 \end{bmatrix} < \begin{bmatrix} 0.1 \\ 0.2 \\ 0.1 \\ 0.1 \\ 0.1 \\ 0.1 \end{bmatrix}. \qquad (2.24)
$$

This means the bottom identification uses WPS. The identification shows almost same result as the third figure. However, the issue is the parameter distances.

Figure 2.20 shows the parameter distance between the identification results and the nominal parameters. The parameter distance of the identification without WPS, that is the constraint of the numerical optimization, is the biggest and the one without WPS is considerably small. When we adopted the criterion (18) for the unconstrained optimization with the L_1 norm of $\|\Theta\left(\theta\right)\|_{L_1}$, the result was similar to the one using the constraint (23). However, the result was similar to in the case of the L_2 norm of $\|\Theta\left(\theta\right)\|_{L_2}$. However, this result can't be general because the weighting factor λ_i highly affects the identification results. Therefore, more investigations are necessary.

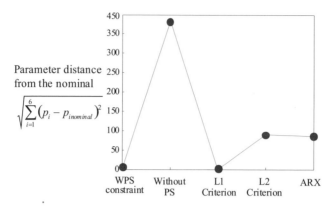

Fig. 2.20 Effect of WPS on parameter distances

2.6 Summary

Rapid and systematic modeling is the urgent issue to realize Model-Based Developments (MBD). However, there is no shared modeling environment among engineers and researchers to timely develop required models. To resolve the issue, a desired modeling environment has been proposed in this paper. The integration of physical and empirical model is the main challenge to construct the proposed environment. The following items are the major technical contributions of this paper.

- Physical model is defined as the one satisfying considered conservation laws. According to the definition, a physical modeling tool based on the conservation laws and the constraints is introduced.
- Empirical mode is defined as the one having adjusted parameters. Empirical models tend to cause the redundancy of model parameter caused by loosing physical structure. That causes the exponential increase of the parameters.
- Three types of approximated physical model have been defined. The combination is possible to make component and system models. A few examples were demonstrated in this paper.
- The key of physical modeling is to make the model error simple from the view of integrating physical and empirical models.
- The key of the integration is to make the parameter space small including the increase of parameter sparsity and binding the parameters to the nominal values.

There have been a lot of remaining studies and the collaboration among researchers and engineers is highly required.

References

[1] Ohata, A., Yamakita, M.: An Application of MPC Starting Automotive Spark Ignition Engine in SICE Benchmark Problem. In: Automotive Model Predictive Control: Models, Methods and Applications. LNCIS, vol. 402. Springer, Heidelberg, ISBN 9781849960700

[2] Ohata, A., Butts, K.: Improving Model-Based Design for Automotive Control Systems Development. In: Proceeding of IFAC WC (2008)

[3] Jones, D.M., Watton, J., Browm, K.J.: Comparison of black-, white-, and grey-box models to predict ultimate tensile strength of high-strength hot rolled coils at the Port Talbot hot strip mill. In: Proceedings of the Institution of Mechanical Engineers, Part L: Journal of Materials

[4] Murakami, K., Rao, N., Oho, P., Shimada, S.: Toward the Certain of an ECU Model Exchange Market. In: ICCAS-SICE (2009)

[5] Oho, S.: Model-Based Implementation Design of Automotive Controllers. In: Proceeding of IFAC WC 2008 (2008)

[6] Hanselmann, H.: Hardware-in-the loop Simulation Testing and Its Integration into CACSD toolset. In: Proceeding of IEEE CAIC/CACSD (1996); Design and Applications 221(1). Professional Engineering Publishing (2007)

[7] Heywood, J.B.: Internal Combustion Engine Fundamentals. McGraw-Hill, New York (1998)

[8] Kiecke, U., Nielsen, L.: Automotive Control Systems for Engine, Driveline and Vehicle, 2nd edn. Springer, Heidelberg (2004)

[9] Guzzela, L., Onder, C.H.: Introduction to Modeling and Control of Internal Combustion Engine Systems, 2nd edn., pp. 178–179. Springer, Berlin (2004)

[10] Rowell, D., Wormley, D.N.: System Dynamics – An Introduction. Prince-Hall (1997)

[11] Thoma, J., Bouamama, B.O.: Modeling and Simulation and Chemical Engineering - A Bond Graph Approach. Springer, Heidelberg (2000), ISBN 3-540-66388-6

[12] Ohata, A., Ito, H., Gopalwamy, S., Furuta, K.: A Plant Modeling Environment Based on Conversation Laws and Projection Method. SICE JCMSI 1(3) (2008)

[13] Hairer, E., Wanner, G.: Solving Ordinary Differential Equation 2 stiff and Differential-Algebraic Problems. Springer, Heidelberg (2002), ISBN 3-540-60452-9

[14] Kowalska, J., Leger, M., Wittkopt, A.: High-Level Modeling Description and Symbolic Computing. In: Proceeding of IFAC WC 2008 (2008)

[15] Röpke, K.: DoE-Design of Experiments Method and Applications in Engine Development, SV Corporate Media, 889503 (2005)

[16] Elrady, E.A., Gan, L.: Identification of Hammerstein and Wiener Models Using Spectral Magnitude Matching. In: Proceeding of IFAC WC (2008)

[17] Kashima, H., Idé, T., Kato, T., Sugiyama, M.: Recent advances and trends in large-scale kernel methods. Transactions on Information and Systems E92-D(7), 1338–1353 (2009)

[18] Ohata, A., Shen, T., Ito, K., Kako, J.: Introduction to the Benchmark Challenge on SICE Engine Start Control Problem. In: IFAC WC (2008)

[19] Ohata, A., Furuta, K.: Integration of Physical and Statistical Models for Automotive Engine Control. In: Proceeding of IEEE CCA/ISIC/CACSD (2006)

[20] Bishop, C.M.: Pattern Recognition and Machine Learning. Springer Science + Business Media, LLC, Heidelberg (2007), ISBN-10: 0-387-31073-8

[21] Akaike, H.: Fitting Autoregressive Model for Prediction. Annals of the Institute of Statistical Mathematics 21, 243–247 (1969)

[22] Kukreja, S.L., Löfberg, J., Brenner, M.J.: A least absolute shrinkage and selection operator (LASSO) for nonlinear system identification. In: Proceedings of the 14th IFAC Symposium on System Identification, vol.14

Chapter 3
An Overview on System-Identification Problems in Vehicle Chassis Control

Simone Formentin and Sergio M. Savaresi

Abstract. This paper provides a brief survey on some identification issues in vehicle chassis control design. Data-driven techniques appear as the most suitable for the control-oriented modeling task, since in this framework physical laws are strongly affected by unmeasurable parameters, *e.g.* the road conditions.

Four examples are then presented to underline which are the advantages and the drawbacks of identification approaches in every subproblem of chassis control, *i.e.* stability control, traction control, suspension control and braking control. Each single task is challenging from both theoretical and practical point of view. At the end, it becomes evident that in this kind of control application, the identification phase is critical at least as much as the control design itself.

3.1 Introduction

In recent years, the automotive market has witnessed an increasing diffusion of control electronics in the whole vehicle architecture. IEEE Spectrum, an American technical publication, reported that electronics, as a percentage of vehicle costs, climbed from 5% in the late 1970s to 15% in 2005 — and this number has undoubtedly increased since then. Among all things, a privileged place is occupied by the electronic systems devoted to the chassis control, for mainly two reasons. First of all, road safety is an essential element of sustainability in mobility and transport. Secondly, chassis control systems can be used as a performance enhancement; as an example, traction control

Simone Formentin · Sergio M. Savaresi
Dipartimento di Elettronica e Informazione, Politecnico di Milano,
Piazza Leonardo da Vinci 32, Milano, Italy
e-mail: {formentin,savaresi}@elet.polimi.it

D. Alberer et al. (Eds.): Identification for Automotive Systems, LNCIS 418, pp. 35–49.
springerlink.com   © Springer-Verlag London Limited 2012

systems for racing vehicles allow drivers to get maximum traction force, by keeping the tire at the optimum slip ratio.

In this field of automotive control, one of the most challenging tasks is the control-oriented modeling of the plant dynamics. As a matter of fact, physical and analytical models often involve road-tire descriptions or unmeasurable parameters; therefore, they are typically not easy to derive and sometimes even unusable in practice. Identification techniques appear then as the most suitable way to model such complex dynamics for control design purposes.

During last years, much effort has been dedicated to the development of automatic identification procedure for chassis control. These studies have also benefited from the recent improvement of theory of control-oriented identification, or "identification for control", that aims at designing models whose quality is assessed depending on the final control application, see [1], [2] and references therein. To cite just few recent contributions in the area of identification for chassis control, see instead [3], [4], [5], [6], [7]. For a detailed survey on modeling and control problems in vehicle dynamics, the reader is instead referred to [12].

In this chapter, four application examples will be presented to show how identification methods can be used for filling the lack of knowledge of plant dynamics in road vehicles (two on motorcycles and two on cars). In order to provide both an overview on application problems and an overview on available methods that can be useful in this particular field, each paragraph deals with a different sub-problem in the chassis control framework, and each one is solved via a different approach.

Specifically, from a methodological point of view, the approaches are grey-box identification, black-box identification, estimation-oriented identification and direct controller identification. From an application point of view, the different themes are stability control, traction control, suspension control and anti-lock braking systems. Notice that from a vehicle dynamics viewpoint, all the (six) degrees of freedom of the rigid chassis are taken into account by this class of applications: traction and braking control are devoted to the longitudinal dynamics, stability control is strictly connected to the lateral movement and the yaw angle, and finally suspension systems aim at regulating roll, pitch and vertical dynamics.

3.2 *Example 1:* Grey-Box Identification for Yaw Control of Four-Wheeled Vehicles ([8])

Yaw control systems can be used for different applications, *e.g.* to correct under- or over-steering behaviour or to compensate the lateral drift that can occur during braking in certain conditions. In the second case, the basic idea to cope with this control problem is to rely only on brake actuation; in fact all modern cars are equipped with ABS systems that independently modulate the braking pressure at each wheel: by applying a differential braking force it is possible to generate a yawing moment to control the drift.

Such a control strategy requires to model the dynamics from the steering wheel and differential braking to the lateral dynamics of the vehicle. A grey-box approach will be followed, by first writing the first-principles equations that handle the physical phenomena and then identifying the unknown parameters. Figure 3.1 schematically represents the main parameters and conventions employed in the model of the vehicle.

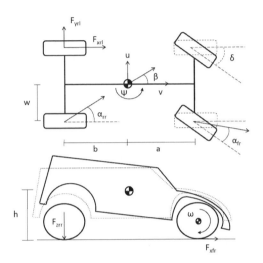

Fig. 3.1 Double-track vehicle model nomenclature.

In the figure: $\alpha_{..}$'s are the wheel side-slips β is the vehicle side-slip, $F_{x..},F_{y..},$ $F_z.$ are the longitudinal, lateral and vertical forces, δ is the steering angle at the wheels, ψ and r are the yaw angle and yaw rate, v and u are the longitudinal and lateral velocity in body fixed coordinates; finally the geometric parameters are a, b, w and h standing for the distance from the front axle to the center of mass, the distance from the rear axle to the center of mass, half the track width and the height of the center of mass. Other symbols not defined in figure are J, m, J_w which represent the yaw moment of inertia of the vehicle, the total vehicle mass and wheel moment of inertia, $\omega_{..}$'s are the wheel velocities, $\lambda_{..}$'s are the longitudinal wheel slips, r_w is the wheel radius and $T_{b..}$'s are the braking torque at each wheel.

The vehicle model is obtained by writing the force and momentum balances relative to 7 degrees of freedom: 2 linear degrees of freedom (longitudinal and lateral) and 5 rotational (yaw and the 4 wheels). The balances can be written as follows. Notice that the longitudinal force is modeled according to

Burkhardt model [23] whereas, thanks to the small steering angles hypothesis, a linear model can be employed for the lateral forces.

$$
\begin{cases}
\dot{v} = 1/m\left(F_{xrl} + F_{xrr} + (F_{xfl} + F_{xfr})\cos(\delta) + (F_{yfl} + F_{yfr})\sin(\delta)\right) \\
\dot{u} = 1/m\left(F_{yrl} + F_{yrr} + (F_{yfl} + F_{yfr})\cos(\delta) - (F_{xfl} + F_{xfr})\sin(\delta) + rvm\right) \\
\dot{r} = 1/J((F_{yrl} + F_{yrl})b + (-F_{xrl} + F_{xrl})w + (F_{yfl} + F_{yfr})\cos(\delta)a + \\
\quad -(F_{xfl} + F_{xfr})\sin(\delta)a + (-F_{xfl} + F_{xfr})\cos(\delta)w + (-F_{yfl} + F_{yfr})\sin(\delta)w) \\
\dot{\omega}_{..} = \frac{1}{J_w}(-T_{b..} + r_w F_{x..}) \\
F_{x..} = -F_{z..}\left(c_1(1 - e^{-c_2\lambda_{..}}) - c_3\lambda_{..}\right) \\
F_{y..} = F_{z..}C_\alpha(\lambda_{..})\alpha_{..}
\end{cases}
$$

$$(3.1)$$

In the model, c_1, c_2 and c_3 are the tire characteristic parameters and $\lambda_{..}$ is the longitudinal wheel slip, defined as the normalized difference between the vehicle speed v and the wheel speed $\omega_{..}r$, that is: $\lambda_{..} = (v - \omega_{..}r)/v$.

The tire side slip can be written as a function of the state space variable according to the following expressions:

$$
\begin{cases}
\alpha_{fl} = (v\sin(\beta + \delta) - ra\cos(\delta))/(v\cos(\beta - \delta) + ra\sin(\delta)) \\
\alpha_{fr} = (v\sin(\beta + \delta) - ra\cos(\delta))/(v\cos(\beta - \delta) + ra\sin(\delta)) \\
\alpha_{fl} = (v\sin(\beta) + ra)/(v\cos(\beta) - rw) - \delta \\
\alpha_{fr} = (v\sin(\beta) + ra)/(v\cos(\beta) + rw) - \delta \\
\alpha_{rl} = (v\sin(\beta) - rb)/(v\cos(\beta) - rw) \\
\alpha_{rr} = (v\sin(\beta) - rb)/(v\cos(\beta) + rw)
\end{cases}
$$

$$(3.2)$$

where the vehicle side slip β and the cornering stiffness $C_\alpha(\lambda)$ are defined as $\beta = atan(u/v)$ and $C_\alpha(\lambda) = C_{\alpha,0} - k\lambda$ (with k suitably chosen as in [8]). The vertical load, which strongly influences the traction force, is constituted by two main terms: the static load and the load transfer.

The measurable parameters such as lengths, widths, masses are assumed to be known, whereas the unknown parameters, i.e. the cornering stiffness, the yaw inertia and the pressure-to-torque gain K_b, are to be identified from data, following a grey-box approach. Ideally also the longitudinal characteristic of the tire should be identified, but in this case it has been assumed known and set equal to the one of a high grip highway road. The pressure-to-torque gain can be identified independently from the other parameters by means of a simple sufficiently exciting experiment (e.g. a step input) and prediction error methods (see [13]). Notice that the identification procedure can be performed off-line. The two remaining parameters required a more elaborate protocol since the identification should not be biased by the drift caused by braking. The identification has been therefore performed on off-line data collected on runs executed at constant velocity, where a square wave excitation of the rear brakes pressure was being applied. Moreover, even though the driver was asked to keep the steering wheel at 0°, the steering wheel angle cannot be assumed constant and thus it has been treated as a measured disturbance. In [8], the presented modeling strategy has been implemented in a

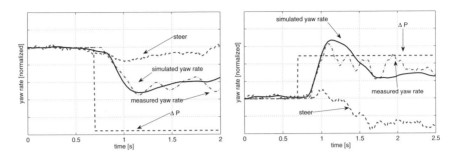

Fig. 3.2 Model Validation. Measured and simulated yaw rate, steering wheel position and pressure gradient request at constant velocity; 100 km/h (left plot) and 120 km/h (right plot).

real experimental setup, by using a car equipped with all necessary sensors. In Figure 3.2 the obtained validation tests at constant velocity are shown. The two figures prove that the model captures even very minute variation of the yaw rate due to small steer action and differential rear wheel braking.

3.3   *Example 2:* Black-Box Identification of Engine-to-Slip Dynamics for Motorcycle Traction Control ([9])

This example considers the problem of modeling the engine-to-slip dynamics in two-wheeled vehicles, by using the data collected with some experiments specifically designed for this purpose. The proposed identification protocol allows to derive dynamic models suited for the design of traction control systems for two-wheeled vehicles. The effects of two different control variables are studied: throttle position and spark-advance. All the analysis is performed on experimental data collected on a real motorbike around a single well-specified working condition, that is in-plane conditions (null roll angle), 2nd-gear, 14000 rpm, dry asphalt.

In order to identify the slip dynamics, sinusoidal sweep signals are used, whereas step variations are employed for validation purpose. As expected, the overall I/O behavior is significantly nonlinear; this is particularly true when using the spark-advance. The nonlinearity of the response is evident in Figure 3.3 where the spectrograms of the two responses (throttle and spark-advance) are plotted. From the spectrograms the nonlinearity of the response is manifest under the form of high order harmonics. If the dynamics were linear, only the 1st harmonic would be present; in this case it is possible to identify three zones:

- **Zone I**, in the range [0 Hz, 2.5 Hz] where the response is essentially linear, only the 1st harmonic is present in the signal.

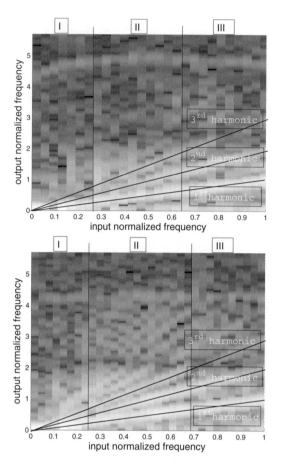

Fig. 3.3 Spectrograms of the rear wheel slip. Top: throttle sweep experiment; Bottom: spark-advance sweep experiment.

- **Zone II**, in the range [2.5 Hz, 7 Hz] where the response is determined by three harmonics. This is the range where the dynamics are mostly nonlinear.
- **Zone III**, in the range [7 Hz, 10 Hz] where the power of the third harmonic diminishes and the response is determined by two harmonics.

Consequently, if an input signal $u(t) = sin(\omega\, t)$ is considered, then the output signal can be written as:

$$y(t) \approx \sum_{i=1}^{N} A_i(\omega) \sin(i\,\omega t + \psi_i(\omega)) \qquad (3.3)$$

where $A_i(\omega)$ and $\psi_i(\omega)$ are the amplitude amplification and phase shift of the i-th harmonic and N is the number of harmonics that are taken into account. If $N = 1$, then a classical describing function is obtained.

The first harmonic generator is a classical linear system with frequency response $G(j\omega)$ and in [9] it has been demonstrated that it is the dominant one in describing slip dynamics. Therefore, the system to be identified will be characterized as a linear system and it will be assumed that it has the following form:

$$G(j\omega) = \frac{B(j\omega)}{A(j\omega)} = \frac{b_1(j\omega)^n + b_2(j\omega)^{n+1} + \ldots b_{n+1}(j\omega)}{a_1(j\omega)^m + a_2(j\omega)^{m-1} + \ldots a_{m+1}(j\omega)}, \tag{3.4}$$

where $\omega \in \mathcal{R}$. The parameters a_i and b_i are determined by solving the following optimization problem:

$$\min_{b,a} \sum_{k=1}^{l} W_f(k) \left| h(k) - \frac{B(\omega(k))}{A(\omega(k))} \right|^2, \tag{3.5}$$

where l is the number of available frequencies, $W_f(k)$ is a weight that can be used to drive the fitting toward certain frequencies, $h(k)$ is the experimental frequency response. In the present work, the optimization problem has been solved via an iterative approach based on the damped Gauss-Newton method (see [14]). The order of the numerator and denominator is determined with a method adapted from the classical Finite Prediction Error method (see [13]), which allows to find a trade-off between model complexity and accuracy.

Moreover, a $10ms$ pure delay has been introduced to model the air-box dynamics. This hypothesis is confirmed also by the black-box approach; if the delay is not introduced, the optimization problem (3.5) yields a non-minimum phase transfer function, which cannot be physically explained.

The first harmonics approximation for the throttle-to-slip and spark-to-slip dynamics are reported in Figure 3.4: From Figure it is possible to draw the following conclusions:

- Both dynamics show a resonance around 8 Hz, due to the transmission. It is interesting to note that the resonance in the throttle-to-slip dynamics is less damped than in the spark-to-slip; this might be due to the throttle servo.
- Spark advance is shown to be faster than throttle action. At 10 Hz, there is a 60° difference in phase. A bandwidth of 7-8 Hz can be anticipated with throttle actuation and a bandwidth of 8-9 Hz with spark advance actuation, but it is only a marginal difference.

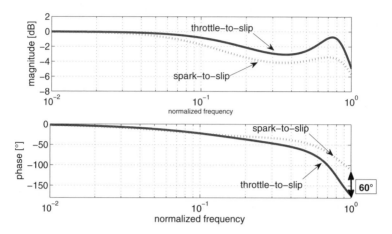

Fig. 3.4 Spark-advance-to-slip and throttle-to-slip dynamics. The low frequency gains are normalized to 1.

- Although the spark advance allows a slightly faster actuation, the response of the system to the spark advance variation is generally less linear, and therefore more difficult to model and control.

The two obtained linear models are thus very useful for designing traction control systems by using actuation via throttle and/or via spark advance.

3.4 *Example 3:* Estimation-Oriented Identification for Sensor Reduction in Semi-active Suspension Systems for Cars ([10])

This example focuses on the semi-active suspension control for a four-wheeled vehicle (see [18] for a survey), aiming at employing a reduced sensor layout. As damping control, the Mix-1-Sensor algorithm presented in [15] [17] will be considered. The goal is then to provide a model to estimate the body accelerations of the four corners on the basis of three sensors. The motivation of this work is that, even if pricing and setup of three sensors is nearly the same as four for a single vehicle, the fourth sensor might represent a huge additional cost for series production.

The model employed for estimation is the following:

$$\widehat{\ddot{z}_4}(t) = a\ddot{z}_1(t) + b\ddot{z}_2(t) + c\ddot{z}_3(t). \tag{3.6}$$

where $\widehat{\ddot{z}_4}(t)$ is the estimated value of the vertical acceleration $\ddot{z}_4$ of the fourth corner, $\ddot{z}_i(t)$ is acceleration measured by the i-th accelerometer and a, b and c are parameters to be identified. Note that model 3.6 is a linear and algebraic model and it is based on the assumption that the accelerometers 1, 2 and 3 define a unique plane π. The accelerometer 4 is eventually translated from its ideal position on plane π. Moreover, the plane is assumed to be rigid (see Figure 3.5).

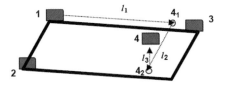

Fig. 3.5 Graphical representation of the sensor layout.

In order to identify the parameters in (3.6), the following cost function is employed:

$$J(a, b, c) = \frac{\sum_{t=1}^{N} \left(\widehat{\ddot{z}}_4(t) - \ddot{z}_4(t) \right)^2}{\sum_{t=1}^{N} \left(\ddot{z}_4(t) \right)^2}. \tag{3.7}$$

The minimization of (3.7) can be performed by means of simple least squares techniques (see [16]). Nevertheless, the estimation of the fourth acceleration suffers from high frequency noise that characterizes the other measurements. Further, at high frequencies, the hypothesis of rigid chassis no longer holds and resonant modes appear. However, in [10] it is demonstrated by experimental results that spurious contributions at high frequencies are suitably filtered by the closed loop control, without adding pre-filters with undesirable phase shifting.

In Figure 3.6, the real acceleration signal collected with the 4th accelerometer is compared with the estimation (3.6) obtained by using the parameters resulting from the identification procedure. Figure 3.7 reports instead the approximate transfer function from the road profile to the driver acceleration obtained if different semi-active strategies are applied to a multibody simulator. From inspecting the figure, it is clear that the degrades of performances are negligible with respect to the advantage of a sensor reduction in the system layout.

3.5 *Example 4:* Direct Braking Control Design for Two-Wheeled Vehicles ([11])

Designing effective braking controllers for two-wheeled vehicles is a very challenging task, due to the complex vehicle dynamics and the highly nonlinear nature of the road-tire interactions . These facts make it difficult to devise control-oriented dynamical models capable of describing the dynamics of interest in all working conditions (for a complete overview of the problem,

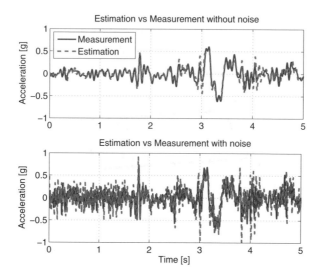

Fig. 3.6 Estimation of the 4th acceleration signal and comparison with real experimental data: noiseless and noisy case.

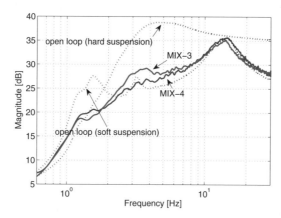

Fig. 3.7 Approximate transfer function from the road vertical profile to driver vertical acceleration. Comparison between over-damped suspensions, under-damped suspensions, vehicle controlled with Mix-1-sensor (Mix-4) and vehicle controlled with reduced sensor layout (Mix-3).

see [20]). This example proposes a model-free approach to solve the problem (see [11] for further details).

Let the control variable and the output signal be, respectively, the braking torque and a convex combination of the normalized wheel deceleration η and λ, as in the standard Mixed-Slip-Deceleration (MSD) braking control framework (see [19]), *i.e.*,

$$\varepsilon = \alpha\lambda + (1 - \alpha)\eta, \qquad \alpha \in [0, 1], \tag{3.8}$$

where $\alpha = 0.9$. By choosing to control ε, instead of λ, it is possible to guarantee that there always exists a unique equilibrium for all road conditions and that output measurements are not too noisy (see [19]).

The overall control scheme is illustrated in Figure 3.8. The intelligent action is built up by following the idea in [21]. Assume that the following ultra-local model of the plant is valid for each time instant:

$$\dot{y}(t) = F(t) + \beta u(t), \tag{3.9}$$

where β is a (non-physical) constant parameter, such that $\beta u(t)$ and $\dot{y}(t)$ are of the same magnitude. Starting from this description, a data-driven feedforward compensation of the nonlinear dynamics can be derived as:

$$u_i(t) = \frac{\dot{y}^o(t)}{\beta} - \frac{\widehat{F}(t)}{\beta}$$
$$\widehat{F}(t) = \widehat{\dot{y}}(t) - \beta u(t),$$

where $\dot{y}^o(t)$ is the time derivative of the reference trajectory and $\widehat{F}(t)$ is an approximation of $F(t)$ at time instant t, computed by using an estimate $\widehat{\dot{y}}(t)$ of the derivative of the output signal. According to [21], the complete control law needs a Proportional-Integral (PI) action in order to guarantee robustness

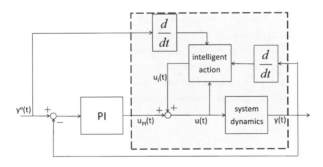

Fig. 3.8 Closed-loop control scheme with nonlinear data-driven compensator and VRFT-synthesized PI.

with respect to errors in the dynamics compensation. Defining the tracking error $e(t)$ as $e(t) = y^o(t) - y(t)$, the final expression of $u(t)$ is given by

$$u(t) = u_i(t) + u_{PI}(t) = u_i(t) + K_P e(t) + K_I \int_{t_0}^{t} e(t)dt, \qquad (3.10)$$

where K_P and K_I are the proportional and integral gains. Thanks to the Virtual Reference Feedback Tuning (VRFT) approach (see [22]), this PI controller can be easily tuned by translating the control design into an identification problem.

As explained in [22], the controller can be identified by a double set of noisy data collected in an open-loop setting. In detail, the VRFT synthesis solves a model-reference problem by minimizing from data the cost function:

$$J_{MR}(\theta) = \left\| \frac{P(z)C(z,\theta)}{1 + P(z)C(z,\theta)} - M(z) \right\|_2^2 \qquad (3.11)$$

where $P(z)$ is the process under control, $\mathcal{C} = \{C(z,\theta) , \theta \in \mathcal{R}^n\}$ is the considered class of controllers, and $M(z)$ a given target closed-loop behaviour. In view of the VRFT approach, the best controller with the requested structure is the one that minimizes the variance of the error between the input signal $u_{PI}(t) = u(t) - u_i(t)$ and the input that the controller generates when fed by $e_V(t) = (M(z)^{-1} - 1) y(t)$. Formally, the actual cost criterion minimized by the VRFT algorithm is the following:

$$J_{VR}^N(\theta) = \frac{1}{N} \sum_{t=1}^{N} (u_{PI}(t) - C(z,\theta)e_V(t))^2, \qquad (3.12)$$

and, in [22], it is proved that (3.11) and (3.12) are the same if some assumptions hold and data are suitably pre-filtered.

Performing a closed-loop simulation on a multibody motorcycle software package, the closed-loop step response shown on the left side of Figure 3.9 can be obtained if the same conditions of the identification experiment are considered. The right side of Figure 3.9 shows instead the closed-loop responses obtained on different road conditions (different longitudinal friction coefficient μ). As could be expected, it ensures repeatable tracking performance on all conditions, while the action of a simple PI (tuned via VRFT) cannot effectively handle this model variation.

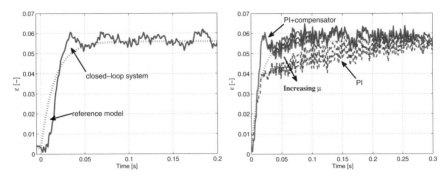

Fig. 3.9 Left: step response of ε with the applied nonlinear control structure. Right: sensitivity to the longitudinal friction μ: reference signal (dotted lines), PI (dashed lines) and data-driven control (solid lines).

3.6   Conclusions

Designing a chassis control system is in practice a rather intricate problem, as vehicle dynamics are complex to model and some important physical parameters are unknown. Data-driven techniques allows the control engineer to overcome the foregoing problems with mild experimental and computational effort.

This contribution did not aim to be (and is far from being) an exhaustive or complete account of all these problems. On the other hand, the proposed examples have been chosen as they are illustrative of the great potential and the challenging aspects of data-driven techniques when applied to modeling, estimation and control design in vehicle chassis control.

Specifically, four different sub-problems in this research area have been reviewed: yaw control, traction control, semi-active suspension control and braking control. In all cases, system identification theory allows one to fill the lack of knowledge about the system with simple and reliable numerical techniques.

Acknowledgements. This work has been partially supported by the Austrian Center of Competence in Mechatronics (ACCM), the MIUR project "New methods for Identification and Adaptive Control for Industrial Systems" and the FIRB project "Highly innovative motorbikes with ultra-low emission engine, active suspensions, electronic brakes and new materials".

The authors would also like to thank all the people who worked on the contributions presented in this note: Matteo Corno (examples 1 and 2), Mara Tanelli (examples 1 and 4), Pierpaolo De Filippi (example 4), Cristiano Spelta, Diego Delvecchio, Gabriele Bonaccorso and Fabio Ghirardo (example 3).

References

[1] Hjalmarsson, H.: From experiment design to closed-loop control. Automatica 41(3), 393–438 (2005)

[2] Gevers, M.: Identification for control. Annual Reviews in Control 20, 95–106 (1996)

[3] Daily, R., Bevly, D.M.: The use of GPS for vehicle stability control systems. IEEE Transactions on Industrial Electronics 51(2), 170–177 (2004)

[4] Rajamani, R., Hedrick, J.K.: Adaptive observers for active automotive suspensions: Theory and experiment. IEEE Transactions on Control Systems Technology 4(3), 86–93 (1995)

[5] Wesemeier, D., Isermann, R.: Identification of vehicle parameters using stationary driving maneuvers. In: 5th IFAC Symposium on Advances in Automotive Control, Seascape Resort, USA (2007)

[6] Fischer, D., Kaus, E., Isermann, R.: Fault detection for an active vehicle suspension. In: American Control Conference, Denver, Colorado, USA (2003)

[7] Ryu, J., Rossetter, E.J., Gerdes, J.C.: Vehicle sideslip and roll parameter estimation using GPS. In: International Symposium on Advanced Vehicle Control (AVEC), Hiroshima, Japan (2002)

[8] Corno, M., Tanelli, M., Boniolo, I., Savaresi, S.M.: Advanced Yaw Control of Four-wheeled Vehicles via Rear Active Differential Braking. In: 28th IEEE Conference on Decision and Control, Shanghai, P.R. China (2009)

[9] Corno, M., Savaresi, S.M.: Experimental Identification of Engine-to-Slip Dynamics for Traction Control Applications in a Sport Motorbike. European Journal of Control 16(1), 88–108 (2010)

[10] Spelta, C., Delvecchio, D., Savaresi, S.M., Bonaccorso, G., Ghirardo, F.: Analysis of a sensor reduction in a semi-active suspension system for a 4-wheels vehicle. In: 3rd ASME Annual Dynamic Systems and Control Conference (DSCC), Boston, MA, USA

[11] Formentin, S., De Filippi, P., Tanelli, M., Savaresi, S.M.: Model-free control for active braking systems in sport motorcycles. In: 8th IFAC Symposium on Nonlinear Control Systems (NOLCOS), Bologna, Italy (2010)

[12] Rajamani, R.: Vehicle dynamics and control. Springer, Heidelberg (2006)

[13] Ljung, L.: System identification: theory for the user. Prentice-Hall, Englewood Cliffs (1987)

[14] Dennis, J.E., Schnabel, R.B.: Numerical methods for unconstrained optimization and nonlinear equations. Prentice-Hall, Englewood Cliffs (1983)

[15] Savaresi, S.M., Spelta, C.: Mixed Sky-Hook and ADD: Approaching the Filtering Limits of a Semi-Active Suspension. ASME Transactions: Journal of Dynamic Systems, Measurement and Control 129(4), 382 (2007)

[16] Boyd, S., Vandenberghe, L.: Convex Optimization. Cambridge University Press, Cambridge (2004)

[17] Savaresi, S.M., Spelta, C.: A Single Sensor Control Strategy for Semi-Active Suspension. IEEE Transaction on Control System Technology 17(1), 143–152 (2009)

[18] Savaresi, S.M., Poussot-Vassal, C., Spelta, C., Sename, O., Dugard, L.: Semi-active suspension control for vehicles. Elsevier (Butterworth-Heinemann), UK (2010)

[19] Savaresi, S.M., Tanelli, M., Cantoni, C.: Mixed slip-deceleration control in automotive braking systems. ASME Transactions: Journal of Dynamic Systems, Measurement and Control 129(1), 20–31 (2007)

[20] Savaresi, S.M., Tanelli, M.: Active braking control systems for vehicles. Springer, London (2010)

[21] Fliess, M., Join, C.: Model-free control and intelligent PID controllers: towards a possible trivialization of nonlinear control? In: IFAC Symposium on System Identification (SYSID), Saint-Malo, France (2009)

[22] Campi, M.C., Lecchini, A., Savaresi, S.M.: Virtual Reference Feedback Tuning: a direct method for the design of feedback controllers. Automatica 38, 1337–1346 (2002)

[23] Burckhardt, M.: Fahrwerktechnik: Radschlupfregelsysteme. Vogel-Verlag (1993)

Part II
Suitable Identification Methods

Chapter 4
Linear Parameter-Varying System Identification: The Subspace Approach

M. Corno, J.-W. van Wingerden, and M. Verhaegen

Abstract. In the past two decades, a significant amount of research has been carried out on Linear Parameter-Varying (LPV) systems. It has been shown that LTI control synthesis techniques like optimal and robust control can be extended to the LPV case. Notwithstanding the advances in LPV control, the identification of such systems is still not completely developed. The scope of this paper is to present the problem of LPV system identification. Recent results on subspace identification of LPV systems are presented and applied to the identification of a car lateral dynamics.

4.1 Introduction

In recent years the Linear Parameter Varying (LPV) framework received considerable attention from the systems and control community. The success of the LPV framework is mainly due to its capacity of describing nonlinear systems while maintaining some of the linear systems properties. Thanks to these properties, it has been possible to extend control synthesis techniques originally developed for Linear Time-Invariant systems (such as optimal and robust control) to LPV systems [12, 2]. LPV models, roughly speaking, can be defined as linear systems where, either the matrices of the state equations of the coefficients of the input-output relation, depend on one or more time varying parameters. LPV control synthesis techniques have been successfully applied to a number of application domains: aerospace [1], wind industry [4] industrial processes, and automotive control [5].

One of the challenges involved in LPV control is the derivation of the LPV models themselves. Most of the times, they are derived via first principle (FP)

M. Corno · J.-W. van Wingerden · M. Verhaegen
Delft Center for Systems and Control (DCSC), Delft University of Technology,
Mekelweg 2, 2628 CD, Delft, The Netherlands
e-mail: {m.corno,j.w.vanwingerden,m.verhaegen}@tudelft.nl

D. Alberer et al. (Eds.): Identification for Automotive Systems, LNCIS 418, pp. 53–65.
springerlink.com   © Springer-Verlag London Limited 2012

modeling. However, deriving FP models requires a considerable effort; and in some cases a system identification approach is to be preferred. In system identification, measured input and output data are used to obtain a mathematical description of the system. Because this approach uses measured data, it has the potential to considerably simplify the modeling of complex systems. The scope of the document is that of describing the available approaches to the LPV identification problem, in particular attention will be given to recent advances in the field of subspace identification.

The outline of the paper is as follows; in Section 4.2 an overview of LPV identification is given. In Section 4.3 a predictor based algorithm suitable for subspace identification of MIMO LPV systems in open or closed loop is detailed. The usage and potentials of the above algorithm are shown in Section 4.3 where the lateral dynamics of a four wheeled vehicle is identified. The paper is closed with some remarks in Section 4.5.

4.2 LPV System Identification Overview

The goal of LPV system identification in that of obtaining an LPV model directly from input output data. The available approaches can be classified along two directions: structure of the model and the scope of the experiment.

Contrary to the LTI case where the input-output (I/O) and state space (SS) representations of a system are equivalent, in the LPV case [15] there exist no input-output equivalent statically scheduling parameter dependent transformation between the I/O and the SS representations. The choice of the LPV model structure is therefore important. Often the decision is made on the basis of the planned use of the LPV model.

LPV input-output Models. A discrete time Linear Parameter-Varying Input-Output model (LPV-I/O) is defined as:

$$y_k = -\sum_{i=1}^{n} a_i(\mu_k)y_{k-i} + \sum_{i=1}^{n} b_i(\mu_k)u_{k-i}$$

where k is the discrete time index, $\mu_k \in \mathbb{R}^M$ is the scheduling parameter vector and $u_k \in \mathbb{R}^m$, $y_k \in \mathbb{R}^l$ are the input and output signals. The coeffcients $\{a_i(\mu_k)\}$, $\{b_i(\mu_k)\}$ are statically dependent on the scheduling parameter μ_k.

LPV State Space Models. A discrete time Linear Parameter-Varying State-Space model (LPV-SS) is usually defined as:

$$x_{k+1} = A(\mu_k)x_k + B(\mu_k)u_k$$
$$y_k = C(\mu_k)x_k + D(\mu_k)u_k$$

where $\mu \in \mathbb{R}^M$ is the scheduling parameter vector, $x_k \in \mathbb{R}^n$, $u_k \in \mathbb{R}^m$, $y_k \in \mathbb{R}^l$ are the state, input and output vectors. System matrices may depend on the scheduling parameter in different ways. One of the most common structures is the Affine Parameter Dependence (LPV-A). In this representation the time-varying system matrices depend linearly on the scheduling:

$$A(\mu_k) = \sum_{i=1}^{n} \mu_k^i A_i$$

and similarly for B, C, D. The scheduling parameter vector is assumed to have the form: $\mu_k = \begin{bmatrix} 1 & \mu_k^2 & ... & \mu_k^M \end{bmatrix}$

More rencently a third structure has been proposed under the name of Orthonormal Basis Functions Models [14]. Although being theoretically well founded, the LPV-OBF approach is rather complex and practical applications are not yet found in literature.

The second axis of classification is the scope of the experiment. Depending on the trajectories of the scheduling parameter, it is possible to classify the identification technique into Local or Global Methods.

Local Methods. In the Local Approach, multiple series of input-output measurements are carried out keeping the scheduling parameters constant. This yields a family of Linear Models that then can be interpolated. As shown in [14], the interpolation could give rise to errors due to the fact that the local models may be in different bases. Different methods have been devised to address this problem, see for example [13, 11]. In local methods there is no information on what happens in between the chosen set of constant operating conditions. As a consequence, the obtained LPV model does not represent the dynamic behaviour of the real system during fast scheduling variations.

Global Methods. Global methods can provide more reliable and accurate models at the cost of complex persistency of excitation conditions on both the input and the scheduling parameter sequences. This may complicate the practical application of these methods, as in many applications it is not possible to freely choose all of the scheduling parameter signals. Several global methods have been devised both in the Input-Output approach [3, 20] and in the State Space approach [10, 9]. As it will be shown, among the available methods, Global LPV-SS subspace identification offers considerable advantages.

4.3   LPV Subspace Identification

Subspace identification techniques are well-known in the LTI setting, see for example [19]. The advantages of the subspace approach are multiple. There is no need to parameterize the state-space model; they provide a simple and effective way to estimate the system order and are easily adapted to closed-loop identification. Early attempts to carry out LPV identification,

[17], suffered from a very high numerical complexity; this limitation has been reduced thanks for some recent developments [16].

To understand the above methods, refer to the innovation form representation of an LPV-A system:

$$x_{k+1} = A(\mu_k)x_k + B(\mu_k)u_k + K(\mu_k)e_k$$
$$y_k = C(\mu_k)x_k + D(\mu_k)u_k + e_k$$

where $x_k \in \mathbb{R}^n$, $u_k \in \mathbb{R}^r$, $y_k \in \mathbb{R}^l$ are the state, input and output vectors and $e_k \in \mathbb{R}^l$ is the zero mean white noise innovation process. The matrices $A(\mu_k)$, $B(\mu_k)$, $C(\mu_k)$, $D(\mu_k)$ and $K(\mu_k)$ are assumed to be affinely dependent on μ_k as defined in Section 4.2. In the following, for sake of brevity, it will be assumed that C and D are parameter independent.

The innovation form can be easily transformed into the predictor form:

$$x_{k+1} = \tilde{A}(\mu_k)x_k + \tilde{B}(\mu_k)u_k + K(\mu_k)y_k \qquad (4.1)$$
$$y_k = C(\mu_k)x_k + D(\mu_k)u_k + e_k \qquad (4.2)$$

where

$$\tilde{A}(\mu_k) = A(\mu_k) - K(\mu_k)C(\mu_k) \quad \text{and} \quad \tilde{B}(\mu_k) = B(\mu_k) - K(\mu_k)D(\mu_k)$$

The objective is that of estimating the matrices $\{A_i, B_i, K_i\}_{i=1}^M$, C and D from the input, output and scheduling sequences up to a global similarity transformation. By defining $z_k = \begin{bmatrix} u_k^T & y_k^T \end{bmatrix}^T$ and $\breve{B}(\mu_k) = \begin{bmatrix} \tilde{B}(\mu_k) & K(\mu_k) \end{bmatrix}$, the state equation (4.1) can be written as:

$$x_{k+1} = \tilde{A}(\mu_k)x_k + \breve{B}(\mu_k)z_k$$

Hence, with a past window p, the state predictor x_{k+p} is given by:

$$x_{k+p} = \underbrace{\tilde{A}(\mu_{k+p-1})\tilde{A}(\mu_{k+p-2})...\tilde{A}(\mu_k)}_{\phi_{p,k}} x_k +$$

$$+ \underbrace{\begin{bmatrix} \phi_{p-1,k+1}\breve{B}_k & \cdots & \phi_{1,k+p-1}\breve{B}_{k+p-2} & \breve{B}_{k+p-1} \end{bmatrix}}_{\overline{\mathcal{K}}_{k,p}} \underbrace{\begin{bmatrix} z_k \\ \vdots \\ z_{k+p-2} \\ z_{k+p-1} \end{bmatrix}}_{\bar{z}_{k,p}}.$$

The time-varying extended controllability matrix $\overline{\mathcal{K}}_{k,p}$ can be factorized in a time-invariant extended controllability matrix $\mathcal{K}_p$ depending only on $\{A_i, B_i, K_i\}_{i=1}^M$ and a matrix $N_{k,p}$ depending only on μ_k:

$$\bar{B}_i = \begin{bmatrix} \tilde{B}_i \ K_i \end{bmatrix}$$

$$\mathcal{L}_i = \begin{cases} \begin{bmatrix} \bar{B}_1 \ \bar{B}_2 \ \ldots \ \bar{B}_M \end{bmatrix} & \text{for} \quad i = 1 \\ \begin{bmatrix} \tilde{A}_1 \mathcal{L}_{i-1} \ \tilde{A}_2 \mathcal{L}_{i-1} \ \ldots \ \tilde{A}_M \mathcal{L}_{i-1} \end{bmatrix} & \text{for } i = 2, \ldots, p \end{cases} \tag{4.3}$$

$$\overline{\mathcal{K}}_p = \begin{bmatrix} \mathcal{L}_p \ \mathcal{L}_{p-1} \ \ldots \ \mathcal{L}_1 \end{bmatrix}$$

$$P_{p,k} = \mu_{k+p-1} \otimes \mu_{k+p-2} \otimes \ldots \otimes \mu_k \otimes I_{m+l}$$

$$N_{k,p} = diag(P_{p,k}, P_{p-1,k+1}, \ldots, P_{1,k+p-1})$$

$$\overline{\mathcal{K}}_{k,p} = \mathcal{K}_p N_{k,p}$$

where $\otimes$ represents the Kronecker product. The key approximation in this algorithm is that we assume that $\phi_{p,k} \approx 0$ for $j \geq p$. It can be shown that if the system is uniformly exponentially stable the approximation error can be made arbitrarily small by making p large, [17]. With this assumption the state x_{k+p} is given by:

$$x_{k+p} \approx \mathcal{K}_p N_{k,p} \bar{z}_{k,p}. \tag{4.4}$$

From here the input-output behavior can be approximated by:

$$y_{k+p} \approx C\mathcal{K}_p N_k^p \bar{z}_{k,p} + e_{k+p} := \hat{y}_{k+p} \tag{4.5}$$

The following matrices can be constructed from the measured data:

$$U = \begin{bmatrix} u_{p+1} \ \ldots \ u_N \end{bmatrix} \ Y = \begin{bmatrix} y_{p+1} \ \ldots \ y_N \end{bmatrix}$$
$$Z = \begin{bmatrix} N_{0,p}\bar{z}_{0,p} \ \ldots \ N_{N-p+1,p}\bar{z}_{N-p-f+1,p} \end{bmatrix}$$

Note that the number of rows of Z_i is $(r+l)\sum_{j=1}^{p} M^j$, it grows more than exponentially with the past window p.

Let us now define the matrix Γ_p as the extended observability matrix of the first local model, and the state sequence matrix X

$$\Gamma_p = \begin{bmatrix} C \\ C\tilde{A}_1 \\ \vdots \\ C\left(\tilde{A}_1\right)^{p-1} \end{bmatrix}, \quad X = \begin{bmatrix} x_p \ \cdots \ x_N \end{bmatrix}, \tag{4.6}$$

then from (4.4) it follows that

$$\Gamma_p X \approx \Gamma_p \mathcal{K}_p Z. \tag{4.7}$$

The matrix $\Gamma_p \mathcal{K}_p Z$ can be estimated from data either with a linear regression method or with the more efficient kernel method.

The linear regression method estimates the matrices $C\mathcal{K}_p$ and D by minimizing the prediction error $\varepsilon = y_k - \hat{y}_k$. From (4.5) it can be done by solving the linear regression problem:

$$\min_{CK_p \, D} \|Y - CK_p Z - DU\|_F^2 . \tag{4.8}$$

For finite p, approximation (4.4) does not perfectly hold and thus the estimate is affected by a bias. This bias can be made arbitrarily small by increasing p. For large p, the matrix $\begin{bmatrix} Z^T & U^T \end{bmatrix}$ tends to lose row ranks, and the optimization problem (4.8) will not have a unique solution. One way to circumvent the problem is to choose the solution with minimum Frobenius norm, *i.e.* $\min \|CK_p \, D\|_F^2$. It is found by using a Singular Value Decomposition (SVD):

$$\begin{bmatrix} Z \\ U \end{bmatrix} = \begin{bmatrix} \mathcal{U} \; \mathcal{U}_\perp \end{bmatrix} \begin{bmatrix} \Sigma & 0 \\ 0 & 0 \end{bmatrix} \begin{bmatrix} \mathcal{V}^T \\ \mathcal{V}_\perp^T \end{bmatrix}, \quad \begin{bmatrix} \widehat{CK_p} \\ \widehat{D} \end{bmatrix} = Y \mathcal{V} \Sigma^{-1} \mathcal{U}^T. \tag{4.9}$$

Once CK_p has been estimated, it can be used to construct $\widehat{\Gamma_p K_p} Z$. From (4.3) and (4.6) it follows that:

$$\Gamma_p K_p \approx \begin{bmatrix} C\mathcal{L}_p & C\mathcal{L}_{p-1} & \cdots & C\mathcal{L}_1 \\ 0 & C\tilde{A}_1 \mathcal{L}_{p-1} & \cdots & C\tilde{A}_1 \mathcal{L}_1 \\ & & \ddots & \vdots \\ 0 & & \cdots & C\left(\tilde{A}_1\right)^{p-1} \mathcal{L}_1 \end{bmatrix} := \widehat{\Gamma_p K_p}$$

$\widehat{\Gamma_p K_p}$ is constructed form $\widehat{CK_p}$ by noting that the first row of $\Gamma_f K_p$ is CK_p

$$CK_p = \begin{bmatrix} C\mathcal{L}_p & C\mathcal{L}_{p-1} & \cdots & C\mathcal{L}_1 \end{bmatrix},$$

moreover it can be observed that for each combination of i, j

$$C\left(\tilde{A}_1\right)^{i-1} \mathcal{L}_{j+1} = \begin{bmatrix} C\left(\tilde{A}_1\right)^i \mathcal{L}_j & \cdots & C\left(\tilde{A}_1\right)^{i-1} \tilde{A}_M \mathcal{L}_j \end{bmatrix}$$

holds.

The main issue of this method is the size of the data matrix Z, in fact the number of rows of Z grows exponentially with p but on the other hand a large p is needed to reduce the bias. The complexity can be reduced by introducing the kernel method. The kernel method was first introduced in [18]; it assumes that the solution to (4.8) can be written as:

$$[CK_p \; D] = \alpha \begin{bmatrix} Z^T & U^T \end{bmatrix}$$

If this holds then the dual optimization problem can be written as:

$$\min_{\alpha} \|\alpha\|_F^2 \text{ with } Y - \begin{bmatrix} Z^T Z + U^T U \end{bmatrix} = 0.$$

If $\begin{bmatrix} Z^T & U^T \end{bmatrix}$ has full rank, then the solution of the dual problem is

$$\alpha = Y \left(Z^T Z + U^T U \right)^{-1} = Y \mathcal{V} \Sigma^{-2} \mathcal{V}^T$$

The above expression confirms the correctness of the initial assumption as

$$\begin{bmatrix} C\mathcal{K}_p & D \end{bmatrix} = \alpha \begin{bmatrix} Z^T & U^T \end{bmatrix} = \left(Y \mathcal{V} \Sigma^{-2} \mathcal{V}^T \right) \left(\mathcal{V} \Sigma \mathcal{U}^T \right) = Y \mathcal{V} \Sigma^{-1} \mathcal{U}^T$$

thus yielding the same results as (4.9). Note that to solve the dual problem only $Z^T Z$ is needed, which can be computed without computing Z:

$$Z^T Z = \sum_{j=0}^{p} \left(\left(\prod_{v=0}^{p-j} \mu_{N+v+j-1}^T \mu_{N+v+j-1} \right) \left(z_{N+j-1}^T z_{N+j-1} \right) \right)$$

It is interesting to note that for $N \gg p$ and $N \gg m$, the computational complexity of the indirect computation of $Z^T Z$ is of order $\mathcal{O}(N^2 \sum_{j=1}^{p} m^j)$ whereas the direct method yields an overall complexity of $\mathcal{O}(N^2)$, thus rendering the problem computationally tractable. The kernel matrix $\begin{bmatrix} Z^T Z + U^T U \end{bmatrix}$ $\in \mathbb{R}^{N \times N}$ is likely to be ill-conditioned, the issue can be solved by regularization as shown in [16].

Finally, given the predictor form (4.1), (4.2) and the definition of $P_{p,k}$ we have that

$$C \left(\tilde{A} \right)^i \mathcal{K}_p = \alpha_i Z_0^T.$$

The estimate of $\Gamma_p \mathcal{K}_p Z_0$ can now be constructed as follows:

$$\widehat{\Gamma_p \mathcal{K}_p Z_0} = \alpha Z_0^T Z_0$$

Once $\widehat{\Gamma_p \mathcal{K}_p Z}$ is obtained, the state sequence can be finally estimated using an SVD and approximation (4.7) in a similar way as it is done in the LTI subspace identification. The state sequence is

$$\widehat{X} = \Sigma_n \mathcal{V}$$

where

$$\widehat{\Gamma_p \mathcal{K}_p Z} = \begin{bmatrix} \mathcal{U} & \mathcal{U}_\perp \end{bmatrix} \begin{bmatrix} \Sigma_n & 0 \\ 0 & \Sigma_0 \end{bmatrix} \begin{bmatrix} \mathcal{V} \\ \mathcal{V}_\perp \end{bmatrix}.$$

In the above expression matrix Σ_n contains the n largest singular values, where n is the order of the system. The order of the system can be determined by searching for gaps in the singular values sequence.

Once the state, input and scheduling sequences are known the system matrices are determined by solving two more linear least squares problems.

In this section an algorithm for LPV subspace identification has been detailed. The use of the kernel method and the dual optimization problem reduces the numerical complexity of the method. The proposed approach has two main advantages: (1) as it does not require any hypothesis on the correlation properties of the noise, it can be directly used for closed-loop system identification and (2) it does not need any assumption on the model order; an accurate estimation of the model order is indeed provided by the method itself.

4.4 Simulation Example

In this Section the LPV subspace identification approach is applied to the lateral dynamics of a car. Recently, several model-based yaw stability control systems have been proposed [7, 6]; the development and tuning of such controllers require accurate models. Deriving an FP model of the car often requires elaborate measurements on the inertia properties of the vehicle and tire characteristics. The possibility of directly deriving those models from measurements would render the design of such controllers easier.

4.4.1 Analytical LPV Modeling

The most popular model for lateral dynamics [8] is the so called single-track model. Despite its simplicity the model is able to describe the vehicle lateral behavior under many conditions. The equations of motion are:

$$mV_x \left(\dot{\beta} + r \right) = F_{yf} + F_{yr}$$
$$I_z \dot{r} = l_f F_{yf} - l_r F_{yr} + M_z$$

where m is the vehicle mass, I_z is the yaw moment of inertia, V_x is the vehicle longitudinal velocity, M_z is the yaw moment input (representing differential braking), F_{yr} is the rear lateral force, F_{yr} is the front lateral force and l_f and l_r are the distances between the vehicle center of gravity and the front and rear tires.

Under the assumption of small tire slip angles, front and rear lateral forces can be described by:

$$F_{yf} = c_f \alpha_f, \quad F_{yf} = c_r \alpha_r$$

where c_f and c_r are the cornering stiffnesses. The front and rear side slip angles (α_f, α_r) are defined as:

$$\alpha_f = \delta - \beta - r l_f / V_x, \quad \alpha_r = -\beta + r l_r / V_x,$$

with δ the steering angle. By defining the state vector $x = \begin{bmatrix} r & \beta \end{bmatrix}^T$, the state space model can be written as

$$\begin{bmatrix} \dot{r} \\ \dot{\beta} \end{bmatrix} = \begin{bmatrix} -\frac{l_f^2 c_f + l_r^2 c_r}{I_z V_x} & \frac{l_r c_r + l_f c_f}{I_z} \\ -1 + \frac{l_r c_r + l_f c_f}{m V_x^2} & -\frac{c_f + c_r}{m V_x} \end{bmatrix} \begin{bmatrix} r \\ \beta \end{bmatrix} + \begin{bmatrix} \frac{l_f c_f}{I_z} & -\frac{1}{I_z} \\ \frac{c_y}{m V_x} & 0 \end{bmatrix} \begin{bmatrix} \delta \\ M_z \end{bmatrix}. \qquad (4.10)$$

The above model can be written as a continuous time LPV-A model by considering the scheduling vector $\mu = \begin{bmatrix} 1/V_x & 1/V_x^2 \end{bmatrix}^T$. In light of this scheduling vector augmentation the model can be written as:

$$\begin{bmatrix} \dot{r} \\ \dot{\beta} \end{bmatrix} = \left\{ A_0 + \frac{1}{V_x} A_1 + \frac{1}{V_x^2} A_2 \right\} \begin{bmatrix} r \\ \beta \end{bmatrix} + \left\{ B_0 + \frac{1}{V_x} B_1 \right\} \begin{bmatrix} \delta \\ M_z \end{bmatrix}.$$

The LPV system given in (4.10) is used to obtain the input, output, and scheduling sequences for the identification algorithm. We assume that the scheduling variable V_x can be chosen arbitrarily (with realistic limitations on the acceleration) and that only yaw rate measurements are available.

Note that the subspace LPV identification algorithm deals with discrete time systems, when the system at hand is discretized the affine LPV structure is lost; nevertheless it is still possible to approximate the discrete time model with an affine LPV structure. To evaluate the effects of sampling, a simulation study is carried out. The system has been discretized via Tustin for various sampling periods and then approximated with two different types of affine dependence. In the *linear case* the same dependence on $1/V_x$ and $1/V_x^2$ as in the continuous case is assumed; in the *quadratic case* the scheduling parameter vector is augmented to $1/V_x$, $1/V_x^2$, $1/V_x^3$ and $1/V_x^4$ so to achieve a better approximation of the discrete time parameter dependence. The results, in terms of Variance Accounted For (VAF), are plotted in Fig. 4.1. The VAF values is defined as:

$$VAF(y_k, \widehat{y}_k) = \max \left\{ 1 - \frac{\text{var}(y_k - \widehat{y}_k)}{\text{var}(y_k)}, 0 \right\} \cdot 100$$

where y and $\hat{y}$ are respectively the measured output and the simulated one. In the simulation the vehicle longitudinal velocity is varied sinusoidally with a frequency of 0.07 Hz between 10 m/s and 30 m/s (corresponding to a maximum acceleration of 0.5g); the steering input is a frequency sweep from 0.01 Hz to 10 Hz in 100 s. The analysis confirms that a sampling time of $T = 0.01$s guarantees a reasonable level of approximation.

4.4.2 Simulation Results

In order to provide indications on the usage of the algorithm, in the following the effect of three parameters will be evaluated: the past window p, the noise affecting the measurement and the number of available samples. In order to provide comparable results the same identification protocol will be employed in all the analyses. The continuous time system is simulated with a sinusoidally varying speed as before, in the meantime the inputs are excited with two frequency sweeps. The steering angle is varied between -4° and 4° (at the wheel) with a frequency sweep; similarly, the yaw moment input is excited with a frequency sweep; by choosing a slightly different frequency range and introducing a phase shift it is possible to guarantee the persistency of excitation requirements.

Figure 4.2 plots the position of the identified frozen velocity poles for three different values of the past window for noiseless data. As pointed out in the previous section, a small p will lead in general to biased estimates while for large p, even without noise, the variance will increase due to the curse of dimensionality. This analysis indicates that for the system at hand a choice of $p = 4$ is good.

A similar effect on the estimation bias can be seen in the experiment length. To evaluate the effects of the number of available samples, the identification measurements were repeated for different number of velocity cycles; the results are plotted in Fig. 4.3. The higher the number of cycles the more accurate results one can get. Note that as many as 10 cycles are enough to obtain an accurate estimation.

The final considerations are on the effect of noise. In the following tests a white noise zero mean signal is added to the simulated output. The performances of the identification algorithm are assessed on a noiseless validation set with a scheduling vector with a different frequency and band limited white

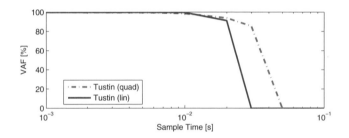

Fig. 4.1 VAF between the continuous time model and the model obtained via Tustin linearization with two different augmented scheduling vectors.

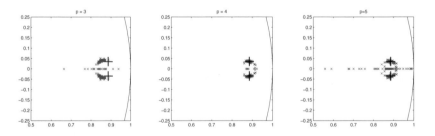

Fig. 4.2 Eigenvalues of the estimated A for $V_x = 30$ m/s in one plot for 100 experiments for different values of the past window p. The big crosses correspond to the real values of the eigenvalues of the matrices.

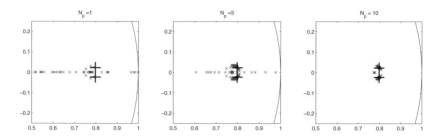

Fig. 4.3 Eigenvalues of the estimated A for $V_x = 30$ m/s in one plot for 100 experiments for different values of the number of periods N_p. The big crosses correspond to the real values of the eigenvalues of the matrices.

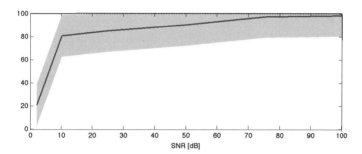

Fig. 4.4 The mean VAF of a fresh data set for 100 Monte Carlo simulations for different level of noise.

noise as inputs. Fig. 4.4 plots the mean VAF of a fresh validation data set for 100 Monte Carlo simulations for different levels of noise, showing that the method is robust to level noise up to 30dB.

4.5   Conclusions

This paper presented recent advances in the field of LPV subspace identification algorithms. The approach has been described and successfully tested on a simulated automotive application. The identification of the vehicle lateral dynamics was used to underline the main features of the algorithm. It was shown that subspace LPV identification methods have the potential to simplify the modeling and identification phase in the development of advanced vehicle dynamics control systems.

References

[1] Balas, G.J.: Linear, Parameter-Varying Control and its Application to a Turbofan Engine. International Journal of Robust and Nonlinear Control 12, 763–796 (2002)

[2] Balas, G.J., Lind, R., Packard, A.K.: Optimally Scaled H_inf Full Information Control with Real Uncertainty: Theory and Application. AIAA Journal of Guidance, Dynamics and Control 19, 854–862 (1998)

[3] Bamieh, B., Giarre, L.: Identification of linear parameter varying models. International Journal of Robust and Nonlinear Control 12(9), 841–853 (2002)

[4] Bianchi, F.D., De Battista, H., Mantz, R.J.: Wind Turbine Control Systems Principles, Modelling and Gain Scheduling Design. Springer, Heidelberg (2006)

[5] Corno, M., Savaresi, S.M., Balas, G.J.: On Linear Parameter Varying (LPV) Slip-Controller Design for Two-Wheeled Vehicles. International Journal of Robust and Nonlinear Control 19(12), 1313–1336 (2009)

[6] Corno, M., Tanelli, M., Boniolo, I., Savaresi, S.M.: Advanced Yaw Control of Four-wheeled Vehicles via Rear Active Differential Braking. In: Proceedings of the 48th IEEE Conference on Decision and Control (2009)

[7] Jia, Y.: Robust control with decoupling performance for steering and traction of 4WS vehicles under velocity-varying motion. IEEE Transactions on Control Systems Technology 8(3), 554–569 (2000)

[8] Kiencke, U., Nielsen, L.: Automotive Control Systems. Springer, Berlin (2000)

[9] Lee, L.H., Poolla, K.: Identification of linear parameter-varying systems via LFTs. In: Proceedings of the 35th IEEE Decision and Control Conference (1996)

[10] Lee, L.H., Poolla, K.: Identification of linear parameter-varying systems using nonlinear programming. ASME Journal of Dynamic Systems Measurement and Control 121, 71–78 (1999)

[11] Lovera, M., Mercere, G.: Identification for Gain-Scheduling: a Balanced Subspace Approach. In: Proceedings of the American Control Conference, New York, USA (2007)

[12] Souza, C.E., Trofino, A.: Gain-scheduled H2 controller synthesis for linear parameter varying systems via parameter-dependent lyapunov functions. International Journal of Robust and Nonlinear Control 16(5), 243–257 (2005)

[13] Steinbuch, M., Van De Molengraft, R., Van Der Voort, A.J.: Experimental Modelling and LPV Control of a Motion System. In: Proceedings of the American Control Conference (2003)

[14] Tóth, R.: Modeling and Identification of Linear Parameter-Varying Systems. PhD thesis, TU Delft (2008)

[15] Tóth, R., Felici, F., Heuberger, P.S.C., Van den Hof, P.M.J.: Discrete time LPV I/O and state space representations, differences of behavior and pitfalls of interpolation. In: Procodeedings of the European Control Conference, pp. 5418–5425 (2007)

[16] van Wingerden, J.W., Verhaegen, M.: Subspace identification of Bilinear and LPV systems for open-and closed-loop data. Automatica 45(2), 372–381 (2009)

[17] Verdult, V., Verhaegen, M.: Subspace identification of multivariable linear parameter-varying systems. Automatica 38(5), 805–814 (2002)

[18] Verdult, V., Verhaegen, M.: Kernel methods for subspace identification of multivariable LPV and bilinear systems. Automatica 41(9), 1557–1565 (2005)

[19] Verhaegen, M., Verdult, V.: Filtering and system identification: a least squares approach. Cambridge Univ. Press, Cambridge (2007)

[20] Wei, X.: Adaptive LPV techniques for Diesel engines. PhD thesis, Johannes Kepler University (2006)

Chapter 5
A Tutorial on Numerical Methods for State and Parameter Estimation in Nonlinear Dynamic Systems

Boris Houska*, Filip Logist, Moritz Diehl, and Jan Van Impe

Abstract. In this chapter we provide a tutorial on state of the art numerical methods for state and parameter estimation in nonlinear dynamic systems. Here, we concentrate on the case that the underlying models are based on first-principles, giving rise to systems of ordinary differential equations (ODEs). As a general introduction the different dynamic model types, the generic modeling cycle and several approaches for dynamic optimization, i.e., optimization problems with dynamic systems as constraints, are briefly mentioned. Then, the estimation problem is posed as a maximum likelihood dynamic optimization problem. Afterwards, we review Multiple Shooting techniques and generalized Gauss-Newton methods for general least-squares and L1-norm optimization problems and discuss the benefits of the recently developed Lifted Newton Method in the context of state and parameter estimation. Finally, we present an illustrative example involving the estimation of the states and parameters of a pendulum using the freely available software environment ACADO Toolkit in which many of the discussed algorithms are implemented.

Boris Houska · Moritz Diehl
SCD & OPTEC, Electrical Engineering Department (ESAT),
K.U. Leuven, Kasteelpark Arenberg 10, 3001 Leuven, Belgium
e-mail: {boris.houska,moritz.diehl}@esat.kuleuven.be

Filip Logist · Jan van Impe
BioTeC & OPTEC, Chemical Engineering Department (CIT),
K.U. Leuven, W. de Croylaan 46, 3001 Leuven, Belgium
e-mail: {filip.logist,jan.vanimpe}@cit.kuleuven.be

* Corresponding author.

D. Alberer et al. (Eds.): Identification for Automotive Systems, LNCIS 418, pp. 67–88.
springerlink.com   © Springer-Verlag London Limited 2012

5.1 Introduction

Mathematical models and simulations are nowadays indispensable tools for the analysis, design, operation and optimization of a huge number of engineering processes. Many of these processes are intrinsically *dynamic* in nature, i.e., properties and variables vary over time, giving rise to dynamic models. However, before the models can be employed in practice for the above mentioned purposes, a *model calibration* is required, i.e., estimating the unknown parameters and unmeasured states based on experimental data.

5.1.1 What Are the Different Classes of Dynamic Models?

In general, dynamic models can be classified according to the amount of *a-priori* knowledge about the dynamic process that is incorporated [16]. *White-box* or *first-principles* models start from the physical, chemical, biological, ... mechanisms and principles underlying the process. Due to the conservation laws in nature, typically balance type of equations are involved giving rise to systems of differential equations. Alternatively, *black-box* or *data driven* models are based on generic mathematical relations which can flexibly be adapted to predict the observed process input output behavior, without looking into the underlying process. First-principles models allow an easier interpretation and insight in the process and have a validity domain that often largely surpasses the region from which the experimental data have been taken. In contrast, data driven models are often preferred whenever a first-principles approach would require too much experimental and modeling effort or when the underlying process mechanisms are too complex or unclear. However, there exists no strict separation between the model classes. In practice often *grey box* models are employed, as they combine the best of both worlds, i.e., a mechanistic backbone for the process fundamentals with empirical correlations for the complex side phenomena. In the current chapter, we focus on *dynamic models described by differential equations*, which in practice can include both white-box and grey-box models. Nevertheless, the estimation of the unknown parameters and unmeasured states for these dynamic models can be challenging, not only due to the dynamic nature but also due to nonlinearities which may often be present [43].

5.1.2 How to Calibrate Models? The Modeling Cycle

Dynamic models exhibit to two main characteristics: a *model structure* and *model parameters*. Under the assumption that a correct model structure is selected for the underlying dynamic process, only the model parameters need to be estimated, which can be done according the general modeling procedure or so-called *modeling cycle* [4, 25].

A schematic view is given in Figure 5.1. First, an initial experiment is performed. Based on these experimental data the model parameters can be estimated. This model calibration involves an optimization procedure in order to select the parameter values which minimize the differences between the model predictions and the measurements. Then, it is evaluated whether or not the calibrated model performs satisfactorily. If so, the model is ready to be employed, otherwise the collection of additional data and a re-estimation of the parameters is needed. Moreover, the novel experiments can be carefully designed using Optimum Experimental Design techniques in order to increase the information content of the experiments and, hence, limit the experimental burden as much as possible [3, 4, 17]. If the re-estimated model performs satisfactorily, the modeling cycle is exited, otherwise the loop has to be re-run until a satisfactory result is obtained.

For dynamic models, it should be noted that not only the parameters are estimated, but that also estimates for the continuous states are returned (in contrast to the experimental data which are only measured at specific time instants). The described modeling cycle is mainly applicable for processes where the estimation can be performed *off-line*, i.e., a specific experimental run is used to derive a process model, which will be used afterwards (e.g., modeling the growth of micro-organisms in a bioreactor [5, 39]). However, as it is not always possible to perform a dedicated experimental run, parameter re-estimation may have to be carried out *on-line*, i.e., during the regular process operation (e.g., in Model Predictive Control applications [44, 45]). In the latter case, the procedures and algorithms are intrinsically highly similar. However, the possibilities for iterating on the modeling cycle and designing optimal experiments are much fewer, and also real-time constraints, demanding a completed re-estimation within a fixed and short time interval may be present.

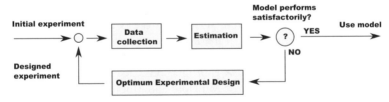

Fig. 5.1 The modeling cycle [25].

5.1.3 Approaches for the Optimization of Dynamic Systems

As mentioned in the previous section, model calibration involves optimization. The presence of the dynamic system yields so-called dynamic optimization or optimal control problems. Numerical methods for solving optimal control problems are generally classified into two categories, *direct* and *indirect* approaches (see, e.g., [38]). Methods of the latter class try to find solutions according to Pontryagin's Minimum Principle [30], i.e., the first-order necessary conditions for optimality, while techniques of the former class convert the original infinite dimensional optimal control problem into a finite dimensional Nonlinear Programming problem (NLP) via discretization. However, most dynamic optimization problems are nowadays solved by direct approaches.

5.1.3.1 Sequential Approaches: Single Shooting

In Single Shooting [34, 40, 41] only the time-varying parameters are discretized, most often using piecewise polynomials. For each parametrization the differential equations are solved using a standard integration algorithm, and the objective function is evaluated. The parameter values are then updated employing a standard optimization algorithm. Hereto, most often a deterministic, sequential quadratic programming (SQP) routine is employed. Consequently, the solution of the differential equations and the minimization are decoupled, and proceed sequentially within one iteration step of the optimizer. As these methods provide a correct solution to the differential equations during all iterations, they are also known as *feasible path* methods.

The most important advantage of the sequential approach is its straightforward implementation resulting in rather small-scale NLPs. However, the sequential approach may be slow, especially in the presence of path constraints on the states, since these constraints cannot be enforced directly.

5.1.3.2 Simultaneous Approaches: Multiple Shooting and Orthogonal Collocation

Simultaneous approaches discretize both the control and the states. Consequently, the simulation and optimization are performed simultaneously in the space of both the discretized controls and states, yielding a large-scale NLP which requires tailored numerical methods. Since in this case the differential equations are only satisfied at the solution of the optimization problem, these methods are also called *infeasible path* methods.

Two different approaches exist: Multiple Shooting [11, 23] and Orthogonal Collocation [6, 7]. In the former case, the total integration range is split into a finite number of intervals on which integration of the states is continuous. The value of the control and the initial value of the states in each interval are chosen by the NLP-solver in each iteration while trying to ensure the continuity of the states between the different intervals. In the latter case, also the states are fully discretized based on (orthogonal) polynomials. Hence, the minimal objective value has to be found while satisfying the discretized differential equations. Clearly, Multiple Shooting and Orthogonal Collocation allow a more direct enforcement of the state constraints. However, the size of the NLPs significantly increases, requiring tailored optimization algorithms that exploit the problem's structure and sparsity. Obviously, the largest but also sparsest NLPs are encountered with Orthogonal Collocation. Consequently, most often Interior Point [42] and partially reduced SQP methods [23, 24] are employed for Orthogonal Collocation and Multiple Shooting, respectively. In view of parameter estimation, simultaneous approaches form a natural choice because the measurements of the states can directly be used for initializing the discretized problem.

This chapter is organized as follows: In Section 5.2 we start with a general introduction on the modeling with differential equation systems while the generalized Gauss-Newton method is discussed in Section 5.3. In Section 5.4 we discuss lifted Newton techniques for parameter estimation. Finally, Section 5.5 introduces the software ACADO toolkit in which many of the discussed algorithms are implemented.

5.2   Maximum Likelihood Estimation for Differential Equation Models

In this section, we introduce the class of parameter estimation problems, we are concerned with. As outlined in the introduction, we are interested in models describing the feasible behavior B by a dynamic system:

$$
B := \left\{ (x(\cdot), w(\cdot), p) \; \left| \; \begin{array}{ll} \forall t \in [0, T] : \\ \dot{x}(t) & = f(t, x(t), w(t), p) \\ 0 & = r(x(0), x(T), p) \\ 0 & \geq s(t, x(t), w(t), p) \end{array} \right. \right\}.
$$

In this notation, $x(\cdot)$ is the state of the dynamic system, while $w(\cdot)$ and p are unknowns which influence the dynamics. Here, we distinguish between the time varying parameters $w(\cdot)$ and the time-constant parameters p. The aim is to discuss strategies for choosing a $\xi \in B$ which maximizes the likelihood function of a family of given probability distributions $\varphi(\xi, \cdot)$ evaluated at a given measurement $\eta \in \mathbb{R}^{n_\eta}$. Note that the measurements η are affected by random measurement errors, whose probability distributions $\varphi : \mathbb{R}^{n_\eta} \times B \to \mathbb{R}$ are assumed to be given. All the other variables $\xi \in B$ are unknown and have to be estimated. As no explicit expression for the likelihood function can be worked out, numerical optimization routines have to be employed to find a maximizer of this function. In practice, it is often more convenient to work with the logarithm of φ leading to maximum likelihood estimation problems of the form

$$
\min_{\xi} \; -\log\left(\varphi(\xi, \eta)\right) \quad \text{s.t.} \quad \xi \in B . \tag{5.1}
$$

Let us remark on some practical aspects of the above formulation.

5.2.1 Interpretation of Given Prior Information

Note that the set B can be chosen in such a way that it contains most of our prior information on the states $x(\cdot)$, time-varying parameters $w(\cdot)$, and time-constant parameters p which we would like to estimate. Typically, the physics of the system is modeled in the right-hand-side function $f : \mathbb{R} \times \mathbb{R}^{n_x} \times \mathbb{R}^{n_w} \times \mathbb{R}^{n_p} \to \mathbb{R}^{n_x}$ containing the dynamic equations. In fact, the first principles which we are using to model our system can be interpreted as a prior information which we have obtained from someone else who has identified these physical laws before. However, note that almost all models come along with a feasible domain to be included in the equality and inequality functions $r : \mathbb{R}^{n_x} \times \mathbb{R}^{n_x} \times \mathbb{R}^{n_p} \to \mathbb{R}^{n_r}$, and $s : \mathbb{R} \times \mathbb{R}^{n_x} \times \mathbb{R}^{n_w} \times \mathbb{R}^{n_p} \to \mathbb{R}^{n_h}$. Here, the equality $0 = r(x(0), x(T), p)$ can not only express given information on the initial state but also coupled boundary information. For example, if the initial state x_0 of a process is at time $t = 0$ exactly known, we might include this information by formulating a constraint function of the form

$$r(x(0), x(T), p) := x(0) - x_0 .$$

In another case, if an open loop stable system is measured while being in its periodic steady state, we might include the information that the system is periodic in order to improve our estimate, i.e. we would formulate a constraint function of the form

$$r(x(0), x(T), p) := x(0) - x(T)$$

in this case. Summarizing this argumentation, a first principle model is not only described by the dynamic system f. Rather, the behavior B is for us a synonym for a first principle model.

5.2.2 Smoothing Heuristics

So far, we have not required any further assumptions on the functions f, r, and h. However, for the algorithm which we discuss in this article f, r, and h are assumed to be sufficiently often differentiable in their arguments. Here, the crucial point is that most algorithm will require that f is smooth - with the only exception that non-smoothness of f in its first argument, the time t, can be reformulated into a smooth problem by re-defining time-intervals provided that the non-smoothness is explicitly known. In practice, the main source of non-smoothness seems to arise from "look-up tables" which are common in certain branches of engineering. In this case we recommend to apply smoothing heuristics. One way is to fit the look-up table with a smooth function. However, from the perspective of the numerical algorithm, only

$f : \Omega \to \mathbb{R}^{n_x}$ and its derivatives are evaluated which suggests to apply the smoothing to the original non-smooth function $\widetilde{f}$ directly, i.e.

$$\forall (\omega_1, \omega_2) \in \Omega_1 \times \Omega_2 : \quad f(\omega_1, \omega_2) := \int_{\Omega_2} \widetilde{f}(\omega_1, \omega_2') \tau(\omega_2, \omega_2') \, d\omega_2' .$$

Here, we have summarized the arguments of f in one variable $\omega \in \Omega$ dividing the domain Ω into the subspaces Ω_1 and Ω_2 which summarize the spaces of variables which enter f smoothly and non-smoothly respectively. Here, τ is a suitable smooth relaxation of the Dirac-distribution acting as a filter function such that f is smooth and its derivatives are known as

$$\partial_{\omega_2}^{\alpha} f = \int_{\Omega_2} \widetilde{f}(\omega_1, \omega_2') \, \partial_{\omega_2}^{\alpha} \tau(\omega_2, \omega_2') \, d\omega_2' .$$

From a numerical point of view, the above filter approximation might be recommended for the case that a non-smooth look-up table enters f but numerical errors should be kept small. On the other hand, if efficiency is more important than the numerical, the above smoothing heuristic might be too expensive for differentiable $\widetilde{f}$ as higher order quadrature rules can not be applied in this case leading to many evaluations of $\widetilde{f}$ for one evaluation of f.

In other cases absolute values or ramp functions are a source of non-smoothness. In this case approximation tricks like

$$\sqrt{x^2 + \epsilon} \approx \mathrm{abs}(x)$$

for some small $\epsilon > 0$ are applied. Clearly, all these smoothing strategies are of a rather heuristic nature but the influence of the corresponding approximation errors can often be estimated in a linear approximation. Finally, we note that switching model behavior, when essential, can and should be treated by suitable numerical methods for switched dynamic systems [15, 21].

5.2.3 The Importance of Convexity

We say that a model is convex if the associated set B is convex. Unfortunately, for estimation problems with dynamic systems convexity can often only be obtained if the function f is linear in all its variables. However, once the convexity of the set B is established, the optimization problem (5.1) is typically also convex - at least if we think of common probability distributions such as Gaussian distributions. For an overview of convex parameter estimation formulations we refer to [14]. In this chapter, we are rather concerned about general nonlinear and consequently most often non-convex problems. For the case that the parameters enter affinely there exist still global optimization approaches [12, 13]. However, for completely non linear problems an appropriate initialization strategy for the numerical algorithms should be used - e.g. by applying a lifted Newton method as we will discuss in Section 5.4.2.

5.2.4   Estimation of Time-Varying Parameters

In many practical situations the physical control input which is applied to the system is not exactly known. This might be due to the fact that the physical input quantity might not exactly coincide with our input, or because the input is only measured and must be estimated, too. Here, it is advisable to distinguish between the physical control input u and the measured input $u + w$ which might be affected by the input noise w to be regarded as a time-varying parameter. However, the noise in the control input u might not be the only uncertainty which is not included in the plain first-principles model. One common approach to robustify an estimate with respect to such a knowledge about uncertainties is to assume that the model is modified by an additional time dependent offset

$$f(t, x(t), w(t), p) := f_{\text{first\_principle}}(t, x(t), p) + w(t) \qquad (5.2)$$

which would model an additive uncertainty on the given first principle function $f_{\text{first\_principle}}$. Analogously, the equality and inequality constraints could be robustified as well. Unfortunately, it is in practice often not clear how to weight between model uncertainty and measurement errors. In a heuristic formulation the objective

$$-\log\left(\varphi(x(\cdot), p, \eta)\right) + \gamma \Phi(w)$$

is minimized where Φ is typically a convex penalty weighted with a heuristic scaling factor γ. Another formulation uses multi-objective optimization [26, 27] to discuss the weighting between robustness and nominal optimality in a more systematic way. Note that the above nonlinear formulation can be interpreted as a generalization of Kalman filters, where typically an additive white noise to a linear system is taken into account.

5.2.5   Least Squares Terms versus l1-Norms

The most common assumption about the probability distribution φ is that the measurements have Gaussian distribution, which leads to least squares objectives of the form

$$-\log\left(\varphi(\xi, \eta)\right) = \sum_{i=1}^{N} \left\| \Sigma^{-\frac{1}{2}} \left(h(t_i, x(t_i), p) - \eta_i \right) \right\|_2^2 - \log(C).$$

Here, C is only a scaling constant which can be set to 1 in our context as it does not affect the minimization. Moreover, $h : \mathbb{R} \times \mathbb{R}^{n_x} \times \mathbb{R}^{n_p}$ denotes a smooth measurement function. However, for different settings another common assumption is that

$$- \log\left(\varphi(\xi, \eta)\right) = \sum_{i=1}^{N} \left\| \Sigma^{-\frac{1}{2}} \left(h(t_i, x(t_i), p) - \eta_i\right) \right\|_1 - \log(C)$$

which leads to l1-norm estimation problems. Note that in both cases the objective is convex if h is linear in its arguments. Although the estimation problem might still become non-convex if B is non-convex, a local minimizer can be guaranteed to be globally optimal if the optimal value is 0 for $C = 1$, i.e. if we have a local minimizer which leads to a perfect fit. The other way round, 0 is a lower bound on the objective which might help us to assess optimality of a local minimizer. However, we recognize already at this point that global optimality is not the only thing we are interested in as we might also ask the question how good our estimate for the states, controls and parameters is. This question will be addressed in the next section.

Note that in practice, the choice of the norm is not always motivated by a stochastic argumentation. For example if information about the probability distribution of the errors at the sensors is completely lacking, we can only use an empirical choice. Here, the main motivation for l1-norms is typically that it does not weight outliers strongly as the squared l2-norm does. For good survey of this and other practical heuristics we refer to [10, 14].

5.3   Generalized Gauss-Newton Methods

In order to solve general nonlinear parameter estimation problems of the form (5.1) most algorithms pass through two stages. First, the continuous problem is discretized and transformed into a nonlinear program. And second, the discrete nonlinear optimization problem is solved locally. This strategy is known as the direct method which is in contrast to indirect methods based on the Pontryagin principle, which is nearly never used for parameter estimation (a notable exeption is [28]). As mentioned before, examples for direct discretization methods are Single- and Multiple Shooting [11] as well as Collocation techniques [6]. While this comment holds for general optimal control problems, we are in this chapter mainly interested in least squares problems, i.e. in the case that the discrete problem takes the form

$$\min_{y} \| F(y) \|_2^2 \quad \text{s.t.} \quad \begin{cases} G(y) = 0 \\ H(y) \leq 0 \end{cases},$$

where $y \in \mathbb{R}^{n_y}$ summarizes the degrees of freedom in the discretized version of the problem (5.1). For the moment, we assume here that we work with Gaussian distributions.

The generalized Gauss-Newton method, as originally proposed in [8], starts from an initial guess y_0 and generates iterates of the form $y^+ = y + \alpha \Delta y$ where Δy solves the QP

$$\min_{\Delta y} \| F_y \Delta y + F \|_2^2 \quad \text{s.t.} \quad \begin{cases} G_y \Delta y + G = 0 \\ H_y \Delta y + H \leq 0 \end{cases}.$$

Here, we have used the short hands $F := F(y)$, $F_y := \partial_y F(y)$, $G := G(y)$, $G_y := \partial_y G(y)$, $H := H(y)$ and $H_y := \partial_y H(y)$.

Note that the above generalized Gauss-Newton method will in general converge with a linear rate only. However, if either the non-linearity of the functions F, G, and H is small or if the objective value in the optimal solution is close to zero, we can expect a reasonable convergence behavior [8]. Here, the main advantage of the Gauss-Newton method is that no second order derivatives are needed while the subproblems are convex by construction.

Finally, we outline that the Gauss-Newton method can be embedded into a more general class of methods which might be called sequential convex optimization techniques [31]. For example, if the above method which is transferred for the l1-norm case:

$$\min_{y} \| F(y) \|_1 \quad \text{s.t.} \quad \begin{cases} G(y) = 0 \\ H(y) \leq 0 \end{cases},$$

This problem is solved iteratively as above but by generating the steps Δy by solving sub-problems of the form

$$\min_{\Delta y} \| F_y \Delta y + F \|_1 \quad \text{s.t.} \quad \begin{cases} G_y \Delta y + G = 0 \\ H_y \Delta y + H \leq 0 \end{cases}.$$

The convergence properties of such 1-norm methods have intensively been studied in [10]. In [31] this approach is generalized further for other convex objective and constraint functions.

Let us assume that the above methods are successful and that we have found the optimal solution

$$y^*(\epsilon) := \operatorname*{argmin}_{y} \| F(y) + \epsilon \| \quad \text{s.t.} \quad \begin{cases} G(y) = 0 \\ H(y) \leq 0 \end{cases} \tag{5.3}$$

for $\epsilon = 0$. The question that arises next is how we can assess the quality of the estimate. This a posteriori analysis can for nonlinear problems become very expensive as the optimal estimate $y^*(\epsilon)$ depends nonlinearly on the measurement noise ϵ. However, under the additional assumption that the measurement error is small, we may analyze the map y^* in a linear approximation. Here, we need to check first that y^* is differentiable with respect to ϵ. Fortunately, this differentiability can be guaranteed under some mild regularity conditions which are known from Robinson's generalization of the implicit function theorem [32, 33]. Thus, we may assume that y^* is differentiable at the given measurement $\epsilon = 0$ and we denote its derivative at this point by y_ϵ^*. Now, we can formally compute the covariance matrix of the estimate in a linear approximation:

$$\mathbb{E}\left\{(y^* - y)(y^* - y)^T\right\} \doteq y_\epsilon^* \Sigma (y_\epsilon^*)^T =: C . \qquad (5.4)$$

Note that C can easily be computed from the solution of the QP (5.3) in the optimal solution as this QP can already be interpreted as a linear approximation of the original nonlinear program. For details of this approach, we refer to [8].

In the field of Optimum Experimental Design [2] for nonlinear systems, the aim is to optimize the covariance of the estimate provided that we have an input $u \in U$ which can be designed in order to obtain more information from the measurements. In this case, the covariance matrix C can be regarded as a function in u, such that we can ask the input which minimizes a certain scalar quality measure $\Phi(C)$ e.g. the determinant, trace, or the maximum eigenvalue of C. Again, the resulting optimization problems are non-convex, but in some cases it is enough to find locally optimal inputs which can lead to solutions that work well in applications [3, 4, 17, 22].

5.4 Schlöder's Trick or the Lifted Newton Type Method for Parameter Estimation

In this section, we would like to come back to two aspects of nonlinear parameter and state estimation: first, we have to address the question how to discretize the dynamic system and second how to initialize the Gauss-Newton or other methods which find local minimizers. For both questions Lifted Newton Methods can be a suitable tool. The idea of lifted Newton methods, which we refer to in this context, has originally been developed by Schlöder in [36]. More recently, a generalization of this idea was published in [1] under the name "Lifted Newton Method". In this article, we present the Schlöder's method from the perspective of inexact SQP methods. The method is tailored for parameter and state estimation problems where the state is measured while the initial value condition $r(x(0), x(T)) = x(0) - x_0 = 0$ is given.

5.4.1   Modular Forward Lifting Techniques

Let us first briefly recall the idea of Multiple Shooting: we assume that we divide the time horizon $[0, T]$ into N intervals $0 = t_0 < t_1 < \ldots < t_N$ with associated state discretization point $s_0, \ldots, s_N$ and a piecewise constant discretization $w_0, \ldots, w_{N-1}$ of the time-varying parameters. Regarding the solution of the differential system

$$\forall t \in [t_i, t_{i+1}]: \quad \dot{x}(t) = f(t, x(t), w_i, p) \quad \text{with} \quad x(t_i) = s_i$$

at the time t_{i+1} as a function $X_i(s_i, w_i, p) := x(t_{i+1})$ in the discrete variables s_i, w_i, p, we require the matching conditions

$$G(y) := \begin{pmatrix} s_0 - x_0 \\ s_1 - X_0(s_0, w_0, p) \\ s_2 - X_1(s_1, w_1, p) \\ \vdots \\ s_N - X_{N-1}(s_{N-1}, w_{N-1}, p) \end{pmatrix} \tag{5.5}$$

to be satisfied. Here, we assume that f is chosen in such a way that the maps X_i are unique and sufficiently often differentiable. Note that these functions can numerically be evaluated by using a suitable integrator. In the Multiple Shooting technique the variables $s_0, s_1, \ldots s_N$ are not eliminated from the NLP which is in contrast to Single Shooting methods. One major advantage of this formulation is on the one hand that the non-linearity of the problem might be reduced as observed in [29] and in [9] and theoretically investigated in [1]. On the other hand, a second advantage in the context of parameter and state estimation is that in the case of state measurements very natural initialization points for $s_0, \ldots, s_N$ are available which are ideally set to the measurements for the states. In practice, this initialization can be the main trick such that the algorithm converges safely to a local minimizer which could be global or gives at least a good fit. Even if we can not prove anything for general nonlinear problems, it can be shown that for parameter affine, but nonlinear problems, one step convergence can be achieved in the absence of measurement errors, independent of the parameter initialization [12, 13, 37].

Note that the lifted Newton method in the context of Multiple Shooting amounts to a different way of computing the derivatives known as Schlöder's trick [1, 36]. In [1] another advantage of lifted Newton methods is shown for the case that only a few free parameters or controls should be estimated while the state dimension is large. We shall see below that the lifting approach will typically lead to significant savings in terms of sensitivity computation time if the condition

$$\frac{N+1}{2} n_w + n_p + 1 \ll n_x + n_w + n_p \tag{5.6}$$

is satisfied as has been first shown by [35]. Indeed, for a Gauss-Newton method we can expect that we do not need to compute all derivative directions of the functions $X_0, \ldots, X_{N-1}$ with respect to the variables $s_0, \ldots, s_N$ as these variables could even be eliminated in a Single Shooting approach. The interpretation of lifted Newton methods in this paper assumes that the computation of the derivatives of $X_0, \ldots, X_{N-1}$ is much more expensive than solving the sparse QP of the form (5.3) in each iteration of the Gauss-Newton method.

Recall that inexact SQP type methods would replace the matrix $G_y = \partial_y G(y)$ in the QP (5.3) with an approximation $\hat{G}_y \approx G_y$. Here, the general motivation for approximating the matrix $\partial_y G(y)$ is that this matrix is expensive to compute. If only a few parameters are unknown while the state dimension is large, then the derivatives of G_y with respect to the initial states are the most expensive blocks. Thus, we are in our context especially interested in a particular approximation of the form

$$
\hat{G}_y := \begin{pmatrix}
1 & 0 & \ldots & 0 & 0 & \ldots & 0 & 0 \\
-\hat{G}_x^0 & 1 & \ldots & 0 & G_w^0 & \ldots & 0 & G_p^0 \\
\vdots & \ddots & \ddots & \vdots & \vdots & \ddots & \vdots & \vdots \\
0 & \ldots & -\hat{G}_x^{N-1} & 1 & 0 & \ldots & G_w^{N-1} & G_p^{N-1}
\end{pmatrix}.
$$

Here, $G_w^i := \partial_w X_i(s_i, w_i, p)$ and $G_p^i := \partial_p X_i(s_i, w_i, p)$ (for $i \in \{0, \ldots, N-1\}$) are exact derivatives but

$$
\hat{G}_x^0 := 0 \quad \text{and} \quad \hat{G}_x^i := \left(\partial_s X_i(s_i, w_i, p) A_i \right) A_i^\dagger \tag{5.7}
$$

(for $i \in \{1, \ldots, N-1\}$) uses only some directional derivatives of X_i which are stored in the matrix $A_i \in \mathbb{R}^{n_x \times m_i}$. Note that for a small number of directions $m_i \ll n_x$ the pseudo inverse $A_i^\dagger$ can cheaply be computed. For general inexact SQP methods we need at least one adjoint derivative to repair the error in the stationarity conditions which is made by the approximation. However, we will see that this is not needed if we choose the directions A_i properly. For example if A_i is the unit matrix the above approximation would be exact, but we are interested in the question whether we can also choose $m_i < n_x$.

Using the notation $G_x^i := \partial_s X_i(s_i, w_i, p)$ and $b_i := s_{i+1} - X_i(s_i, w_i, p)$ we choose for A_i the matrix

$$
A_i := \left(E_{i-1}^0, \ldots, E_{i-1}^{i-1}, F_{i-1}, d_{i-1} \right) \tag{5.8}
$$

where the matrices are defined via the condensing recursion $d_0 = 0$, $d_{i+1} = G_x^i d_i + b_i$, $F_0 = 0$, $F_{i+1} = G_x^i F_i + G_p^i$ and $E_{i+1}^j = G_x^i E_i^j$ with $E^{i,i} = G_w^i$. Computing the entries of A_i recursively by means of automatic differentiation in forward mode, we need to compute

$$
m_i = i\, n_w + n_p + 1
$$

derivative directions of X_i on the i-th Multiple Shooting interval. Assuming that the time for computing a derivative direction of X_i is approximately the same for all intervals i, we may estimate the total cost of computing sensitivities as

$$\sum_{i=1}^{N} m_i = \frac{N(N+1)}{2} n_w + N n_p + N . \tag{5.9}$$

This cost needs to be compared to the cost $N(n_x + n_w + n_p)$ which we would have to invest if we had to compute the matrices G_x^i completely. This comparison leads to the condition (5.6) for which Schlöder's trick can be expected to lead to savings in the time for computing sensitivities.

Note that due to the construction of $\hat{G}_y$ we can simply replace G_y in the QP (5.3) by its approximation $\hat{G}_y$ obtaining a completely equivalent QP. This can easily be checked using

$$\hat{G}_x^i A_i = \left(G_x^i A_i \right) A_i^\dagger A_i = G_x^i A_i \tag{5.10}$$

as we may use that $A_i A_i^\dagger A_i = A_i$. Finally, we should mention that in [1] a programming trick was suggested which avoids the explicit construction of the sparse QP (5.3) obtaining directly a condensed version. However, in this paper, as in [35, 36], we prefer to construct the matrix $\hat{G}_y$ as it recovers modularity. Indeed, an implementation of the modularly lifted Newton method in the version presented above requires only to replace the matrix G_y by $\hat{G}_y$ without touching the actual implementation of the Gauss-Newton method. Now, it does not matter which QP solver is used in behind, i.e. the implementation of the sparse QP solver and the derivative generation are completely independent. The price to be paid for this modularity is only some linear-algebra overhead which is usually negligible compared to the cost of derivative generation for applications which satisfy the condition (5.6).

5.4.2 Automatic Backward Lifting Techniques

In the last section we have mainly concentrated on the case that the derivatives for the lifting variables are computed in forward mode. It is of course also possible to transfer the idea of lifting by computing one backward and one forward direction in order to construct a condensed version of the QP. As we shall see the backward lifting strategy will require one additional forward derivative to expand the primal solution of the dense QP - similar to forward lifting techniques where we would need an additional backward sweep if we are interested in an expansion of the multipliers.

In this section, we simplify our notation slightly in order to work out the main idea while generalization can later be derived: first we are in our consideration concerned about the case $n_p \gg 1$, possibly even $n_p \gg n_x$, i.e. we want

to estimate many parameters. Thus, it not so crucial to distinguish between
controls and parameters - we can simply regard the controls as parameters,
too, as soon as we have discretized the system as it does not matter for the
backward differentiation how many parameters our functions depend on. The
backward method is especially beneficial if we have only a few constraints or
even no constraints beside the dynamic residuum $G(y) = 0$. Here, G is defined in equation (5.5). In this simplified setting without constraints and for
smooth objectives (which excludes l1-type cost functions), the optimization
can be we written as

$$\min_{y} \Phi(y) \quad \text{s.t.} \quad G(y) = 0 , \tag{5.11}$$

with Φ summarizing the objective and $y := (s^T, p^T)^T$, $s := (s_0^T, \ldots, s_m^T)^T$.
Recall that for Single Shooting the variable s is eliminated or regarded as
a function in p, such that the derivatives of the Single Shooting objective
$\Phi(s(p), p)$ can efficiently be computed by automatic differentiation in backward mode. We are asking the question whether this efficient differentiation
technique can also be rescued for the case that we like to keep the Multiple
Shooting lifting nodes s in the NLP.

   The key idea is to recognize that the linearized condition $G_y \Delta y + G = 0$
can be used to eliminate Δs in dependence on Δp, i.e. we get a linear relation
of the form

$$\Delta s = Z \Delta p + z \quad \Leftrightarrow \quad G_y \Delta y + G = 0$$

where the matrix $Z \in \mathbb{R}^{(N+1) n_x \times n_p}$ and the vector $z \in \mathbb{R}^{(N+1) n_x}$ are defined
such that the above equivalence holds. We are interested in the condensed
objective gradient

$$g := \left[\partial_s \Phi(s, p) + z^T H_{sp} \right] Z + \partial_p \Phi(s, p)$$

with the aim to solve the unconstrained minimization problem

$$\min_{\Delta p} \frac{1}{2} \Delta p^T K \Delta p + g \Delta p \tag{5.12}$$

with $K := H_{pp} + Z^T H_{ss} Z$ and a suitable choice of the Hessian matrix

$$\mathcal{H} = \begin{pmatrix} H_{ss} & H_{sp} \\ H_{sp}^T & H_{pp} \end{pmatrix} ,$$

which we will discuss later. Thus, we would like to compute the condensed
backward derivative $\mu^T Z$ where we use the definition

$$\mu := (\mu_0, \ldots, \mu_N) := \nabla_s \Phi(s, p) + z^T H_{sp}$$

to denote the seed. Here, z can be computed iteratively, by computing one
forward direction on each of the Multiple Shooting intervals. Clearly, $\mu^T Z$ can

also be computed iteratively once z is known, requiring only one backward sensitivity direction on each interval:

$$\sigma_N := \mu_N \qquad\qquad\qquad \sigma_i := \sigma_{i+1} G_x^i + \mu_i$$
$$\nu_i := \sigma_{i+1} G_p^i \quad \text{and finally} \quad \mu^T Z = \sum_i \nu_i \, .$$

Note that the choice of the Hessian matrix $\mathcal{H}$ is of course crucial for the performance of the method. Unfortunately, Gauss-Newton Hessian approximations are in this case not advisable, as we would have to compute an approximation of the projected Hessian term $Z^T H_{ss} Z$. However, there are various other possibilities: for example a direct approximation of K (or its inverse) with BFGS updates is perfectly applicable [18, 35]. Also an computation of the exact Hessian is possible. In this case, a projection is required to ensure that K is always positive definite. Finally, we remark that the solution Δp of the unconstrained condensed problem must be expanded to apply the step $s^+ = s + \alpha \Delta s$. Here, one additional forward direction of Z must be computed to obtain $\Delta s = Z \Delta p + z$. Thus, we need 3 sensitivity directions in total in each lifted Newton iteration.

5.5   State and Parameter Estimation with ACADO Toolkit

In this section we review the software ACADO [20] which implements tools for automatic control and dynamic optimization. It provides a general framework for using a great variety of algorithms for direct optimal control, including model predictive control and, in particular, state and parameter estimation. ACADO Toolkit is implemented as self-contained C++ code, while the object-oriented design allows for convenient coupling of existing optimization packages and for extending it with user-written optimization routines. Note that the package can be downloaded from [19].

In ACADO direct methods, as explained in the previous sections, are implemented. Of the three algorithms discussed in this paper, "Multiple Shooting, forward lifting, backward lfitng", only the first two are implemented so far. Here, several software levels are available: first ACADO comes along with a symbolic syntax which allows convenient modeling using first principles techniques. This symbolic syntax is based on the C++ operator overloading. The benefit of this way of implementing right-hand side functions of differential equation, constraints or objectives is that e.g. automatic detection of dependencies and dimensions, automatic as well as symbolic differentiation, convexity detection etc. are available. An example for setting up a simple toy parameter estimation problem is shown in the code Listing 5.1. The corresponding problem can mathematically be notated as a least squares problem

Listing 5.1 An implementation of a parameter estimation in ACADO

```
int main( ){

        DifferentialState         phi, dphi; // the states of the pendulum
        Parameter                  l, alpha ; // its length and the friction
        const double               g = 9.81 ; // the gravitational constant
        DifferentialEquation       f         ; // the model equations
        Function                   h         ; // the measurement function
//  ————————————————————————————————————————
        OCP ocp( 0.0, 2.0 )                   ; // construct an OCP
        h << phi                              ; // the state phi is measured
        ocp.minimizeLSQ( h, "data.txt" )      ; // fit h to the data

        f << dot(phi ) == dphi                ; // a symbolic implementation
        f << dot(dphi) == -(g/l) * sin( phi ) // of the model
                         -alpha * dphi        ; // equations

        ocp.subjectTo( f                  ); // solve OCP s.t. the model,
        ocp.subjectTo( 0.0 <= alpha <= 4.0 ); // the bounds on alpha
        ocp.subjectTo( 0.0 <=   l    <= 2.0 ); // and the bounds on l.
//  ————————————————————————————————————————

        ParameterEstimationAlgorithm algorithm(ocp);
        algorithm.solve ();

        return 0;
}
```

$$
\begin{array}{l}
\underset{\phi(\cdot),\alpha,l}{\text{minimize}} \quad \sum_{i=1}^{10} \left(\phi(t_i) - \eta_i \right)^2 \\[2mm]
\text{subject to:} \\[2mm]
\forall t \in [0,T] : \ddot{\phi}(t) = -\frac{g}{l}\phi(t) - \alpha\dot{\phi}(t) \\[1mm]
\qquad\qquad 0 \le \alpha \le 4 \\[1mm]
\qquad\qquad 0 \le l \le 2
\end{array}
\tag{5.13}
$$

Here, a simple pendulum is regarded, which consists of the state ϕ representing the excitation angle and the state $\dot{\phi}$ denoting the angular velocity. The constant $g = 9.81$ is the gravitational constant while the friction coefficient α and the length l of the cable are only known to lie between certain bounds. We assume that the state ϕ has been measured at several times.

The problem can be solved with the data file shown in the left part of Figure 5.2 while the result for the estimated state is shown in the upper right part of this figure. Note that the parameter estimation algorithm chooses by default a Gauss-Newton SQP method using the structure of the least-squares objective. The result for the parameter estimation is shown in the lower right part of Figure 5.2. Note that the computation of the standard deviations $\sqrt{C_{i,i}}$ of the parameter estimates is based on a linear approximation of in

ACSII file "data.txt" containing the
measurements:

TIME POINTS	MEASUREMENTS
0.00000e+00	1.00000e+00
2.72321e−01	nan
3.72821e−01	5.75146e−01
7.25752e−01	−5.91794e−02
9.06107e−01	−3.54347e−01
1.23651e+00	−3.03056e−01
1.42619e+00	nan
1.59469e+00	−9.64208e−02
1.72029e+00	−1.97671e−02
2.00000e+00	9.35138e−02

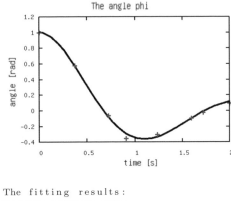

The fitting results:

l
= 1.001e+00 +/− 1.734e−01
alpha = 1.847e+00 +/− 4.059e−01

Fig. 5.2 Data file containing the measurements as well as the fitting results obtained by the Gauss-Newton method applied to problem (5.13). The "nan" values in the data file are automatically ignored.

the optimal solution as discussed in Section 5.3, i.e. the matrix C is defined by equation (5.4).

Note that in ACADO operators such as +,-,*,sin etc. are overwritten and the whole function tree of the dynamic equations, objective and constraints is stored. Now, the user does not need to provide any dimensions or other dependency information about the functions as this can be detected inside.

The second level of ACADO tools are integration routines. Here, several Runge-Kutta as well as BDF integrators are available in order to also deal with stiff systems. Note that the ACADO integrators come along with internal automatic differentiation and are also available as a stand-alone package. The third level of tools implements discretization schemes such as Multiple Shooting while the fourth level implements optimization algorithm such as the above discussed Gauss-Newton method. However, there are also other sequential quadratic programming (SQP) methods with exact Hessians or BFGS updates available. For more details on the structure of ACADO and more tutorial examples we refer to [20].

5.6 Conclusions

In this paper, we have provided an overview on the problem classes and numerical algorithm which are employed for non-linear and non-convex state and parameter estimation tasks. Starting with a general formulation of a maximum likelihood estimation problem for first-principles models we have discussed practical aspects of the formulation. Here, we have introduced smoothing techniques and commented on the difference between least-squares optimization and 1-norm formulations. When it comes to the numerical algorithm, generalized Gauss-Newton methods have turned out to be very efficient in practice. Here, the main problem for non-convex formulations is that we need a suitable initial guess to start the iterations. This problem can be addressed by lifting techniques in order to allow both: making the problem less non-linear and being able to use the measurements to improve the initialization. In the last part, we have discussed the software environment ACADO.

References

[1] Albersmeyer, J., Diehl, M.: The Lifted Newton Method and its Application in Optimization. SIAM Journal on Optimization 20(3), 1655–1684 (2010)

[2] Atkinson, A., Donev, A.: Optimum Experimental Designs. Oxford Statistical Sciences Series, vol. 8. Oxford University Press, Oxford (1992)

[3] Balsa-Canto, E., Alonso, A.A., Banga, J.R.: Computing optimal dynamic experiments for model calibration in predictive microbiology. Journal of Food Process Engineering 31, 186–206 (2008)

[4] Balsa-Canto, E., Alonso, A.A., Banga, J.R.: An iterative identification procedure for dynamic modeling of biochemical networks. BMC Systems Biology 4 (2010)

[5] Bernaerts, K., Servaes, R.D., Kooyman, S., Versyck, K.J., Van Impe, J.F.: Optimal temperature input design for estimation of the Ratkowsky Square Root model parameters: parameter accuracy and model validity restrictions. International Journal of Food Microbiology 73(2-3), 145–157 (2002)

[6] Biegler, L.T.: Solution of dynamic optimization problems by successive quadratic programming and orthogonal collocation. Computers and Chemical Engineering 8, 243–248 (1984)

[7] Biegler, L.T.: An overview of simultaneous strategies for dynamic optimization. Chemical Engineering and Processing 46, 1043–1053 (2007)

[8] Bock, H.G.: Recent advances in parameter identification techniques for ODE. In: Deuflhard, P., Hairer, E. (eds.) Numerical Treatment of Inverse Problems in Differential and Integral Equations. Birkhäuser, Boston (1983)

[9] Bock, H.G.: Randwertproblemmethoden zur Parameteridentifizierung in Systemen nichtlinearer Differentialgleichungen. Bonner Mathematische Schriften, vol. 183. Universität Bonn, Bonn (1987)

[10] Bock, H.G., Kostina, E., Schlöder, J.P., Gienger, G., Pallaschke, S., Ziegler, G.: Robust Parameter Estimation for Identifying Satellite Injection Orbits. In: Modeling, Simulation and Optimization of Complex Processes, pp. 37–46. Springer, Heidelberg (2003)

[11] Bock, H.G., Plitt, K.J.: A multiple shooting algorithm for direct solution of optimal control problems. In: Proceedings 9th IFAC World Congress Budapest, pp. 243–247. Pergamon Press, Oxford (1984)

[12] Bonilla, J., Diehl, M., Logist, F., De Moor, B., Van Impe, J.: A Convex Approximation for Parameter Estimation Involving Parameter-Affine Dynamic Models. In: Proceedings of the 48th IEEE Conference on Decision and Control, pp. 4670–4675 (2009)

[13] Bonilla, J., Diehl, M., Logist, F., De Moor, B., Van Impe, J.: An automatic initialization procedure in parameter estimation problems with parameter-affine dynamic models. Computers & Chemical Engineering 34(6), 953–964 (2010)

[14] Boyd, S., Vandenberghe, L.: Convex Optimization. University Press, Cambridge (2004)

[15] Brandt-Pollmann, U.: Optimization of discontinuous dynamical models. Technical report, First International Conference on Optimization and Software, Hangzhou, China (2002)

[16] Edgar, T.F., Himmelblau, D.M., Lasdon, L.S.: Optimization of Chemical Processes. McGraw-Hill, New York (2001)

[17] Franceschini, G., Macchietto, S.: Model-based design of experiments for parameter precision: State of the art. Chemical Engineering Science 63(19), 4846–4872 (2008)

[18] Gill, P.E., Murray, W., Saunders M.A.: SNOPT: An SQP algorithm for large-scale constrained optimization.Technical report, Numerical Analysis Report 97-2, Department of Mathematics, University of California, San Diego, La Jolla, CA, (1997)

[19] Houska, B., Ferreau, H.J.: ACADO Toolkit User's Manual (2009), http://www.acadotoolkit.org

[20] Houska, B., Ferreau, H.J., Diehl, M.: ACADO Toolkit – An Open Source Framework for Automatic Control and Dynamic Optimization. In: Optimal Control Applications and Methods (2010) (in print), doi:10.1002/oca.939

[21] Kirches, C.: A Numerical Method for Nonlinear Robust Optimal Control with Implicit Discontinuities and an Application to Powertrain Oscillations. Diploma thesis, University of Heidelberg (October 2006)

[22] Körkel, S., Kostina, E.: Numerical Methods for Nonlinear Experimental Design. In: Bock, H.G., Kostina, E., Phu, H.X., Rannacher, R. (eds.) Modelling, Simulation and Optimization of Complex Processes, Proceedings of the International Conference on High Performance Scientific Computing, Hanoi, Vietnam, pp. 255–272. Springer, Heidelberg (2004)

[23] Leineweber, D.B., Bauer, I., Bock, H.G., Schlöder, J.P.: An Efficient Multiple Shooting Based Reduced SQP Strategy for Large-Scale Dynamic Process Optimization. Part I: Theoretical Aspects. Computers and Chemical Engineering 27, 157–166 (2003)

[24] Leineweber, D.B., Schäfer, A.A.S., Bock, H.G., Schlöder, J.P.: An Efficient Multiple Shooting Based Reduced SQP Strategy for Large-Scale Dynamic Process Optimization. Part II: Software Aspects and Applications. Computers and Chemical Engineering 27, 167–174 (2003)

[25] Ljung, L.: System identification: Theory for the User. Prentice Hall, Upper Saddle River (1999)

[26] Logist, F., Houska, B., Diehl, M., Van Impe, J.: Fast Pareto set generation for nonlinear optimal control problems with multiple objectives. In: Structural and Multidisciplinary Optimization (2010) (in press)

[27] Miettinen, K.: Nonlinear Multiobjective Optimization. Kluwer Academic Publisher, Boston (1999)

[28] Ohtsuka, T.: Nonlinear receding-horizon state estimation with unknown disturbances. Trans. of the Society of Instrument and Control Engineers 35(10), 1253–1260 (1999)

[29] Osborne, M.R.: On shooting methods for boundary value problems. Journal of Mathematical Analysis and Applications 27, 417–433 (1969)

[30] Pontryagin, L.S., Boltyanski, V.G., Gamkrelidze, R.V., Miscenko, E.F.: The Mathematical Theory of Optimal Processes. Wiley, Chichester (1962)

[31] Quoc, T.D., Diehl, M.: Local Convergence of Sequential Convex Programming for Nonconvex Optimization. In: Recent Advances in Optimization, BFG Conference, Leuven, Belgium (2009) (in press)

[32] Robinson, S.M.: Strongly Regular Generalized Equations. Mathematics of Operations Research 5(1), 43–62 (1980); 5, 43–62 (1980)

[33] Robinson, S.M.: Perturbed Kuhn-Tucker points and rates of convergence for a class of nonlinear programming algorithms. Mathematical Programming 7, 1–16 (1974)

[34] Sargent, R.W.H., Sullivan, G.R.: The development of an efficient optimal control package. In: Stoer, J. (ed.) Proceedings of the 8th IFIP Conference on Optimization Techniques (1977), Part 2. Springer, Heidelberg (1978)

[35] Schäfer, A., Kühl, P., Diehl, M., Schlöder, J.P., Bock, H.G.: Fast reduced multiple shooting methods for Nonlinear Model Predictive Control. Chemical Engineering and Processing 46(11), 1200–1214 (2007)

[36] Schlöder, J.P.: Numerische Methoden zur Behandlung hochdimensionaler Aufgaben der Parameteridentifizierung. Bonner Mathematische Schriften, vol. 187. Universität Bonn, Bonn (1988)

[37] Schulz, V.H.: Ein effizientes Kollokationsverfahren zur numerischen Behandlung von Mehrpunktrandwertaufgaben in der Parameteridentifizierung und Optimalen Steuerung. Master's thesis, Universität Augsburg (1990)

[38] Srinivasan, B., Bonvin, D., Visser, E., Palanki, S.: Dynamic Optimization of Batch Processes: II. Role of Measurements in Handling Uncertainty. Computers and Chemical Engineering 27(1), 27–44 (2003)

[39] Van Derlinden, E., Bernaerts, K., Van Impe, J.: Accurate estimation of cardinal growth temperatures of Escherichia coli from optimal dynamic experiments. International Journal of Food Microbiology 128(1), 89–100 (2008)

[40] Vassiliadis, V.S., Sargent, R.W.H., Pantelides, C.C.: Solution of a class of multistage dynamic optimization problems. 1. Problems without path constraints. Industrial and Engineering Chemistry Research 10(33), 2111–2122 (1994)

[41] Vassiliadis, V.S., Sargent, R.W.H., Pantelides, C.C.: Solution of a class of multistage dynamic optimization problems. 2. Problems with path constraints. Industrial and Engineering Chemistry Research 10(33), 2122–2133 (1994)

[42] Wächter, A., Biegler, L.T.: On the Implementation of a Primal-Dual Interior Point Filter Line Search Algorithm for Large-Scale Nonlinear Programming. Mathematical Programming 106, 25–57 (2006)

[43] Walter, E., Pronzato, L., Norton, J.: Identification of parametric models: from experimental data. Springer, Heidelberg (1997)

[44] Zavala, V.M., Biegler, L.T.: Optimization-based strategies for the operation of low-density polyethylene tubular reactors: Moving horizon estimation. Computers & Chemical Engineering 33(1), 379–390 (2009)

[45] Zavala, V.M., Biegler, L.T.: Optimization-based strategies for the operation of low-density polyethylene tubular reactors: Nonlinear model predictive control. Computers & Chemical Engineering 33(10), 1735–1746 (2009)

Chapter 6
Using Genetic Programming in Nonlinear Model Identification

Stephan Winkler*, Michael Affenzeller, Stefan Wagner, Gabriel Kronberger, and Michael Kommenda

Abstract. In this paper we summarize the use of genetic programming (GP) in nonlinear system identification: After giving a short introduction to evolutionary computation and genetic algorithms, we describe the basic principles of genetic programming and how it is used for data based identification of nonlinear mathematical models. Furthermore, we summarize projects in which we have successfully applied GP in R&D projects in the last years; we also give a summary of several algorithmic enhancements that have been successfully researched in the last years (including offspring selection, on-line and sliding window GP, operators for monitoring genetic process dynamics, and the design of cooperative evolutionary data mining agents). A short description of HeuristicLab (HL), the optimization framework developed by the HEAL research group, and the use of the GP implementations in HL are given in the appendix of this paper.

6.1 Evolutionary Computation and Genetic Algorithms

Work on what is nowadays called evolutionary computation started in the sixties of the 20th century, and there are two basic approaches in computer science that copy evolutionary mechanisms: evolution strategies (ES) and genetic algorithms (GA). Genetic algorithms go back to Holland [6], an

Stephan Winkler · Michael Affenzeller · Stefan Wagner · Gabriel Kronberger · Michael Kommenda
Heuristic and Evolutionary Algorithms Laboratory,
Upper Austria University of Applied Sciences, School of Informatics,
Communications and Media, Softwarepark 11, 4232 Hagenberg, Austria
e-mail: {Stephan.Winkler,Michael.Affenzeller,Stefan.Wagner,
Gabriel.Kronberger,Michael.Kommenda}@fh-hagenberg.at
http://www.heuristiclab.com

* Corresponding author.

D. Alberer et al. (Eds.): Identification for Automotive Systems, LNCIS 418, pp. 89–109.
springerlink.com   © Springer-Verlag London Limited 2012

American computer scientist and psychologist who developed his theory not only under the aspect of solving optimization problems but also to study self-adaptiveness in biological processes; the theoretical foundations of evolution strategies were formed by Rechenberg and Schwefel (see for example [17] or [18]). Both attempts work with a population model whereby the genetic information of each individual of a population is in general different. Among other things this genotype includes a parameter vector which contains all necessary information about the properties of a certain individual. Before the intrinsic evolutionary process takes place, the population is initialized arbitrarily; evolution, i.e., replacement of the old generation by a new generation, proceeds until a certain termination criterion is fulfilled.

Concerning its internal functioning, a genetic algorithm is an iterative procedure which usually operates on a population of constant size and is basically executed in the following way: An initial population of individuals (also called "solution candidates" or "chromosomes") is generated randomly or heuristically. During each iteration step, also called a "generation," the individuals of the current population are evaluated and assigned a certain fitness value. In order to form a new population, individuals are first selected (usually with a probability proportional to their relative fitness values), and then produce offspring candidates which in turn form the next generation of parents. This ensures that the expected number of times an individual is chosen is approximately proportional to its relative performance in the population. For producing new solution candidates genetic algorithms use two operators, namely crossover and mutation:

- Crossover is the primary genetic operator: It takes two individuals, called parents, and produces one or two new individuals, called offspring, by combining parts of the parents. In its simplest form, the operator works by swapping (exchanging) parts of the parents' genetic make-up before and after a randomly selected crossover point.
- The second genetic operator, mutation, is essentially an arbitrary modification which helps to prevent premature convergence by randomly sampling new points in the search space. In the case of bit strings, mutation is applied by simply flipping bits randomly in a string with a certain probability called mutation rate.

Genetic algorithms are stochastic iterative algorithms, which cannot guarantee convergence; termination is hereby commonly triggered by reaching a maximum number of generations or by finding an acceptable solution or more sophisticated termination criteria indicating premature convergence.

Overviews and detailed descriptions of GAs and adequate genetic operators for selected optimization problems can for example be found in [6], [15] and the authors' book [3].

6.2   Genetic Programming and Its Use in System Identification

6.2.1   Basics of Genetic Programming

Genetic programming (GP, [9]), an extension of the genetic algorithm, is a domain-independent, biologically inspired method that is able to create computer programs from a high-level problem statement. GP was first explored in depth in 1992 by John R. Koza who pointed out that virtually all problems in artificial intelligence, machine learning, adaptive systems, and automated learning can be recast as a search for computer programs, and that genetic programming provides a way to successfully conduct the search in the space of computer programs. Similar to GAs, GP works by imitating aspects of natural evolution to generate a solution that maximizes (or minimizes) some fitness function; a population of solution candidates evolves through many generations towards a solution using evolutionary operators and a "survival of the fittest" selection scheme. Whereas GAs are intended to find an array of characters or integers representing the solution of a given problem, the goal of a GP process is to produce a computer program solving the optimization problem at hand. Typically, a GP population contains a few hundred or thousand individuals and evolves through the action of operators known as crossover, mutation and selection; as in every evolutionary process, new individuals (in GP's case, new programs) are created, tested, and the fitter ones in the population succeed in creating children of their own, whereas unfit ones are removed from the population. Theory and application scenarios have been widely discussed [9], [13], [25].

As genetic programming is an extension to the genetic algorithm, GP also uses two main operators for producing new solution candidates in the search space, namely crossover and mutation: In the case of genetic programming, crossover is seen as the exchange of parts of programs resulting in new program structures, and mutation is applied by modifying a randomly chosen node of the respective structure tree: A sub-tree could be deleted or replaced by a randomly re-initialized sub-tree, or a function node could for example change its function type or turn into a terminal node. Of course, numerous other mutation variants are possible, many of them depending on the problem and chromosome representation chosen. Figure 6.1 shows exemplary operations on structure trees representing formulas in GP.

Figure 6.2 illustrates the main components of the GP process as also given for example in [13]; please note that this chart shows an enhanced version of the GP cycle also including offspring selection as described in Section 6.4.1.2.

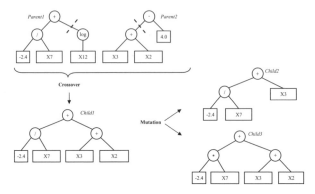

Fig. 6.1 Exemplary operations on structure trees representing formulas in GP: The crossover of programs *Parent1* and *Parent2* can for example lead to *Child1*; by mutation this model could be transformed to *Child2* or *Child3*.

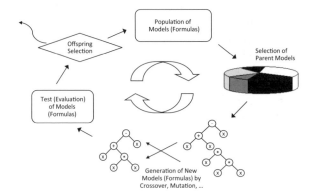

Fig. 6.2 The genetic programming cycle including offspring selection.

6.2.2 Evolutionary Structure Identification Using Genetic Programming

One of the ways how GP can be used in data based modeling is its application in systems analysis: A given system is to be analyzed and its behavior modeled by a mathematical model; this process is (especially in the context of modeling dynamic physical systems) called system identification [14]. In this context we use data including a set of variables (features), of which one (or more) is (are) specified as the target variable(s); target variables are to be described using other variables and a mathematical model. Modeling numerical data consisting of values of a target variable and of one or more independent variables (also denoted as explanatory variables) is called regression.

In principle, the main goal of regression is to determine the relationship of a dependent (target) variable t to a set of specified independent (input) variables x. Thus, what we want to get is a function f that uses x and a set of coefficients w such that

$$t = f(x, w) + \epsilon \tag{6.1}$$

where ϵ represents the error (noise) term.

Applying this procedure we assume that a model can be created with which it will also be able to predict correct outputs for other data examples (test samples). From the training data we want to generalize to situations not known (or allowed to analyze) during the training phase. Using genetic programming for data-based modeling has the advantage that we are able to design an identification process that automatically incorporates variables selection, structural identification and parameters optimization in one process.

When it comes to evaluating a model (i.e., a solution candidate in a GP based modeling algorithm), the formula has to be evaluated on a certain set of evaluation (training) data X yielding the estimated values E. These estimated target values are compared to the original values T, i.e. those which are known from experiments or calculated applying the original formula to X. This comparison is done by calculating the error between original and calculated target values; there are several ways how to measure this error, one of the simplest and probably most frequently used ones is the mean squared error (mse) function.

Accuracy on training data is not the only requirement for the result of the modeling process, compact (and if possible minimal) models are preferred as they can be used in other applications easier. One of the major problems of in data-based modeling is *overfitting*: It is, of course, not easy to find that models that ignore unimportant details and capture the behavior of the system that is analyzed. Due to this challenging character of the task of system identification, modeling has been considered as "an art" [16].

When applying genetic programming to data-based modeling the function f, which is searched for, is not of any pre-specified form; low-level functions are during the GP process combined to more complex formulas. Given a set of functions $f_1, \ldots, f_u$, the overall function induced by genetic programming can take a variety of forms. Usually, standard arithmetical functions such as addition, subtraction, multiplication, and division are considered, but also trigonometric, logical, and more complex functions can be included. Thus, the key feature of this technique is that the object of search is a symbolic description of a model, not just a set of coefficients in a pre-specified model. Of course, the models created by GP also include coefficients and constants, these are also optimized by the evolutionary process (i.e., by the interplay of random creation, mutation, crossover, and selection). This is in sharp contrast with other methods of regression, including linear regression, polynomial approaches, or also artificial neural networks, where a specific structure is assumed and often only the complexity of this model can be varied.

6.3 Application Examples

We have in recent years successfully applied GP-based system identification, especially in the context of identifying models for technical, mechatronical systems as well as in the analysis of medical data sets. In the two following sections we give a compact introduction to these application fields and some examples; an extensive overview of these application fields, a lot of application examples and empirical analysis of algorithmic enhancements of the GP-based system identification process can for example be found in [25].

6.3.1 Designing Virtual Sensors for Emissions (NO_x, Soot) of Motor Engines

The first research work of members of the Heuristic and Evolutionary Algorithms Laboratory (HEAL) in the area of system identification using GP was done in cooperation with the Institute for Design and Control of Mechatronical Systems (DesCon) at JKU Linz, Austria. The framework and the main infrastructure was given by DesCon who maintain a dynamical motor test bench (manufactured by AVL, Graz, Austria). A BMW diesel engine is installed on this test bench, and a lot of parameters of the ECU (engine control unit) as well as engine parameters and emissions are measured; for example, air mass flows, temperatures, and boost pressure values are measured, nitric oxides (NO_x, to be described later) are measured using a Horiba Mexa 7000 combustion analyzer, and an opacimeter is used for estimating the opacity of the engine's emissions (in order to measure the emission of particulate matters, i.e., soot).

In general, being able to predict emissions on-line (i.e., during engine operation) would be very helpful for low emissions engine control. While NO_x formation is widely understood (see for example [5] and the references given therein), the computation of NO_x and soot turns out to be too complex and - at the moment - not easy to be used for control. The reason for this is that in theory it would be possible to calculate the engine's emissions if all relevant parameters (pressures, temperatures, ...) of the combustion chambers were known, but (at least at the moment) we are not able to measure all these values.

In this context, our goal is to use system identification approaches in order to create models that are designed to replace or support physical sensors; we want to have models that can be potentially used instead of these physical sensors (which can be damageable or simply expensive). This is why we are here dealing with the design of so-called *virtual sensors*.

During several years of research on the identification of NO_x and soot emissions, members of DesCon have tried several modeling approaches, some of them being purely data-based as for example those using artificial neural networks (ANNs). These results were not very satisfying, as is for example documented in [5]: Even though modeling quality on training data was very good, the model's ability to predict correct values for operating points not included in the training data was very poor. This is why it was decided at DesCon to initiate research on the ability of GP to produce reasonable virtual sensors for diesel engines; we are here once again thankful to Prof. del Re for initiating these studies.

In the last years, the authors have published numerous publications in which the research results on the evolutionary design of virtual sensors for emissions of diesel engines have been described; details can for example be found in [4], [5], [27], [28], [35], [37], [38], [25], and [3]. As summarized in [38], we have seen that GP has shown to be a viable alternative to much better established methods (as for example the NARX polynomial modeling approach), with validation results absolutely comparable, if not superior. Differently from classes like the polynomial ARX models, GP is able to build new function kernels which can allow a much better insight into the system. Of course, as for every heuristic method, there is no guarantee for it and the computational effort can become rather large, but it can provide a very interesting and probably unique approach to the system model, in this case to the emissions.

Screenshots of the evaluation of a model for NO_x emissions produced by GP implemented in HL3.3 are shown in the appendix of this paper.

6.3.2   Quality Pre-assessment in Steel Industry Using Data Based Estimators

Within a cooperation project of the Institute for Design and Control of Mechatronical Systems at Johannes Kepler University Linz, Austria, and the Industrial Competence Center for Metallurgical Process Engineering, Austria, we have used several variables selection and data based classification methods (linear modeling, k-nearest-neighbor modeling, neural networks, and GP) for designing quality pre-assessment models for a steel production process; the goal of this work was to examine the ability of data based estimators to formulate models that predict the final quality of steel products (as binary output value, i.e., "ok" or "not ok") on the basis of process parameter values.

Quality assessment is a standard and central issue in industrial processes and is usually performed on the basis of an inspection of the final product, either by a human operator or in a computer assisted way. The latter approach includes usually an automatic inspection and/or classification method, very often based on pattern recognition tools. A much more appealing possibility, however, consists in performing an indirect assessment, i.e. without visual inspection of the final product. These methods can include intermediate process data and are therefore not necessarily predictive in a strict sense, but offer the essential advantage of allowing to understand the relationships between process quantities and quality. To this end, different approaches can be used, in particular a classical issue is the choice or combination of model based vs. data based approaches.

Essentially, the results of this case study can be summarized as follows:

- It is indeed possible to implement an automatic quality assessment scheme which reproduces rather well the results of the human inspection of the final product.
- The specific choice of the modeling algorithms, as long as nonlinear modeling methods are used, is not the critical issue.
- The binary decision on the quality by the human operator is the crisp expression of a continuous value, and therefore is the wrong quantity on which to train the algorithms.
- Even though large amounts of data are recorded on typical steel (and other) industrial plants, they might not contain the necessary information.

Detailed descriptions of the results achieved in this project have been published in [36] and [25].

6.3.3 Medical Data Analysis

Apart from applications of GP in the analysis of data of technical systems, we have also done intensive research on the analysis of medical data using heuristic data mining methods. We have described our research results (especially on the data based design of prediction models for the presence of tumors and other diseases) in numerous publications, please see for example [29], [34], [25], [3], or [26] for details.

6.4 Algorithmic Enhancements

6.4.1 Enhanced Selection Concepts

6.4.1.1 Gender Specific Parents Selection

Inspired by the idea of male vigor and female choice as it is considered in the model of sexual selection discussed in the area of population genetics, a new selection paradigm for GAs called SexualGA has been developed [20]. The main idea of this gender specific selection scheme is to use two different selection schemes for the selection of the two parents required for each crossover. So it becomes possible to simulate the concept of male vigor and female choice by using random selection as the first selection scheme and another selection strategy with far more selection pressure as the second one (e.g. roulette wheel selection or linear rank selection).

This gender specific selection concept not only brings the concept of GAs a little bit more towards its biological archetype, but it also has relevant advantages compared to classical GA approaches particularly concerning flexibility.

By using two different selection concepts simultaneously a GA user can influence the selection pressure level of a GA run more precisely. It is thus also possible to control the interplay between genetic diversity supporting and reducing forces in a more directed way and to better tune GA behavior depending on the individual needs of the attacked optimization problem. Further discussions and a comparison of solution qualities achieved using this principle can for example be found in [20].

6.4.1.2   Offspring Selection

Offspring selection (OS, [1], [2], [3]) considers not only the fitness of the parents in order to produce a child for the ongoing evolutionary process. Additionally, the fitness value of the evenly produced offspring is compared with the fitness values of its own parents. Offspring is accepted as a candidate for the further evolutionary process if and only if the reproduction and possibly the mutation operator were able to produce an offspring that could outperform the fitness of its own parents. This strategy guarantees that evolution is presumed mainly with crossover results that were able to mix the properties of their parents in an advantageous way.

As in the case of conventional GAs or GP, offspring are generated by parent selection, crossover, and mutation. In a second (offspring) selection step, the number of offspring to be generated is defined to depend on a predefined ratio parameter giving the quotient of next generation members that have to outperform their own parents (success ratio, $SuccRatio$). As long as this ratio is not fulfilled, further children are created and only the successful offspring will definitely become members of the next generation. When the postulated ratio is reached, the rest of the next generation members are randomly chosen from the children that did not reach the success criterion.

When using this selection model, the selection pressure $selPres$ is a measure for the effort that is necessary to produce a sufficient number of successful solutions; it is defined as the ratio of generated candidates to the population size:

$$selPres = \frac{|virtualPOP| + |POP| \cdot SuccRatio}{|POP|} \tag{6.2}$$

where $virtualPOP$ denotes the virtual population, the pool of solutions that are not considered immediately but might be inserted into the new population as "lucky losers". An upper limit for selection pressure gives a quite intuitive termination heuristics: If it is no more possible to find a sufficient number of offspring that outperform their parents, the algorithm terminates.

In the context of GP applications we have seen that maximally strict settings of OS yield best results, i.e., we have set $SuccRatio$ and the comparison factor (that determines the quality threshold for crossover products before the offspring selection step) to 1.0. Figure 6.3 schematically displays this strict variant of offspring selection.

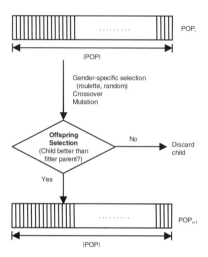

Fig. 6.3 Flowchart for embedding a simplified version of offspring selection into the GP based machine learning process.

6.4.2 On-Line and Sliding Window Genetic Programming

The idea of sliding window behavior in computer science is not novel; in machine learning, drifting concepts are often handled by moving the scope (of either fixed or adaptive size) over the training data (see for example [23] or [7]). The main idea is the following: Instead of considering all training data for training models (in the case of GP, for evaluating the models produced), the algorithm initially only considers a part of the data. Then, after executing learning routines on the basis of this part of the data, the range of samples under consideration is shifted by a certain offset. Thus, the window of samples considered is moved, it slides along the training data; this is why we are talking about sliding window behavior.

When it comes to GP based structure identification, sliding window approaches are not all too common; in general, the method is seen as a global optimization method working on a set of training samples (which are completely considered by the algorithm within the evaluation of solution candidates). On the contrary, GP is often even considered as an explicitly off-line, global optimization technique. Nevertheless, during research activities in the field of on-line systems identification [35], we discovered several surprising aspects. In general, on-line GP was able to identify models describing a diesel engine's NO_x emissions remarkably fast; the even more astonishing fact was that these models were even less prone to overfitting than those created using standard methods. After further test series and reconsidering the basic algorithmic processes, these facts did not seem to be surprising to us anymore: On the one hand, especially the fact that the environment, i.e. the training data currently considered by the algorithm, is not constant but rather changing during the execution of the training process, contributes positively to the models' quality, it obviously decreases the threat of overfitting. On the other hand, the interplay of a changing data basis and models created using different data also seems to be contributing in a positive way. As the on-line algorithm is executed and evaluates models using (new) current training data forgetting samples that were recorded in the beginning, those "old" data are really forgotten from the algorithm's point of view. Still, the models created on the basis of these old data are still present. The behavior that results out of this procedure is more or less that several possible models that explain the first part of the data are created, and as the scope is moved during the algorithm's execution, only those models are successful that are also able to explain "new" training data.

So, the most self-evident conclusion was that these benefits of online training should be transferred to off-line training using GP. The idea for selection pressure driven sliding window GP (described in [30] and [25]) is to initially reduce the amount of data that is available for the algorithm as identification data. As the identification process is executed, better and better models are created which leads to a rise of the selection pressure; as soon as the selection pressure reaches a predefined maximum value, the limits of the identification data are shifted and the algorithm goes on considering another part of the available identification data set. This procedure is then repeated until the actual training data scope has reached the end of the training data set available for the identification algorithm, i.e. when all data have been considered. By doing so, the algorithm is even less exposed to overfitting, and due to the fact that the models created are evaluated on much smaller data sets we also expect a significant decrease of runtime consumption.

6.4.3   Cooperative Evolutionary Data Mining Agents

One of the disadvantages of GP is that it takes a long time until the result of a run is available; especially for non-trivial data sets it is usually necessary to analyze the result of a previous run before a new run can be started for instance to counteract over or under fitting or to exclude dominant input variables. This is an even bigger problem for domain experts who do not fully understand the internal details of GP and thus often have problems to configure the algorithm correctly. Usually a few iterations are necessary until a configuration for the algorithm is found that works for data mining task at hand. However, even when such a configuration has been found there is another aspect that causes friction in the knowledge discovery process.

One approach to improve the discovery of more detailed knowledge is to run many independent GP processes to generate a large number of models for each possible target variable with different complexities and to extend the data mining process to search for implicit dependencies between input

variables. In combination with an interactive user-interface to filter and analyze the generated models and to compose simple models to hierarchic models the user gains new knowledge step by step while investigating the set of
models.

This approach to use cooperative evolutionary data mining agents (CED-
MA) has for example been described in [11]: The authors have there described
the architecture of a distributed data mining system in which genetic programming agents create a large amount of structurally different models which
are stored in a model database. A search engine for models that is connected
to this database allows interactive exploration and analysis of models, and
composition of simple models to hierarchical models. The search engine is
the crucial component of the system in the sense that it supports knowledge
discovery and paves the way for the goal of finding interesting hidden causal
relations.

Figure 6.4 shows how the standard GP process could be improved by
using parallelism to run different GP processes at the same time. Each of the
distributed genetic programming agents has different settings for the target
variable, maximally allowed model complexity and the set of allowed input
variables. Controller agents create new GP jobs and coordinate the running
GP agents. It would be interesting to have more intelligent controller agents
which try to predict which models are more interesting for the user and guide
the GP agents to search especially for such models; this remains a topic for
future research.

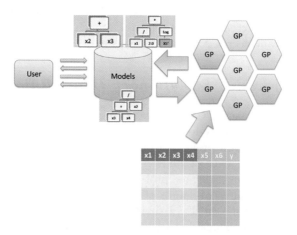

Fig. 6.4 In the context of CEDMA, distributed GP agents continuously analyze
the data set and store new models in the database. These models can be analyzed
separately or combined to models describing the overall system behavior.

An especially interesting aspect is the intelligent coordination and control of GP agents: On the one hand these agents should be used for finding different models that describe the analyzed process in different ways, on the other hand the models found by the agents (that describe different aspects of the system or target variables) should be combined to models that describe the whole system.

6.5 Algorithm Analysis: Population Diversity Dynamics

Of course, any modification of the standard GP procedure can have severe effects on internal genetic processes and population dynamics in GP; especially the use of strict offspring selection influences the GP process significantly. This is why we have used several approaches to describe and analyze GP population dynamics as for example the following ones:

- We have analyzed genetic propagation in GP in order to find out which individuals of the population are more or less able to pass on their genetic material to the next generation. On the one hand, better individuals (providing better genetic material) are supposed to pass on their genetic make-up more than other individuals; still, of course for the sake of genetic diversity also solutions that are not that good should also be able to contribute to the genetic process. In [31] and [25] we have summarized empirical studies using several data sets for analyzing the differences in GP population dynamics of different variants of genetic programming.

- When analyzing genetic diversity in GP we systematically compare the structural components of solutions in a GP population and so calculate how similar the individuals are to each other. Of course, on the one hand the GP optimization process is supposed to converge so that eventually all solutions will be more or less similar to each other, but on the other hand the genetic diversity should be maintained as long as possible (or necessary) in order to keep the genetic process active. In [32], [25], and [33] we have explained methods how to measure structural similarities in models, and we have also analyzed the respective differences discovered in various variants of genetic programming; examples of the graphical representation of the progress of solutions similarity in GP processes is shown in Figure 6.5.

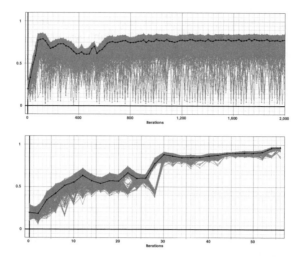

Fig. 6.5 Population diversity analysis: The progress of the mutual solution similarity in an exemplary standard GP process (shown in the upper figure) and in an exemplary GP process with strict offspring selection (shown in the lower figure). These figures are taken from [24].

Acknowledgements

In this paper we have reported on research work that has been done within several research and development projects. In detail, the following institutions have substantially supported the authors' work on GP based modeling over the last years:

- Since 2004, the Linz Center of Mechatronics (LCM) has supported studies in the field of data based identification of technical systems within the strategic project 230100 as well as under grant LCM-001.
- Work done on the identification of models for the quality pre-assessment in steel industry using data based estimators has been conducted at the Johannes Kepler University Linz jointly with the Industrial Competence Center for Metallurgical Process Engineering, Austria and has been partly funded by the Austrian Ministry for Economy and Labor in the frame of its Industrial Competence Center Program K-ind/K-net.
- In the years 2006 – 2009, GP related research work was done by the authors within the Translational Research Project L284-N02 "GP-Based Techniques for the Design of Virtual Sensors" sponsored by the Austrian Science Fund (FWF). The following institutions were involved in the execution of this project:
    - the Department of Software Engineering as well as Research Center Hagenberg, Upper Austria University of Applied Sciences, Campus Hagenberg,
    - the Institute for Design and Control of Mechatronical Systems, Johannes Kepler University Linz, and
    - the Linz Center of Mechatronics (LCM).
- In the years 2007 – 2008, GP related research work on cooperative evolutionary data mining agents done by the authors was sponsored by the basis financing program of the Upper Austria University of Applied Sciences.

- In September 2008 the Josef Ressel Center for Heuristic Optimization, *Heureka!*, was initiated at Campus Hagenberg of the Upper Austria University of Applied Sciences. This research center is funded by the Austrian Research Promotion Agency (FFG) and the center's partner companies voestalpine, Rosenbauer, Carvatech, and the General Hospital Linz; the authors' research on heuristic optimization (including GP based system identification) is continued and applied for solving various real world problems within *Heureka!*.

References

[1] Affenzeller, M., Wagner, S.: Offspring selection: A new self-adaptive selection scheme for genetic algorithms. In: Ribeiro, B., Albrecht, R.F., Dobnikar, A., Pearson, D.W., Steele, N.C. (eds.) Adaptive and Natural Computing Algorithms. Springer Computer Series, pp. 218–221. Springer, Heidelberg (2005)

[2] Affenzeller, M., Wagner, S., Winkler, S.: GA-selection revisited from an ES-driven point of view. In: Mira, J., Álvarez, J.R. (eds.) IWINAC 2005. LNCS, vol. 3562, pp. 262–271. Springer, Heidelberg (2005)

[3] Affenzeller, M., Winkler, S., Wagner, S., Beham, A.: Genetic Algorithms and Genetic Programming - Modern Concepts and Practical Applications. Numerical Insights. CRC Press, Boca Raton (2009)

[4] Alberer, D., del Re, L., Winkler, S., Langthaler, P.: Virtual sensor design of particulate and nitric oxide emissions in a DI diesel engine. In: Proceedings of the 7th International Conference on Engines for Automobile ICE 2005 (2005)

[5] del Re, L., Langthaler, P., Furtmüller, C., Winkler, S., Affenzeller, M.: NOx virtual sensor based on structure identification and global optimization. In: Proceedings of the SAE World Congress 2005 (2005)

[6] Holland, J.H.: Adaption in Natural and Artifical Systems. University of Michigan Press, Ann Arbor (1975)

[7] Hulten, G., Spencer, L., Domingos, P.: Mining time-changing data streams. In: Proceedings of the 7th ACM SIGKDD International Conference on Knowledge Discovery and Data Mining, pp. 97–106 (2001)

[8] Kommenda, M., Kronberger, G., Winkler, S., Affenzeller, M., Wagner, S., Schickmair, L., Lindner, B.: Application of genetic programming on temper mill datasets. In: Proceedings of the 2nd International Symposium on Logistics and Industrial Informatics LINDI 2009. IEEE Publications, Los Alamitos (2009)

[9] Koza, J.R.: Genetic Programming: On the Programming of Computers by Means of Natural Selection. The MIT Press, Cambridge (1992)

[10] Kronberger, G., Winkler, S., Affenzeller, M., Beham, A., Wagner, S.: On the success rate of crossover operators for genetic programming with offspring selection. In: Moreno-Díaz, R., Pichler, F., Quesada-Arencibia, A. (eds.) EUROCAST 2009. LNCS, vol. 5717, pp. 793–800. Springer, Heidelberg (2009)

[11] Kronberger, G., Winkler, S., Affenzeller, M., Wagner, S.: Data mining via distributed genetic programming agents. In: Bruzzone, A., Longo, F., Piera, M.A., Aguilar, R.M., Frydman, C. (eds.) Proceedings of the 20th European Modeling and Simulation Symposium (EMSS 2008), pp. 95–99. DIPTEM University of Genova (2008)

[12] Kronberger, G., Winkler, S., Affenzeller, M., Wagner, S.: On crossover success rate in genetic programming with offspring selection. In: Vanneschi, L., Gustafson, S., Moraglio, A., De Falco, I., Ebner, M. (eds.) EuroGP 2009. LNCS, vol. 5481, pp. 232–243. Springer, Heidelberg (2009)

[13] Langdon, W.B., Poli, R.: Foundations of Genetic Programming. Springer, Heidelberg (2002)

[14] Ljung, L.: System Identification – Theory For the User, 2nd edn. PTR Prentice Hall, Upper Saddle River (1999)

[15] Michalewicz, Z.: Genetic Algorithms + Data Structures = Evolution Programs. Springer, Heidelberg (1992)

[16] Morrison, F.: The Art of Modeling Dynamic Systems: Forecasting for Chaos, Randomness, and Determinism. John Wiley & Sons, Inc., Chichester (1991)

[17] Rechenberg, I.: Evolutionsstrategie. Friedrich Frommann Verlag (1973)

[18] Schwefel, H.-P.: Numerische Optimierung von Computer-Modellen mittels der Evolutionsstrategie. Birkhäuser Verlag, Basel (1994)

[19] Wagner, S.: Heuristic Optimization Software Systems - Modeling of Heuristic Optimization Algorithms in the HeuristicLab Software Environment. PhD thesis, Johannes Kepler University, Linz, Austria, (2009)

[20] Wagner, S., Affenzeller, M.: SexualGA: Gender-specific selection for genetic algorithms. In: Callaos, N., Lesso, W., Hansen, E. (eds.) Proceedings of the 9th World Multi-Conference on Systemics, Cybernetics and Informatics (WMSCI) 2005, vol. 4, pp. 76–81. International Institute of Informatics and Systemics (2005)

[21] Wagner, S., Kronberger, G., Beham, A., Winkler, S., Affenzeller, M.: Modeling of heuristic optimization algorithms. In: Bruzzone, A., Longo, F., Piera, M.A., Aguilar, R.M., Frydman, C. (eds.) Proceedings of the 20th European Modeling and Simulation Symposium, pp. 106–111. DIPTEM University of Genova (2008)

[22] Wagner, S., Winkler, S., Pitzer, E., Kronberger, G., Beham, A., Braune, R., Affenzeller, M.: Benefits of plugin-based heuristic optimization software systems. In: Moreno Díaz, R., Pichler, F., Quesada Arencibia, A. (eds.) EUROCAST 2007. LNCS, vol. 4739, pp. 747–754. Springer, Heidelberg (2007)

[23] Widmer, G., Kubat, M.: Learning in the presence of concept drift and hidden contexts. Machine Learning 23(2), 69–101 (1996)

[24] Winkler, S.: Evolutionary System Identification - Modern Approaches and Practical Applications. PhD thesis, Johannes Kepler University, Linz, Austria (2008)

[25] Winkler, S.: Evolutionary System Identification - Modern Approaches and Practical Applications. Schriften der Johannes Kepler Universität Linz, Reihe C: Technik und Naturwissenschaften, Universitätsverlag Rudolf Trauner (2009)

[26] Winkler, S., Affenzeller, M., Jacak, W., Stekel, H.: Classification of tumor marker values using heuristic data mining methods. In: Proceedings of the GECCO 2010 Workshop on Medical Applications of Genetic and Evolutionary Computation (MedGEC 2010), Portland, OR. Association for Computing Machinery (ACM), New York (2010)

[27] Winkler, S., Affenzeller, M., Wagner, S.: Genetic programming based model structure identification using on-line system data. In: International Mediterranean Modeling Multiconference, Proceedings of Conceptual Modeling and Simulation Conference, CMS 2005, pp. 177–186 (2005)

[28] Winkler, S., Affenzeller, M., Wagner, S.: New methods for the identification of nonlinear model structures based upon genetic programming techniques. Journal of Systems Science 31(1), 5–13 (2005)

[29] Winkler, S., Affenzeller, M., Wagner, S.: Advanced genetic programming based machine learning. Journal of Mathematical Modelling and Algorithms 6(3), 455–480 (2007)

[30] Winkler, S., Affenzeller, M., Wagner, S.: Selection pressure driven sliding window behavior in genetic programming based structure identification. In: Moreno Díaz, R., Pichler, F., Quesada Arencibia, A. (eds.) EUROCAST 2007. LNCS, vol. 4739, pp. 788–795. Springer, Heidelberg (2007)

[31] Winkler, S., Affenzeller, M., Wagner, S.: Off spring selection and its effects on genetic propagation in genetic programming based system identification. In: Trappl, R. (ed.) Cybernetics and Systems 2008, vol. 2, pp. 549–554. Austrian Society for Cybernetic Studies (2008)

[32] Winkler, S., Affenzeller, M., Wagner, S.: Variables diversity in systems identification based on extended genetic programming. Journal of Systems Science 34(2), 27–34 (2008)

[33] Winkler, S., Affenzeller, M., Wagner, S.: On the reliability of nonlinear modeling using enhanced genetic programming techniques. In: Skiadas, C.H., Dimotikalis, I., Skiadas, C. (eds.) Topics on Chaotic Systems: Selected Papers from Chaos 2008 International Conference, pp. 24–31. World Scientific Publishing, Singapore (2009)

[34] Winkler, S., Affenzeller, M., Wagner, S.: Using enhanced genetic programming techniques for evolving classifiers in the context of medical diagnosis. Genetic Programming and Evolvable Machines 10(2), 111–140 (2009)

[35] Winkler, S., Efendic, H., Affenzeller, M., Del Re, L., Wagner, S.: On-line modeling based on genetic programming. International Journal on Intelligent Systems Technologies and Applications 2(2/3), 255–270 (2007)

[36] Winkler, S., Efendic, H., del Re, L.: Quality pre-assesment in steel industry using data based estimators. In: Cierpisz, S., Miskiewicz, K., Heyduk, A. (eds.) Proceedings of the IFAC Workshop MMM 2006 on Automation in Mining, Mineral and Metal Industry, International Federation for Automatic Control (2006)

[37] Winkler, S., Hirsch, M., Affenzeller, M., del Re, L., Wagner, S.: Incorporating physical knowledge about the formation of nitric oxides into evolutionary system identification. In: Bruzzone, A., Longo, F., Piera, M.A., Aguilar, R.M., Frydman, C. (eds.) Proceedings of the 20th European Modeling and Simulation Symposium (EMSS 2008), pp. 69–74. DIPTEM University of Genova (2008)

[38] Winkler, S.M., Hirsch, M., Affenzeller, M., del Re, L., Wagner, S.: Virtual sensors for emissions of a diesel engine produced by evolutionary system identification. In: Moreno-Díaz, R., Pichler, F., Quesada-Arencibia, A. (eds.) EUROCAST 2009. LNCS, vol. 5717, pp. 657–664. Springer, Heidelberg (2009)

Appendix: The HeuristicLab Framework for Heuristic Optimization

HeuristicLab (HL) is an optimization environment which has continuously evolved since 2002. By incorporating beneficial features of existing frameworks as well as several novel concepts, HeuristicLab represents a powerful and mature framework (written in C# using the Microsoft .NET framework) which can be used for the development, analysis, comparison, and productive application of heuristic optimization algorithms. In the course of the development of HeuristicLab, from the first major version to the most recent version 3.3 (HL3), the architecture has gone through several major changes. Similar to many other frameworks the separation of algorithms and problems was one of the initial core concepts; in the most recent version of HL an algorithm is represented by a customizable graph of operations that can be specified and modified even down to very detailed levels in the GUI itself. The experimentation possibilities allow defining, executing, and analyzing runs of algorithms with different parameter settings in one place.

The following list highlights the most important aspects influencing the HeuristicLab development process:

- **Paradigm independence:** HeuristicLab is an environment for heuristic optimization in general and is not limited to any specific optimization paradigm or optimization problem. Due to a very high level of abstraction it is possible to implement very different kinds of optimization algorithms and problems for HeuristicLab. This approach has already proven its worth as lots of different algorithms and problems are already available within the HeuristicLab environment.
- **Plug-in architecture:** HeuristicLab builds upon a plugin architecture so that it can be extended without developers having to touch the original source code [22]. A lightweight plugin concept is implemented in HeuristicLab by keeping the coupling between plugins very simple: Collaboration between plugins is described by interfaces. The plugin management mechanism contains a discovery service that can be used to retrieve all types implementing an interface required by the developer; it takes care of locating all installed plugins, scanning for types, and instantiating objects. As a result, building extensible applications is just as easy as defining appropriate interfaces (contracts) and using the discovery service to retrieve all objects fulfilling a contract. Every HeuristicLab feature, even the Optimizer user interface, is available as a plugin. It is possible to easily derive completely customized GUIs tailored to very specific applications or bundle specific plugins into packages, so that users can be given exactly the functionality they need.
- **Usability:** In spite of the high level of abstraction HeuristicLab is not just another optimization library but should be understood as optimization environment. Usability and user-friendliness are very important as

HeuristicLab is not only used for rapid algorithm and problem development but also as optimization tool in industrial projects and as supporting tool in university courses. HeuristicLab provides a state-of-the-art graphical user interface, is very easy to install, and includes comprehensive documentation. The genericity of HL allows to build almost any kind of algorithm through a click-and-drag user interface.

- **Algorithm model:** The core of HeuristicLab is its algorithm model [21]. It is very easy to understand on an abstract level, but naturally reveals more complexity the deeper one descends into it. However, luckily for most users, understanding more than the abstract level of the core language is not necessary for applying heuristic optimization. The modularity of HL allows to combine any number of operators which can be saved and reused. The algorithms that are created in HL are saved in files called workbenches which can be distributed and modified; thus, algorithm development does not happen in a programming language anymore and parts of the algorithms can be shared among workbenches.
- **Scalability:** The scalability of HL allows to fully use a multi-core machine or even a multicomputer parallel system such as a cluster making use of today's increased parallel processing power.

The architecture of HL has been described in several publications; the most detailed overview is given in the PhD thesis of Stefan Wagner [19]. An overview of the publications of the HEAL research group can be found online on the research group's website[2].

The software framework is released under the GNU General Public License (GPL). The source code as well as a binary version of the environment can be downloaded from the HL website[3]; interested readers may also find screenshots and other development related material there.

Genetic Programming Based System Identification in HeuristicLab

Genetic programming has already been implemented in several versions of HeuristicLab; the architecture of these implementations as well as their application in several research projects have been published in numerous conference papers, journal articles, and one PhD thesis.

For example, the identification of nonlinear models in the context of technical systems using GP in HL1 and HL2 has been described in [35], [29], [32], [34], the authors' book [3] and the PhD thesis of Stephan Winkler [25].

The GP implementation in the most recent version of HL (HL3) has for example been applied for evolving cooperative GP agents (as described in Section 6.4.3 as well as in the analysis of temper mill data sets and blast furnace processes; please see for example [11], [12], [10] and [8] for details. The following Figures 6.6, 6.7, and 6.8 show screenshots of a GP application

---

[2] http://heal.heuristiclab.com/publications
[3] http://dev.heuristiclab.com

implemented in HL3: Data measured at the motor engine test bench of the Institute for Design and Control of Mechatronical Systems at JKU Linz have been used for identifying models that describe the NO_x emissions of a BMW diesel engine.

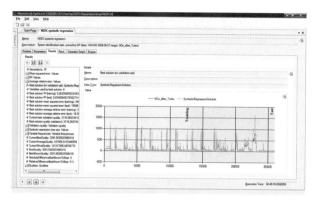

Fig. 6.6 GP in HL3: Visualization of measured vs. modeled target values (NO_x emissions of a BMW diesel engine).

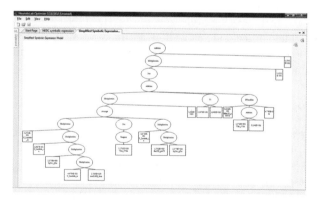

Fig. 6.7 GP in HL3: Visualization of a model for the NO_x emissions of a BMW diesel engine, generated using GP.

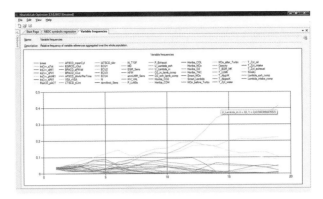

Fig. 6.8 GP in HL3: Variable frequencies in a GP process modeling the NO_x emissions of a BMW diesel engine.

Chapter 7
Markov Chain Modeling and On-Board Identification for Automotive Vehicles

Dimitar P. Filev and Ilya Kolmanovsky

Abstract. The paper considers issues and algorithmic approaches related to modeling and identification of Markov Chain type models for vehicle applications. The use of Markov Chain models in these applications is stimulated by their ability to reflect aggregate vehicle operating conditions and induce "best on average" control policies based on application of stochastic dynamic programming and stochastic Model Predictive Control. A novel fuzzy encoding approach of continuous signals is proposed in which a signal value is simultaneously associated with multiple cells, and it is shown to enhance identification and prediction accuracy of Markov Chain type models. A computationally simple identification algorithm, suitable for on-board applications, is proposed to learn Markov Chain transition probabilities in real-time. Examples of real-time learning models of vehicle speed and road grade are reported to illustrate the overall identification approach.

7.1 Introduction

Traditionally, fuel economy improvement efforts in the automotive industry have focused on the development of improved engines, transmissions, and lighter-weight vehicles. Recently, growing computing power and information connectivity (through sensors, GPS, digital maps, road and traffic information systems, wireless internet, vehicle-to-vehicle communication, etc.)

Dimitar P. Filev
Ford Motor Company, Research and Advanced Engineering,
2101 Village Rd., Dearborn, MI 48121, Phone: 313-248-1652
e-mail: dfilev@ford.com

Ilya Kolmanovsky
Department of Aerospace Engineering,
the University of Michigan, Ann Arbor, MI
e-mail: ilya@umich.edu

D. Alberer et al. (Eds.): Identification for Automotive Systems, LNCIS 418, pp. 111–128.
springerlink.com © Springer-Verlag London Limited 2012

provided further, transformational opportunities to improve vehicle fuel efficiency. Some of the main research and technological challenges to realize these opportunities are in the development of algorithms that can continuously monitor and analyze driving conditions (terrain, traffic, etc.) and driver behavior and their impact on the fuel efficiency, and either offer an intelligent advice to the driver or automatically adjust vehicle control strategy, while minimizing the detrimental effects of disturbing and annoying the driver.

As an example, we have recently considered powertrain control strategies which are optimized for frequently travelling in a given geographic region [8]. A control policy for adjusting vehicle speed has been defined which, on one hand, is responsive only to current operating conditions, as is the case for the conventional powertrain control strategy, but on the other hand, it provides best on-average performance (in terms of fuel economy and travel time) when travelling frequently in this region. Markov Chain type models [2, 3] were used to aggregate and summarize terrain and traffic properties and stochastic dynamic programming was applied to generate an optimal control policy. Several other references (see e.g., [7, 11, 14, 15, 17]) have considered the use of Markov Chain type models to represent driving conditions and obtain control policies using stochastic dynamic programming, stochastic Model Predictive Control and reinforcement learning.

Markov Chains models are typically applied to sampled signals, where a signal is considered to be in one of a finite number of states. To treat signals which vary in a continuous range an encoding scheme needs to be introduced to attribute the signal value to one of the states of the finite-state Markov Chain. The common approach is to use interval partitioning associating each interval with a discrete Markov Chain state.

Since interval type encoding is a special case of a more general process of information granulation [16], alternative granulation paradigms, e.g. those based on the use of fuzzy sets or rough sets, can also be employed as transformations assigning continuous signal values to discrete states. This approach has been suggested in our previous work [5], where it was shown that introducing fuzzy encoding can be advantageous but it involves numerous theoretical and algorithmic issues that are not evident consequences of the conventional Markov Chain theory. In [6] we generalized Chapman-Kolmogorov theorem to the case of Markov chain models with fuzzy encoding and in [5] we have highlighted the use of fuzzy encoding in facilitating Markov Chain aggregation.

The objective of this paper is to further consider issues and algorithmic approaches related to modeling and identification of Markov Chain type models based on fuzzy encoding for real-time vehicle applications. In Section 12.1.1 Markov chains based on interval and fuzzy granulation are introduced and compared, assuming that a signal being modeled is a scalar. A computationally simple identification algorithm (as compared e.g., to Baum-Welch algorithm in [1]), suitable for on-board applications, is proposed to learn Markov Chain transition probabilities in real-time in Section 12.2. In Section 12.3 we consider the case of a vector signal being modeled and present an aggregation approach to signal values to reduce the number of Markov chain states. Finally, concluding remarks are made in Section 7.5.

7.2   Generalization of Conventional Markov Chain through Interval and Fuzzy Granulation

A conventional finite-state Markov chain is a stochastic process with finite number of states satisfying the Markov assumption [2, 3] - *the future is independent of the past given present.* We denote the state space of the finite state Markov chain as $S = \{\bar{x}_j, j = 1, \cdots, M\} \subset X$ where $X \subset R$ is a bounded set, and we introduce transition probabilities as

$$\pi_{ij} = P\left(x^+ = \bar{x}_j | x = \bar{x}_i\right), \tag{7.1}$$

which form a transition probability matrix, Π. For a given $x = \bar{x}_i$ the next state inferred by the Markov chain is the one that maximizes the probability distribution corresponding to $\bar{x}_i$:

$$x^+ = \bar{x}_j, \text{ if } x = \bar{x}_i, j \in \arg\max_k \pi_{ik}.$$

The transition probabilities may be estimated from the frequencies of observed transitions between the states, i.e.,

$$\pi_{ij} \approx \frac{N_{ij}}{N_{oi}}. \tag{7.2}$$

In (7.2), N_{ij} denotes the number of observed transitions from state $\bar{x}_i$ to state $\bar{x}_j$ over a given time interval, and $N_{0i} = \sum_{j=1}^{M} N_{ij}$ is the total number of transitions that are initiated from the state $\bar{x}_i$. The present and one step-ahead state probability vectors (i.e., vectors of probabilities of x taking one of finite values, $\bar{x}_j$) are linked through transition probabilities as

$$(p^+)^T = p^T \Pi,$$

and for $t > 1$ steps ahead by the Chapman-Kolmogorov equation [2] implying that:

$$(p^{+t})^T = p^T \Pi^t. \tag{7.3}$$

In the case when x, x^+ can take arbitrary values in X (e.g., they represent a physical quantity) encoding may be used to transform the continuous domain of x, x^+ to a discrete one so that the main Markov chain concepts may be applied. The main idea is to introduce a finite set of *granules* representing the states of the chain, use the granules to code x, x^+, apply the

Markov approach to the granular representation, and then decode the result
back to the original continuous domain. In the remainder of this section we
consider two granulation techniques – interval and fuzzy encoding. The differences between the conventional Markov chain and the generalized Markov
chain models that are based on interval or fuzzy encoding are illustrated in
Figure 12.1.

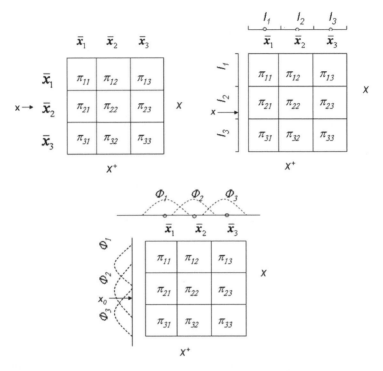

Fig. 7.1 Markov chain models: conventional (top, left), interval encoding-based
(top, right) and fuzzy encoding-based (bottom).

7.2.1   Markov Chains with Interval Encoding

The interval encoding approach is based on partitioning X into a finite set of
disjunct intervals, $I_j, j = 1, \cdots, M$, and assigning to each interval a Markov
chain state, $\bar{x}_j \in I_j$, which is typically the midpoint of the interval I_j. The
state x is considered to be in the discrete state $\bar{x}_j$ if $x \in I_j$.

Based on this interval partitioning, the continuous real valued state $x_0 \in I_j$
may be encoded as an $M-$dimensional *probability vector* with the jth element
equal to 1 and all remaining elements equal to zero:

$$\alpha^T(x_0) = \begin{bmatrix} 0 \cdots 1 \cdots 0 \end{bmatrix}. \tag{7.4}$$

For a given $x_0 \in I_j$ the next state inferred by the Markov chain can be calculated as the one that maximizes the probability distribution corresponding to $\bar{x}_j$:

$$x_0^+ = \bar{x}_i, \text{ if } x_0 \in I_j, i \in \arg\max_k \pi_{jk} \tag{7.5}$$

or, alternatively, as the expectation based on the interval centers:

$$x_0^+ = \sum_{i=1}^M \pi_{ij} \bar{x}_i, \text{ if } x_0 \in I_j.$$

In a more general setting, the above expression of the expectation based inferred value can be formally re-written in a matrix form:

$$x_0^+ = \alpha^T(x_0)\, \Pi\, \bar{x}^+ \tag{7.6}$$

where $\bar{x}^+$ is the vector of the interval midpoints $\bar{x}_j \in I_j$. Therefore, the interval encoding in the example shown in Figure 12.1 infers a deterministic output x^+ that can be expressed by either

$$x^+ = \bar{x}_j^+, j = \arg\max_k \pi_{2k},$$

or

$$x^+ = \pi_{21}\bar{x}_1^+ + \pi_{22}\bar{x}_2^+ + \pi_{23}\bar{x}_3^+.$$

Transition probabilities may be estimated analogously to the case of conventional finite-state Markov chain [10].

7.2.2   *Markov Chains with Fuzzy Encoding*

We introduced fuzzy encoding in [5] motivated by the approximation properties of fuzzy granules. Following the theory of approximate reasoning [12, 18], we consider fuzzy subsets A_j, rather than intervals, defined by their membership functions

$$a_j(x) : X \rightarrow [0,1]; \forall x \in X, \exists j, 1 \le j \le M, a_j(x) \ne 0. \tag{7.7}$$

Fuzzy encoding provides an additional opportunity to interpolate between multiple Markov chain states based on the degree of membership (closeness) of the random input x to the individual states - a property that is not available in the interval encoding [5].

Unlike interval encoding, in the fuzzy encoding case, a single $x \in X$ may be associated with several states $\bar{x}_j$ of the underlying finite-state Markov chain model. As shown in Figure 12.1, in the case of interval encoding the state variable x belongs to one of the intervals, specifically to I_2 in Figure 12.1. The next state probability distribution is the one that is associated with the second interval, i.e. the second row of the transition probability matrix. In the case of fuzzy encoding, the state variable x belongs simultaneously to the fuzzy subsets A_2 and A_3, with different degrees of membership $a_2(x)$ and $a_3(x)$.

From Figure 12.1 one can easily grasp the similarity between the concepts of Markov chains with interval or fuzzy encoding and rule-based systems. The Markov chain with interval encoding resembles a set of M^2 Boolean logic rules which operate on the interval granules I_i while the Markov chain with fuzzy granulation can be expressed as a collection of M^2 rules with fuzzy predicates of the form

$$\text{If } x \text{ is } A_i \text{ Then } x^+ \text{ is } A_j \text{ with weight } \pi_{ij} \tag{7.8}$$

where rule weights π_{ij}, $i, j = 1, \cdots, M$, are the elements of the transition probability matrix Π. The antecedent subsets are defined by their membership functions $a(x)$ on the universe of x. The consequent subsets are defined by the same membership functions $a(x^+)$ but on the universe of x^+. The rule weights represent the credibility (frequency) of the antecedent / consequent relations that are covered by the rules; see e.g. [12] and [18] for the frequency interpretation of the weights. Following this analogy and the fundamentals of the theory of approximate reasoning we introduced some of the basic ideas of using Markov chains with fuzzy encoding as an on-board modeling tool [5]. The main results are summarized below.

Using the membership functions, the continuous real valued state x_0 may be encoded as a possibility vector of the degrees membership of x_0 to each of the subsets A_j:

$$\tilde{\alpha}^T(x_0) = \begin{bmatrix} a_1(x_0) \, a_2(x_0) \, \cdots \, a_M(x_0) \end{bmatrix} \tag{7.9}$$

The possibility vector $\tilde{\alpha}(x_0)$ is analogous to the concepts of the "degree of firing,", "firing level" [18] of a rule and characterizes the measure of similarity (compatibility, matching) between x_0 and the subsets A_j.

For a numerical (crisp) state x_0 each rule with antecedent A_i is fired with firing level $\tilde{\alpha}_i(x_0)$ by (7.8) and infers a fuzzy subset (possibility distribution):

$$\chi_i^+(x_0) = \tilde{\alpha}_i(x_0) \sum_{j=1}^{M} \pi_{ij} a_j(x^+)$$

$$= \tilde{\alpha}_i(x_0) \Pi_i^T a(x^+) \tag{7.10}$$

By aggregating the subsets $\chi_i^+(x_0)$ with weights $\alpha(x_0)$ that are proportional to the firing levels of the rules $\tilde{\alpha}_i(x_0)$ we obtain for the overall possibility distribution $\chi^+(x_0)$ generated by the rules (see [12, 18]):

$$\chi^+(x_0) = \frac{\displaystyle\sum_{i=1}^{M} \tilde{\alpha}_i(x_0) \sum_{j=1}^{M} \pi_{ij} a_j(x^+)}{\displaystyle\sum_{i=1}^{M} \tilde{\alpha}_i(x_0) \sum_{j=1}^{M} \pi_{ij}} = \frac{\displaystyle\sum_{i=1}^{M} \tilde{\alpha}_i(x_0) \Pi_i^T a(x^+)}{\displaystyle\sum_{i=1}^{M} \tilde{\alpha}_i(x_0)} \tag{7.11}$$

$$= \sum_{i=1}^{M} \alpha_i(x_0) \Pi_i^T a(x^+) = \alpha^T(x_0) \, \Pi \, a(x^+).$$

where $\alpha(x_0)$ is the vector of the normalized degrees of firing (matching) the subsets A_j by x_0:

$$\alpha(x_0) = \tilde{\alpha}(x_0)/(\sum_{j=1}^{M} \tilde{\alpha}_j(x_0)) \tag{7.12}$$

For a given crisp state x_0 the deterministic prediction of the next state inferred by the Markov chain with fuzzy encoding is calculated as the defuzzified value (i.e. the *centroid* or "center of gravity" of possibility distribution $\chi^+(x_0)$):

$$x^+ = \int_X \chi^+(y) \cdot y dy / \int_X \chi^+(y) dy.$$

Note that by substituting for $\chi^+(x_0)$ from (7.11) we have:

$$\int_X \chi^+(y) y dy = \sum_{i=1}^{M} \alpha_i(x_0) \sum_{j=1}^{M} \pi_{ij} \int_X y a_j(y) dy,$$

$$\int_X \chi^+(y) dy = \sum_{i=1}^{M} \alpha_i(x_0) \sum_{j=1}^{M} \pi_{ij} \int_X a_j(y) dy.$$

After algebraic manipulations (see also [12]), these expressions imply that the predicted deterministic next state value x_0^+ given x_0 is:

$$x_0^+ = \frac{\displaystyle\sum_{i=1}^{M} \alpha_i(x_0) \sum_{j=1}^{M} \pi_{ij} V_j \bar{x}_j^+}{\displaystyle\sum_{i=1}^{M} \alpha_i(x_0) \sum_{j=1}^{M} \pi_{ij} V_j} \tag{7.13}$$

where $\bar{x}_j^+$ and V_j denote the centroid and the volume, respectively, of the membership function, a_j:

$$\bar{x}_j^+ = \int_X y a_j(y) dy / \int_X a_j(y) dy, \quad V_j = \int_X a_j(y) dy.$$

If membership functions have the same volume (i.e. all V_j's are equal), the expression for x^+ further simplifies to

$$x_0^+ = \frac{\displaystyle\sum_{i=1}^{M} \alpha_i(x_0) \sum_{j=1}^{M} \pi_{ij} \bar{x}_j^+}{\displaystyle\sum_{i=1}^{M} \alpha_i(x_0) \sum_{j=1}^{M} \pi_{ij}} = \alpha^T(x_0) \, \Pi \, \bar{x}^+. \tag{7.14}$$

In derivation of (7.14) we used the fact that the rows of the transition probability matrix, Π, sum up to 1 ($\displaystyle\sum_{j=1}^{M} \pi_{ij} = 1, \ \forall i = 1, \cdots, M$) and that $\displaystyle\sum_{i=1}^{M} \alpha_i(x_0) = 1$. Note that expression (7.14) interpolates the transition probabilities and the centroids $\bar{x}^+$ according to the membership function values.

Referring to the illustrative example in Figure 12.1, the deterministic output x_0^+ is calculated by interpolating between the centroids of the subsets Φ_2 and Φ_3:

$$x_0^+ = \frac{\displaystyle\sum_{i=2}^{3} \tilde{\alpha}_i(x_0) \sum_{j=1}^{3} \pi_{ij} \bar{x}_j^+}{\displaystyle\sum_{i=2}^{3} \tilde{\alpha}_i(x_0) \sum_{j=1}^{3} \pi_{ij}} = \frac{\displaystyle\sum_{i=2}^{3} a_i(x(x_0)) \sum_{j=1}^{3} \pi_{ij} \bar{x}_j^+}{\displaystyle\sum_{i=2}^{3} a_i(x(x_0)) \sum_{j=1}^{3} \pi_{ij}}$$

By comparing expressions (7.14) and (7.6) one can notice the formal similarity between the continuous real valued next state that is inferred based on the interval and fuzzy encoding. The only difference is in the calculation of the vector $\alpha^T(x_0)$ which characterizes compatibility of the continuous real valued current state x_0 to a single interval and multiple fuzzy granules – compare (7.4) and (7.12).

In the next section this similarity will be used to derive a general algorithm for real time learning of the transition probability matrix for interval and fuzzy encoding.

7.3   General Algorithm for On-Line Learning of the Transition Probability Matrix

In the finite-state Markov chain model (7.1), the transition probabilities π_{ij} may be estimated from the total frequencies of transitions, see (7.2). If measurements have been taken up to time k,

$$\pi_{ij} \approx \frac{N_{ij}(k)}{N_{oi}(k)} = \frac{N_{ij}(k)/k}{N_{oi}(k)/k} = \frac{F_{ij}(k)}{F_{oi}(k)}, \tag{7.15}$$

where $F_{ij}(k)$ is the mean frequency of transition events $f_{ij}(k)$ from state $\bar{x}_i$ to state $\bar{x}_j$ and $F_{0i}(k)$ is the mean frequency of the transition events, $f_i(k)$, that are initiated from the state i:

$$F_{ij}(k) = \frac{N_{ij}(k)}{k} = \frac{1}{k} \sum_{t=1}^{k} f_{ij}(t), \tag{7.16}$$

$$F_{oi}(k) = \frac{N_{oi}(k)}{k} = \frac{1}{k} \sum_{t=1}^{k} f_i(t), \tag{7.17}$$

($f_{ij}(k) = 1$ if a transition from $\bar{x}_i$ to $\bar{x}_j$ occurs at time instant k; $f_i(k) = 1$ if a transition is initiated in the state $\bar{x}_i$ at time instant k; otherwise these take zero values.)

From the definition of the transition probabilities one can derive a recursive expression for their calculation:

$$F_{ij}(k) = F_{ij}(k-1) + \varphi\,(f_{ij}(k) - F_{ij}(k-1)), \tag{7.18}$$

$$F_{oi}(k) = F_{oi}(k-1) + \varphi\,(f_i(k) - F_{oi}(k-1)), \tag{7.19}$$

where $\varphi = \frac{1}{k}$. In the case of non-stationary data it is more reasonable to weigh the old data with exponentially decreasing weights, i.e., to consider a constant parameter φ (learning rate) instead of the decaying factor, $\frac{1}{k}$. For a constant parameter φ the recursive expressions (7.18) and (7.19) can be interpreted as AR models implementing the exponential smoothing algorithm (a low pass filter) with forgetting factor φ [4]. By substituting expressions (7.18) and (7.19) into (7.15) we obtain a recursive form of the transition probabilities estimates that is convenient for their on-line learning [10]:

$$\pi_{ij} = \frac{F_{ij}(k)}{F_{oi}(k)} = \frac{F_{ij}(k-1) + \varphi\,(f_{ij}(k) - F_{ij}(k-1))}{F_{oi}(k-1) + \varphi\,(f_i(k) - F_{oi}(k-1))}. \tag{7.20}$$

Expression (7.20) can be rewritten in a matrix form by replacing $f_{ij}(k)$ and $f_i(k)$ by their vector multiplication counterparts, $\tau(k)\gamma(k)^T$ and $\tau(k)\gamma(k)^T e$, where $\tau(k)$ and $\gamma(k)$ are M-dimensional vectors with ones as the ith and jth elements, respectively, and zeros elsewhere and e- an M-dimensional vector of ones). More specifically,

$$\Pi(k) = diag(F_0(k))^{-1} F(k), \tag{7.21}$$

where

$$F(k) = F(k-1) + \varphi \ (\tau(k)\gamma(k)^T - F(k-1)), \tag{7.22}$$

$$F_o(k) = F_o(k-1) + \varphi \ (\tau(k)\gamma(k)^T e - F_o(k-1)). \tag{7.23}$$

In the above, we considered the finite state Markov chain model (7.1). We now show that matrix form of the recursive algorithm for on-line updating the transition probabilities (7.21)-(7.23) can be also used for on-line learning in interval and fuzzy encoding cases. The only change necessary is the calculation of $\tau(k)$ and $\gamma(k)$, which is now discussed.

For interval encoding case, the probability vectors $\alpha(x)$ and $\alpha(x^+)$ reflect, respectively, the occupancy of the ith and jth intervals by x and x^+. If a transition between the corresponding states occurs, the ij-th element of the product matrix $\alpha(x)\alpha^T(x^+)$ is equal to one, while other elements are zero. Therefore, by substituting

$$\tau(k) = \alpha(x(k)), \quad \gamma(k) = \alpha(x(k+1)), \tag{7.24}$$

expressions (7.21) - (7.23) directly apply.

For fuzzy encoding case, the membership of x and x^+ in the fuzzy subsets A_j are reflected in the possibility vectors $\tilde{\alpha}(x)$ and $\tilde{\alpha}(x^+)$. The membership degrees of an individual transition between the fuzzy subsets A_i and A_j correspond to $\tilde{\alpha}_i(x) \cdot \tilde{\alpha}_j(x^+)$, which is the ijth element of the matrix, $\tilde{\alpha}(x)\tilde{\alpha}^T(x^+)$. Therefore, similarly to the interval encoding, we may introduce:

$$\tau(k) = \tilde{\alpha}(x(k)), \quad \gamma(k) = \tilde{\alpha}(x(k+1)), \tag{7.25}$$

where vectors τ and γ are the membership values, i.e. the measures of compatibility, between $x(k)$ and $x(k+1)$ and the current and future states represented by the fuzzy subsets A_i and A_j. The products $\tau(k)\gamma(k)^T$ and $\tau(k)\gamma(k)^T e$ present the transition from A_i to A_j and the transition that is initiated in the state A_i at time instant through the correlation between these vectors so that expressions (7.21)-(7.23) can be formally applied.

Example 1: We apply the continuous state Markov chain model to the problem of modeling stochastically the speed of a vehicle that is randomly accelerating and decelerating (this type of model may be used to describe the dynamics of the lead vehicle in the problem of adaptive cruise control of a vehicle that is following it [10]). We consider the speed range [46 mph - 66 mph] and treat the speed of the vehicle as a continuous variable.

The transition probabilities are recursively estimated as vehicle data become available by applying the interval (7.24) and fuzzy encoding (7.25) version of the learning algorithm (7.21)-(7.23). The on-line algorithm for recursive learning the transition probabilities allows for their continuous adaptation to the changing vehicle, road, and traffic conditions, and style of driving. Predicted speed of the lead vehicle x^+ is inferred from the Markov chain model by expression (7.14).

In Figure 12.3 the speed universe is partitioned into 5 intervals of 4 mph each. The upper plot compares the on-line learning with interval partitioning (Mean Square Error (MSE)=1.93) with the case of fuzzy partitioning into 5 fuzzy subsets with Gaussian membership functions that are centered at the centers of the 5 intervals and standard deviations of 1.43 with no initial learning – the lower plot (MSE=0.96). The better approximation can be explained by the interpolative properties of the fuzzy partitioning (compare (7.6) vs. (7.11)). It can be shown that the impact of the fuzzy encoding on the approximation accuracy is even stronger if a coarser quantization of the speed universe is considered.

The contribution of the fuzzy partitioning can be viewed as means for supplying more information about the Markov chain states compared to the case of interval partitioning. Comparing (7.9) and (7.6), one sees that fuzzy encoding provides an additional mechanism to interpolate between discrete states of the finite-state Markov chain based on the degree of membership (closeness) of the uncertain value of x to the states $\bar{x}_j$ of the underlying finite state Markov chain – a property that is not automatically available in the interval encoding case. This interpolation can be beneficial in state prediction and for identification of transition probabilities.

Interpolative properties of the fuzzy encoding are dependent on the degree of fuzziness of the subsets A_j. Figure 12.4 illustrates the impact of the fuzziness of partitioning the continuous domain assuming Gaussian membership functions of the fuzzy subsets. For small values of the standard deviation of the Gaussians the interpolative ability of the model is rather limited resembling the case of interval partitioning. For larger values the individual fuzzy subsets tend to significantly overlap discounting the effect of the partitioning.

Example 2: We now consider the use of the Markov chain model to approximate the road grade of several typical routes, each of about 30 km, in the Dearborn area, Michigan. Measurements of the road grade are taken with a sampling rate of 30 meters. The plots below represent the road grade changes within the range of [-5.65%, 5.88%] on a 3.5 km segment. Figure 12.10 shows the performance of the Markov chain models with interval RMS and fuzzy encoding for predicting the road grade 1 step (30m) ahead; the road grade is considered a continuous state variable that is uniformly granulated into 6 intervals, resp. 6 fuzzy subsets. Table 1 lists the root mean squared error for 2, 4, 6 step prediction of the road grade. In the case of interval partitioning predicted state is calculated by applying the Chapman-Kolmogorov equation (7.3) and taking the expectation of the predicted probability distribution (7.6). Predicted state by the fuzzy encoded chain is calculated by

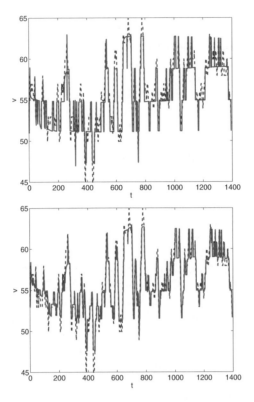

Fig. 7.2 Markov models of vehicle speed (solid) vs. actual speed (dashed): interval encoding into 5 intervals of 4 mph each (upper plot; MSE=1.93) and fuzzy encoding into 5 fuzzy subsets with Gaussian membership functions centered at the center points of the 5 intervals and standard deviations of 1.43 (MSE=0.96).

re-propagating the defuzzified value (7.14) and by applying the fuzzy version of the Chapman-Kolmogorov equation [6] followed by the defuzzification of the inferred possibility distribution.

Table 7.1 Root-mean-square error in predictions of the road grade over different horizons for interval and fuzzy encoding (based on repropagation of the de-fuzzified value and the possibilistic version of the Chapman-Kolmogorov equation in [6].)

Horizon	Int. C-K	Defuzz.	Poss. C-K
2 steps	1.181	1.146	1.105
4 steps	1.212	1.207	1.179
6 steps	1.457	1.235	1.157

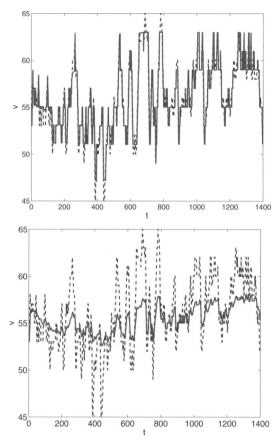

Fig. 7.3 Impact of the shape of the fuzzy subsets on the approximation error (solid:modeled, dashed:actual). Fuzzy encoding with Gaussian membership functions with small standard deviations resembles interval partitioning and lacks the ability for interpolation (upper plot) and vice versa (lower plot).

7.4 Markov Chain Models of Vector-Valued Signals

Previously we have demonstrated that the concepts of granulation (interval or fuzzy) can be applied to stochastic modeling of a scalar signal, x. Suppose now that $x \in R^n$ is a vector-valued signal. For instance, we may be interested to simultaneously represent vehicle speed and road grade values.

One approach in this situation is to treat individual elements of the vector x as statistically independent, apply preceding ideas to each element and then combine the resulting Markov chain models. For the cases when the independence assumption is not justified, the encoding can be performed in terms of granules defined in the Cartesian product space. For example, in

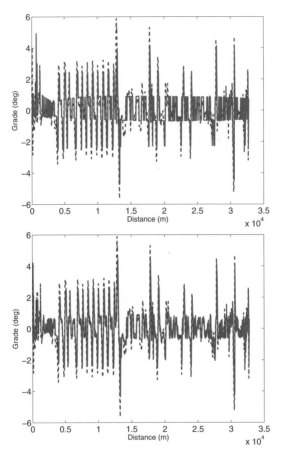

Fig. 7.4 Markov models of road grade (solid) vs. actual road grade (dashed) based on 1 step (30 m) ahead prediction of the road grade on a 3.5 km segment; interval (upper) and fuzzy encoding (lower) plot.

the $n = 2$ case, we can consider granules, $G_{j_1 j_2}$, centered at $x_1 = \bar{x}_{j_1}$ and $x_2 = \bar{x}_{j_2}, j_1 = 1, ..., M_1; j_2 = 1, ..., M_2$ and the corresponding matrix of the transition probabilities between the granules of the dimension $(M_1 M_2) \times (M_1 M_2)$. Similarly, fuzzy encoding can be used by introducing fuzzy sets A_i, $i = 1, \cdots, M$, and membership functions defined in R^n.

Note that the dimensionality of the Markov chain can grow dramatically in the case of vector-valued signals. This issue can be addressed by associating the states of the Markov chain with *clusters* of values of x, where the clusters can be evolved by an algorithm such as the one in [4]. As in the case of scalar signals, the use of fuzzy encoding can greatly enhance the identification and prediction accuracy of the resulting Markov chain.

Example 3: We apply Markov chain model to summarize simultaneous patterns in vehicle speed and road grade recorded when driving an experimental vehicle over eight typical routes, each of about 30 km, around the city of Dearborn, Michigan. To estimate the transition probabilities of the two variables - speed and grade - we divided their ranges [3.05, 34.06] (m/s) and [-9.23, 6.97] (%) into 10 and 8 intervals, respectively. This partitioning of the state variables defines 80 discrete Markov states that are the centers of the 80 cells in the two dimensional Cartesian space. Since some of the cells are not occupied (the specific combination of values corresponding to these states were not observed in the measured data) the number of discrete states of the Markov chain is reduced to 51 (Figure 12.5). Now we can apply the interval or fuzzy encoding concept on the aggregated variable and on-line learn its transition probability matrix similarly to the single state viable case described in the previous sections. Figure 12.6 shows the Markov model based one step ahead predictions of the speed and the grade versus the actual data.

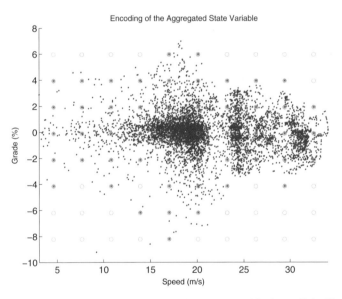

Fig. 7.5 2D cell encoding of the aggregated state variable (80 cells). The centers of the occupied cells (red stars-circles) define the new Markov states.

In the above we considered one special type of granulation of the inputs and the states that was based on assumed rectangular regions (cells) in the Cartesian product space. This approach takes into account the density of the states space by eliminating the centers of the empty cells. A more general type of granulation that accounts for the density distribution of the state variables, and consequent reduction of the number of resulting Markov states, can be obtained through clustering of the state space.

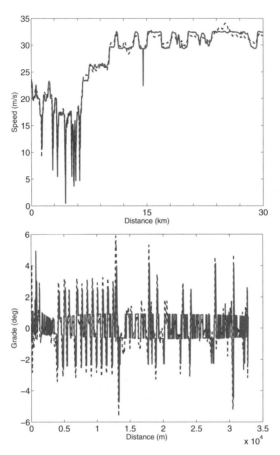

Fig. 7.6 Vehicle speed (upper plot) and road grade (lower plot) predicted 1 step ahead (solid line) versus actual vehicle speed and grade (dashed line) over a 30 km section of the road (30 m segments).

An application of the proposed Markov models of vehicle speed and fuel consumption to stochastic optimal control of the set-point of the vehicle speed as a function of the road grade and reference speed for fuel economy improvement without significant degradation of the travel time is discussed in [8].

7.5 Conclusions

The paper considered modeling and real-time identification of Markov chain models for signals in automotive vehicle applications. The framework of fuzzy encoding has been introduced which facilitates both identification and prediction accuracy. A computationally simple algorithm for real-time identification has been proposed which is applicable to the cases of both interval and fuzzy encoding. Opportunities for using such models for stochastic dynamic programming based optimization of vehicle fuel economy and performance under different road and traffic conditions will be considered in future publications.

References

[1] Fraser, A.M.: Hidden Markov Models and Dynamical Systems. SIAM, Philadelphia (2008)

[2] Chistyakov, V.P.: A Course in Probability Theory, Nauka, Moscow (1987)

[3] Dynkin, E.B., Yushkevich, A.A.: Markov Processes: Theorems and Problems. Plenum, New York (1969)

[4] Filev, D., Georgieva, O.: An extended version of the Gustafson-Kessel algorithm for evolving data stream clustering. In: Angelov, P., Filev, D., Kasabov, N. (eds.) Evolving Intelligent Systems, John Wiley and Sons, New York (2009)

[5] Filev, D., Kolmanovsky, I.: Markov chain modeling approaches for on board applications. In: Proc. of American Control Conference, Baltimore, MD (2010)

[6] Filev, D., Kolmanovsky, I.: A generalized Markov chain modeling approach for on-board applications. In: Proc. of International Journal Conference on Neural Networks, Barcelona, Spain (2010)

[7] Johannesson, L., Ashbogard, M., Egardt, B.: Assessing the potential of predictive control for hybrid vehicle powertrains using stochastic dynamic programming. IEEE Trans. on Intelligent Transportation Systems 8(1) (2007)

[8] Kolmanovsky, I., Filev, D.: Terrain and traffic optimized vehicle speed control. In: Proc. of 6th IFAC Symposium Advances in Automotive Control, Munich, Germany (2010)

[9] Klir, G.: A principle of uncertainty and information invariance. Int. J. of General Systems 17(2), 249–275 (1990)

[10] Kolmanovsky, I., Filev, D.: Stochastic optimal control of systems with soft constraints and opportunities for automotive applications. In: Proc. of IEEE Multi-conference on Systems and Control, St., Petersburg, Russia (2009)

[11] Kolmanovsky, I., Sivergina, I., Lygoe, B.: Optimization of powertrain operation policy for feasibility assessment and calibration: Stochastic dynamic programming approach. In: Proc. of American Control Conference, Anchorage, Alaska (2002)

[12] Kosko, B.: Fuzzy Engineering. Prentice-Hall, Upper Saddle River (1996)

[13] O'Leary, D., Stewart, G., Vandergraft, G.: Estimating the largest eigenvalue of a positive definite matrix. Mathematics of Computation 33, 1289–1292 (1979)

[14] Lin, C., Peng, H., Grizzle, J.: A stochastic control strategy for hybrid electric vehicles. In: Proc. of American Control Conference, Baltimore, MD, pp. 4710–4715 (2004)

[15] Malikopoulos, A.A., Papalambros, P.Y., Assanis, D.N.: Online self-learning
     identification and stochastic control for autonomous internal combustion en-
     gines. ASME J. Dyn. Sys., Meas., Control 132(2) (2010)
[16] Pedrycz, W., Skowron, A., Kreinovich, V.: Handbook of Granular Computing.
     John Wiley and Sons, Chichester (2008)
[17] Ripaccioli, G., Bernardini, D., Di Cairano, S., Bemporad, A., Kolmanovsky,
     I.: A stochastic Model Predictive Control approach for a series hybrid elec-
     tric vehicle power management. In: Proc. of American Control Conference,
     Baltimore, MD (2010)
[18] Yager, R., Filev, D.: Essentials of Fuzzy Modeling and Control. John Wiley
     and Sons, New York (1994)

Chapter 8
Parameter Identification in Dynamic Systems Using the Homotopy Optimization Approach[*]

Chandrika P. Vyasarayani, Thomas Uchida, Ashwin Carvalho, and John McPhee

Abstract. Identifying the parameters in a mathematical model governed by a system of ordinary differential equations is considered in this work. It is assumed that only partial state measurement is available from experiments, and that the parameters appear nonlinearly in the system equations. The problem of parameter identification is often posed as an optimization problem, and when deterministic methods are used for optimization, one often converges to a local minimum rather than the global minimum. To mitigate the problem of converging to local minima, a new approach is proposed for applying the homotopy technique to the problem of parameter identification. Several examples are used to demonstrate the effectiveness of the homotopy method for obtaining global minima, thereby successfully identifying the system parameters.

8.1 Introduction

The problem of identifying the parameters in a mathematical model governed by ordinary differential equations (ODEs), given a set of experimental measurements, is encountered in many fields of physics, chemistry, biology, and engineering [8]. The problem of parameter identification can be posed as an optimization problem [17, 18], where the arguments of the global minimum of

Chandrika P. Vyasarayani (✉) · Thomas Uchida · John McPhee
University of Waterloo, Waterloo, Ontario, N2L 3G1, Canada
e-mail: cpvyasar@uwaterloo.ca, tkuchida@uwaterloo.ca,
    mcphee@real.uwaterloo.ca

Ashwin Carvalho
Indian Institute of Technology Bombay, Powai, Mumbai, 400076, India
e-mail: ashwincarvalho@iitb.ac.in

[*] This manuscript will also appear as an article in *Multibody System Dynamics*, with permission granted by Springer.

D. Alberer et al. (Eds.): Identification for Automotive Systems, LNCIS 418, pp. 129–145.
springerlink.com   © Springer-Verlag London Limited 2012

the objective function are the identified parameters. If the parameters appear linearly in the system equations and full state measurement is available from experimental data, a large class of methods is available for both off-line and on-line parameter identification [17, 18]. In most practical engineering problems, however, it is not possible to obtain measurements for all states, and the parameters often appear nonlinearly in the equations of motion. Off-line identification algorithms are used when the main goal is to develop a mathematical model for system simulation; on-line identification algorithms are more popular in adaptive control applications [15]. The optimization problems are usually solved using deterministic methods, which require the solution of differential equations at each optimization step. The solution of these ODEs can be obtained using initial-value methods [10, 26], shooting methods [1], or collocation methods [2]. When deterministic approaches like the steepest descent [22], Gauss–Newton [22], and Levenberg–Marquardt [20] algorithms are used in the optimization procedure, it is not uncommon to converge to a local minimum rather than the global minimum [7]. Stochastic methods, such as simulated annealing [23] and genetic algorithms [9], can be used to find global minima, but these methods typically require a large number of iterations to converge and, thus, are time-consuming, especially for parameter identification problems where the equations of motion are integrated at every optimization step [10, 11, 26]. An obvious question that arises is whether there exist any non-stochastic algorithms that can find global minima. Although one can never be assured that a deterministic algorithm will be able to find the global minimum in every situation, there are approaches derived from homotopy methods that can find global minima in situations where other deterministic methods cannot.

Homotopy [27] is a powerful technique that is used in several areas of mathematics, including optimization [5, 12] and nonlinear root finding [4]. In homotopy methods, the objective function to be minimized is modified by adding another function whose optimum is known, herein referred to as the known function, and a morphing parameter is used to transform the modified function into the original objective function. A series of optimizations is performed while slowly varying the morphing parameter until the modified function is transformed back into the original objective function [5]. Applying homotopy to algebraic optimization problems is straightforward, but its application to the parameter identification problem is not, since the objective function is, itself, dependent on the solution of differential equations. Homotopy was successfully applied to ARMAX models [13, 14] for the identification of linear parameters. In the work of Abarbanel et al. [1], the authors have coupled the mathematical model to the experimental data for identifying parameters from a chaotic time series for first-order systems, which is related to the homotopy method. The application of homotopy to the general nonlinear parameter identification problem has not been studied in the literature. In this work, we present a methodology to apply homotopy to the problem of parameter identification. We show several examples where the classical deterministic methods fail to find the global minimum while the homotopy method successfully minimizes the objective function.

8.2 Problem Statement

The dynamic equations of the physical system for which the parameters must be identified are assumed to be of the following form:

$$\dot{x}_1 = x_2$$
$$\dot{x}_2 = f(x_1, x_2, p, t) \tag{8.1}$$

For mechanical systems, $x_1(t) = [y_1(t), y_2(t), \ldots, y_n(t)]^{\mathrm{T}}$ are the independent coordinates (displacements) and p is a column vector containing the parameters to be identified. The system is assumed to be nonlinear, and the parameters may also appear nonlinearly in equations (8.1). Experimental data $x_{1e}(t) = [y_{1e}(t), y_{2e}(t), \ldots, y_{ne}(t)]^{\mathrm{T}}$ of all the displacements are assumed to be available over time T; it is assumed that the velocities are not measured. Note that it is possible to identify the system parameters with only a few components of $x_{1e}(t)$ since, in coupled systems, each component of $x_{1e}(t)$ contains information from all the parameters due to the coupling between the system equations. The unknown initial conditions of the states for which measurements are not available can be treated as unknown parameters. In order to simplify the optimization procedure, experiments can be performed starting from rest, such that the velocity initial conditions are zero. The goal is to identify the parameters in the mathematical model such that the solution of the differential equations (8.1) closely matches the experimental data. To identify the parameters, we minimize the integral of the squared difference between the experimental and simulated states:

$$V(p) = \frac{1}{2} \sum_{i=1}^{n} \left\{ \int_0^T \Big(y_{ie}(t) - y_i(t, p) \Big)^2 \, dt \right\} \tag{8.2}$$

Note that a discrete summation can be used in place of the integral in this equation. Given the initial estimates of the parameters, we can minimize equation (8.2) iteratively using the Gauss–Newton method [22] as follows:

$$p^{r+1} = p^r + \sigma d^r \tag{8.3}$$

where σ is the step size and d^r is the search direction, which can be obtained from the following relation [26]:

$$H(p^r) d^r = -G^{\mathrm{T}}(p^r) \tag{8.4}$$

In equation (8.4), G and H are the gradient and the approximate Hessian of the objective function, where the second-order sensitivities have been neglected in the latter:

$$G(p) = \frac{\partial V}{\partial p} = -\sum_{i=1}^{n} \left\{ \int_0^T \left(y_{ie}(t) - y_i(t, p) \right) \frac{\partial y_i}{\partial p} \, dt \right\} \quad (8.5)$$

$$H(p) = \frac{\partial^2 V}{\partial p^2} \approx \sum_{i=1}^{n} \left\{ \int_0^T \frac{\partial y_i}{\partial p}^{\mathrm{T}} \frac{\partial y_i}{\partial p} \, dt \right\} \quad (8.6)$$

The sensitivity data $\frac{\partial y_i}{\partial p} = \left[\frac{\partial y_i}{\partial p_1}, \frac{\partial y_i}{\partial p_2}, \ldots, \frac{\partial y_i}{\partial p_m} \right]$ can be obtained by solving the sensitivity differential equations, which can be derived by the direct differentiation of equations (8.1) with respect to the individual parameters:

$$\frac{\partial \dot{x}_1}{\partial p_j} = \frac{\partial x_2}{\partial p_j}$$
$$\frac{\partial \dot{x}_2}{\partial p_j} = \frac{\partial f(x_1, x_2, p, t)}{\partial p_j} + \frac{\partial f(x_1, x_2, p, t)}{\partial x_1} \frac{\partial x_1}{\partial p_j} + \frac{\partial f(x_1, x_2, p, t)}{\partial x_2} \frac{\partial x_2}{\partial p_j},$$
$$j = 1, 2, \ldots, m \quad (8.7)$$

We will briefly explain how homotopy is applied to a simple algebraic minimization problem. Let $\mathcal{F}(p)$ be the objective function. We are interested in finding parameters p^* at which $\mathcal{F}$ has a global minimum. If we start from an arbitrary point p^0, and if the function has multiple local minima, it is likely that the optimization procedure will converge to a local minimum. In the homotopy method, we first construct the following function:

$$\mathcal{H}(p, \lambda) = (1 - \lambda) \mathcal{F}(p) + \lambda \mathcal{G}(p) \quad (8.8)$$

where $\mathcal{G}(p)$, referred to as the known function, is a chosen function that is convex in the unknown parameters, and for which the arguments of its global minimum are known. We now begin the process by choosing $\lambda_0 = 1$ and minimizing $\mathcal{H}(p, 1) = \mathcal{G}(p)$. In this first stage, the minimum of $\mathcal{H}(p, 1)$ is simply the minimum of $\mathcal{G}(p)$, which is known. Once the minimum has been found, λ is decreased by a small amount $\delta\lambda$ and $\mathcal{H}(p, \lambda_1)$ is minimized, where $\lambda_\ell = 1 - \ell \, \delta\lambda$, using the converged result from the previous stage as the initial guess for p. This process is continued until $\lambda = 0$ and the objective function has been morphed back into $\mathcal{F}(p)$. Provided we are always finding the minimum of $\mathcal{H}(p, \lambda)$ with an initial guess that is close to its global minimum, it is more likely that we will find the global minimum of the function $\mathcal{F}(p)$. A variant of this method has been successfully applied to complex optimization problems involving protein structures [6] and to finding the equilibrium configuration of an elastica [28]. Note that the choice of known function $\mathcal{G}(p)$

is nontrivial. In general, the homotopy method is only capable of finding local minima; however, if the nature of $\mathcal{F}(\boldsymbol{p})$ is known, it may be possible to construct the homotopy transformation in a way that increases the chance of finding the global minimum. It is for this reason that the optimization problem for parameter identification is generally more challenging than it is for purely algebraic problems. In particular, the shape of $\mathcal{F}(\boldsymbol{p}) = V(\boldsymbol{p})$ is unknown in parameter identification problems, since it is dependent on the solution of differential equations (8.1), thereby making the selection of a suitable known function $\mathcal{G}(\boldsymbol{p})$ very difficult. In contrast to algebraic problems, however, where the minimum value of the objective function is unknown, the minimum value of the objective function in parameter identification problems is zero—provided the mathematical model is known exactly. Since the mathematical model is not precisely known in general, the final error may not be exactly zero; however, it is expected to be small. We use this knowledge in developing our homotopy transformation.

We now discuss how the homotopy method can be applied to the problem of parameter identification. To modify the objective function, the experimental data is coupled to the mathematical model as follows:

$$\dot{\boldsymbol{x}}_1 = \boldsymbol{x}_2 + \lambda\, K_1\, (\boldsymbol{x}_{1e} - \boldsymbol{x}_1)$$
$$\dot{\boldsymbol{x}}_2 = \boldsymbol{f}(\boldsymbol{x}_1, \boldsymbol{x}_2, \boldsymbol{p}, t) + \lambda\, K_2\, (\boldsymbol{x}_{1e} - \boldsymbol{x}_1) \tag{8.9}$$

Initially, when $\lambda = 1$, the coupling term acts as a high-gain observer [3, 16], and if sufficiently high values of K_i are used, the experimental data and simulated response will synchronize. Note that λ is introduced to the traditional definition of a high-gain observer so as to construct the homotopy transformation. Also note that the sensitivity equations (8.7) must be modified to account for the added coupling term. For very large K_i, the objective function becomes a flat surface with a very small magnitude, and the experimental data $\boldsymbol{x}_{1e}$ and simulated response $\boldsymbol{x}_1$ will closely match no matter what parameters are used. For the purposes of parameter identification, we choose the lowest values of K_i such that the experimental data and simulated response synchronize to within a desired tolerance ε when the initial parameter estimates are used. We now decrease λ by a small amount $\delta\lambda$ and minimize the objective function (8.2), treating equation (8.9) as the mathematical model. The parameter estimates are refined so as to reduce the error to $\beta\varepsilon$, where $0 < \beta < 1$. We then decrease λ further to $1 - 2\,\delta\lambda$; since the parameter guesses have been refined, the observer gain can be reduced without increasing the error beyond ε. At each stage in this process, we use the converged result from the previous stage as the initial guess for $\boldsymbol{p}$. This process is repeated until $\lambda = 0$, and equation (8.9) has morphed back into equation (8.1). In summary, the homotopy optimization approach follows the path of minimal error as the observer gain is decreased. We ensure that the error is close to zero (no greater than ε) at each value of the observer gain, with the hope that the refined parameter guesses at the final stage are sufficiently close to the global optimum of the original problem. The process of applying the homotopy method to the problem of parameter identification is summarized in Algorithm 1.

Algorithm 1.   Parameter identification using homotopy

Input: Experimental data $(\boldsymbol{x}_{1e})$, objective function tolerance (ϵ)
Output: Identified parameters $(\boldsymbol{p})$

   Initialize

   while $\lambda \geq 0$ **do**

      while $V > \epsilon$ **do**

         Solve ODEs for $\boldsymbol{x}_1$ and $\frac{\partial \boldsymbol{x}_1}{\partial p_j} \, \forall j$

         Minimize $V(\boldsymbol{p}) = \frac{1}{2} \sum_{i=1}^{n} \left\{ \int_0^T \Big(y_{ie}(t) - y_i(t, \boldsymbol{p}) \Big)^2 dt \right\}$

         Solve $\boldsymbol{H}(\boldsymbol{p})\,\boldsymbol{d} = -\boldsymbol{G}^{\mathrm{T}}(\boldsymbol{p})$ for $\boldsymbol{d}$

         $\boldsymbol{p} \leftarrow \boldsymbol{p} + \sigma \boldsymbol{d}$

      end while

      $\lambda \leftarrow \lambda - \delta \lambda$

   end while

   return $\boldsymbol{p}$

8.3   Numerical Examples

In this section, we present numerical examples in which the homotopy method has been used to successfully identify the parameters in mechanical systems.

8.3.1   *Linear Parameters*

Let us begin with the identification of parameters for a simple pendulum. We describe this example in detail to further explain the idea of the homotopy method. The equations of motion for a simple pendulum are given in state-space form as follows:

$$\dot{y}_1 = y_2$$
$$\dot{y}_2 = -p\sin(y_1) \tag{8.10}$$

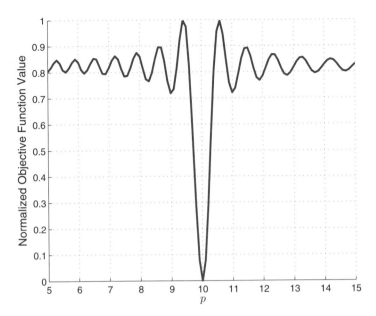

Fig. 8.1 Normalized objective function for simple pendulum

with initial conditions $y_1(0) = \pi/6$ and $y_2(0) = 0$. The solution of this system with $p = 10$ for time $t \in [0, 50]$ is used as experimental data y_{1e}. The goal is to determine p from the experimental data. We define the following minimization problem:

$$p^* = \min_{\text{arg}} \int_0^{50} \left(y_{1e} - y_1(p)\right)^2 dt \qquad (8.11)$$

A direct search has been used to identify the shape of the objective function, as shown in Figure 8.1. It can be seen that the objective function has multiple local minima, and any classical deterministic optimization method will fail to converge to the global minimum unless we choose our initial guess for p to be quite close to 10. To apply the homotopy method, we modify the equations of motion as follows:

$$\dot{y}_1 = y_2 + \lambda K_1 \left(y_{1e} - y_1\right)$$
$$\dot{y}_2 = -p \sin(y_1) + \lambda K_2 \left(y_{1e} - y_1\right) \qquad (8.12)$$

The sensitivity equations are now given as follows:

$$\frac{\partial \dot{y}_1}{\partial p} = \frac{\partial y_2}{\partial p} - \lambda K_1 \frac{\partial y_1}{\partial p}$$

$$\frac{\partial \dot{y}_2}{\partial p} = -\sin(y_1) - p\cos(y_1)\frac{\partial y_1}{\partial p} - \lambda K_2 \frac{\partial y_1}{\partial p} \qquad (8.13)$$

Figure 8.2, again obtained using a direct search, illustrates the shape of the objective function for different values of λ and p, using coupling parameters $K_1 = K_2 = 10$. As can be seen, the modified objective function is convex when $\lambda = 1$; as we decrease λ to 0, the modified objective function slowly morphs into the original objective function shown in Figure 8.1. The line joining the minimum of the modified function for different values of λ is also shown. We have implemented the homotopy method for identifying p as illustrated in Algorithm 1, with $\epsilon = 0.001$ and $\delta\lambda = 0.2$. Using the homotopy method with an initial guess of $p^0 = 15$, we converge to the global minimum at $p = 10$ since we always remain close to the global minimum of the modified objective function. Without using homotopy, again starting with an initial guess of $p^0 = 15$, the parameter converges to a local minimum at $p = 14.6755$.

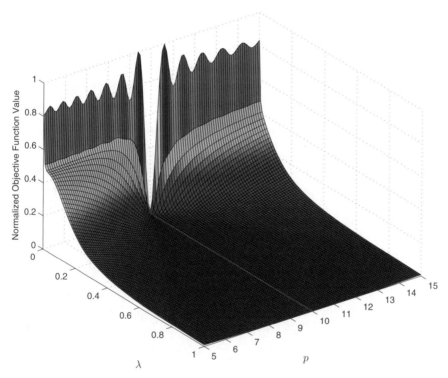

Fig. 8.2 Normalized objective function for simple pendulum with $0 \leq \lambda \leq 1$

8.3.2   Nonlinear Parameters

Consider the following system, which has been studied in [3]:

$$\ddot{u} + 0.1\dot{u} + \tan^{-1}(ku) = \sin\left(\frac{t}{2} + \pi\psi\right) \tag{8.14}$$

The system is nonlinear in both k and ψ. We use $k = 1$ and $\psi = 0.5$ for generating experimental data, and then attempt to identify these parameters using initial guesses $k^0 = 4$ and $\psi^0 = 2$. The identification procedure was first performed without using homotopy, the results of which are shown in Figure 8.3(a). Clearly, the optimization procedure (Gauss–Newton) has converged to a local minimum where $k = 4.5850$ and $\psi = 1.9620$. The homotopy method was then applied using the same initial guesses for k and ψ. In this example, we use coupling parameters $K_1 = K_2 = 10$, step size $\delta\lambda = 0.2$, and objective function tolerance $\epsilon = 0.01$. As shown in Figure 8.3(b), the global minimum was found and both parameters were correctly identified using the homotopy method (we obtain $k = 1.0001$ and $\psi = 2.5000$). Note that the estimate for ψ produces an equivalent response as does $\psi = 0.5$, since $\sin\left(\frac{t}{2} + \pi(\psi + 2)\right) = \sin\left(\frac{t}{2} + \pi\psi\right)$.

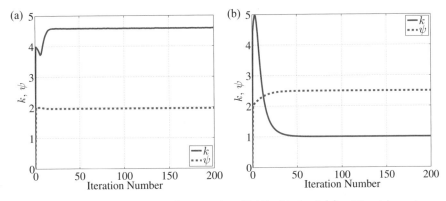

Fig. 8.3 Parameter estimates for equation (8.14) obtained (a) without homotopy, and (b) with homotopy

8.3.3   Noisy Experimental Data

Consider the following two-degree-of-freedom (DOF) system [3]:

$$\ddot{u} + 0.3\dot{u} + u + C_1 u^3 + C_2 v = 0$$
$$\ddot{v} + 0.3\dot{v} + u + C_3 v^3 + \cos(t) = 0 \tag{8.15}$$

where $C_1 = 1$, $C_2 = 2$, and $C_3 = 3$ are used for generating the experimental data for u and v, to which we add 5% white noise. In this example, we assume that the initial conditions for all states are known. Taking as initial guesses $C_1^0 = C_2^0 = C_3^0 = 10$ and without using homotopy, the optimization procedure converges to parameters that are different from the experimental values (we obtain $C_1 = 13.3840$, $C_2 = 7.9444$, and $C_3 = 9.3622$, as shown in Figure 8.4(a)). Although we do not expect to achieve exact convergence due to the noise in the experimental data, these parameter values are significantly different than the actual values. Using the homotopy method with $K_i = 10 \, \forall i$ and $\delta\lambda = 0.2$, we obtain parameter estimates $C_1 = 0.9998$, $C_2 = 2.0008$, and $C_3 = 3.0034$, which are very close to the experimental values (see Figure 8.4(b)). Note that only the experimental data for u is coupled to the equations of motion; it is assumed that the data for v is not measured. This coupling strategy confirms that it is not necessary to measure all the coordinates of a multiple-DOF system in order to identify all the system parameters. The experimental data and the results from the simulated system using the identified parameters are in good agreement, as shown in Figure 8.5.

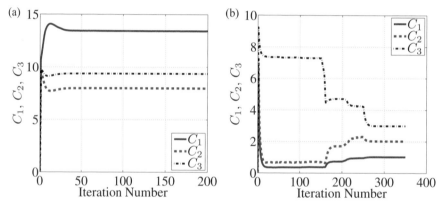

Fig. 8.4 Parameter estimates for equation (8.15) obtained (a) without homotopy, and (b) with homotopy

8.3.4 Reduced-Order Modelling

In this final example, we consider the problem of identifying parameters in a multibody system given experimental data generated using a more complex model. In particular, a 14-DOF vehicle model with a fully independent suspension, shown in Figure 8.6, is used to generate the experimental data. This topology is recommended by Sayers [24] for simulating the handling

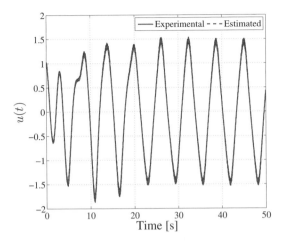

Fig. 8.5 Experimental data and simulated response for equation (8.15) using identified parameters

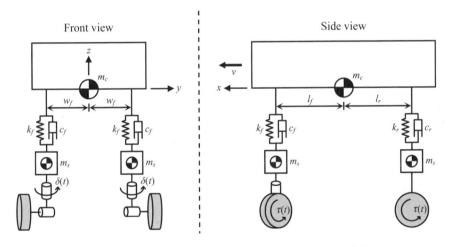

Fig. 8.6 14-DOF vehicle model used for generating experimental data

and braking behaviour of a vehicle, and has been adopted by several commercial software packages. The position and orientation of the vehicle chassis (m_c) together comprise 6 DOF. Four lumped masses (m_s), each representing one-quarter of the suspension components, are connected to the chassis by prismatic joints in parallel with springs (k) and dampers (c), which represent the suspension compliance and together add 4 DOF. Each wheel is connected to its corresponding lumped mass with a horizontally-oriented revolute joint that allows the wheel to spin, collectively accounting for the final 4 DOF; torques $(\tau(t))$ can be applied at these joints to accelerate the vehicle. The

Pacejka2002 Magic Formula tire model [21] is used to model the tire dynamics. The vertically-oriented revolute joints on the front wheels ($\delta(t)$) are used to steer the vehicle on a prescribed trajectory, so do not add any DOF to the system. The system parameters are obtained from [25], some of which are shown in Table 8.1.

Table 8.1 System parameters for 14-DOF vehicle model

Parameter		Value
Mass	Chassis (m_c)	2077 kg
	Quarter of suspension (m_s)	10 kg
Stiffness	Front suspension (k_f)	48.30 kN/m
	Rear suspension (k_r)	30.52 kN/m
Damping	Front suspension (c_f)	3.08 kN-s/m
	Rear suspension (c_r)	2.33 kN-s/m
Dimensions	Front width (w_f)	0.760 m
	Rear width (w_r)	0.795 m
	Front length (l_f)	1.353 m
	Rear length (l_r)	1.487 m

Experimental data generated from the 14-DOF model with Pacejka tires is used to identify parameters in the planar bicycle model shown in Figure 8.7. The planar bicycle model is often used for the simulation of vehicle dynamics and for on-board stability controllers. This simple model has only 3 DOF: the position of the chassis in the x-direction (y_1), the position of the chassis in the y-direction (y_2), and the orientation of the chassis in the plane (y_3). The mass of the chassis is assumed to be $m_b = m_c + 4m_s = 2117$ kg, and the lengths are assumed to be $d_f = l_f = 1.353$ m and $d_r = l_r = 1.487$ m. The yaw inertia of the chassis is also defined to match that of the 14-DOF model. A simple 4-parameter tire model [19] is used in place of the complex 117-parameter Pacejka model used in the 14-DOF system:

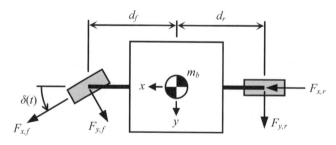

Fig. 8.7 Planar vehicle model for which parameters are sought

$$s_b = 1 - \frac{v_{x,w}}{R\,\omega} \tag{8.16}$$

$$F_{x,w} = A_{long} \left(1 - e^{-B_{long}s_b}\right) F_z \tag{8.17}$$

$$\alpha_b = \arctan\left(\frac{v_{y,w}}{v_{x,w}}\right) \tag{8.18}$$

$$F_{y,w} = A_{lat} \left(1 - e^{-B_{lat}\alpha_b}\right) F_z \tag{8.19}$$

where s_b is the longitudinal slip, α_b is the lateral slip angle, $v_{x,w}$ and $v_{y,w}$ are the longitudinal and lateral components of the wheel velocity in the wheel reference frame, R is the tire radius, and ω is the rotational speed of the wheel. Forces $F_{x,w}$ and $F_{y,w}$ are applied directly to the wheel centers. The tire radius (R) is assumed to be equal to the unloaded radius used in the Pacejka model, and the vertical tire force (F_z) is simply assumed to be half the total static load in the 14-DOF model. We wish to find values for the tire parameters (A_{long}, B_{long}, A_{lat}, and B_{lat}) such that the simulated response of the 3-DOF planar model matches the experimental data obtained from the 14-DOF model as closely as possible. Note that the planar model has no suspension and can neither pitch nor roll.

To generate the experimental data, we must first define the required inputs. In order to adequately capture both the longitudinal and lateral dynamics, we first accelerate the vehicle from 4 m/s to 16 m/s by applying positive torque to each wheel, then perform a lane-change maneuver, and finally slow the vehicle to 6 m/s by applying negative torque to each wheel. The steer angle and wheel torque inputs are shown in Figure 8.8. Using these inputs, the 14-DOF vehicle model is simulated for 22 seconds and the state vector is stored every millisecond. For the purpose of parameter identification, we

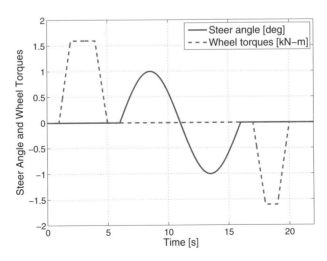

Fig. 8.8 Inputs used to generate experimental data with the 14-DOF model

assume that only five states are known: the position of the chassis along the
x- and y-axes, the orientation of the chassis about the z-axis, and the average
rotational speeds of the two front and two rear wheels. Note that the wheel
speeds of the 14-DOF model are required in order to calculate the tire forces
applied to the 3-DOF model (equations (8.16) to (8.19)). We use the following
objective function:

$$
V = \sum_{i=1}^{3} \left\{ w_i \int_{0}^{22} \left(y_{ie}(t) - y_i(t, A_{long}, B_{long}, A_{lat}, B_{lat}) \right)^2 dt \right\}
\qquad (8.20)
$$

where weights $w_1 = 0.001$, $w_2 = 0.01$, and $w_3 = 1$ are chosen to scale the
longitudinal position, lateral position, and orientation errors to the same
orders of magnitude. We obtain rough initial guesses $A_{long} = B_{long} = 100$
and $A_{lat} = B_{lat} = 1$ by hand, which corresponds to an objective function
value of 1.804×10^{-2} and produces the simulation results shown in Figure
8.9.

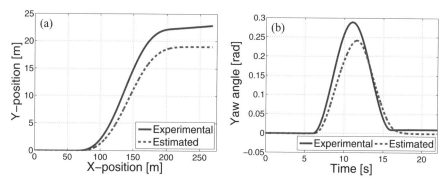

Fig. 8.9 Simulation results for (a) trajectory and (b) yaw angle, obtained using
initial parameter guesses

We perform the optimization procedure using the homotopy method de-
scribed above, with $K_i = 0.5\,\forall i$, $\delta\lambda = 0.1$, and $\epsilon = 10^{-4}$. Convergence is
achieved after a total of 12 iterations, with a final objective function value
of $V_f = 3.3 \times 10^{-5}$, as shown in Figure 8.10. The identified parameters are
$A_{long} = 99.97$, $B_{long} = 99.97$, $A_{lat} = 3.05$, and $B_{lat} = 3.06$; the corresponding simulation results are shown in Figure 8.11.

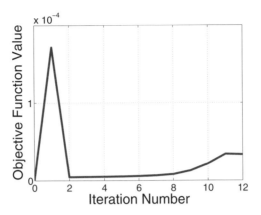

Fig. 8.10 Convergence of the objective function (normalized by the number of data points)

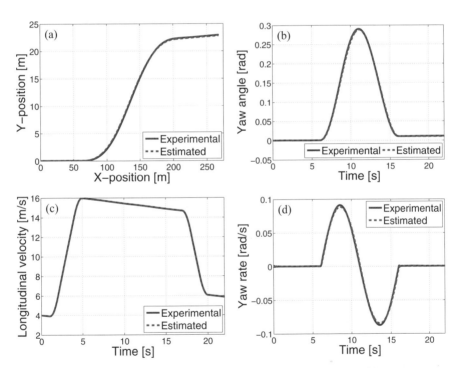

Fig. 8.11 Simulation results for (a) trajectory, (b) yaw angle, (c) longitudinal velocity, and (d) yaw rate, obtained using identified parameters

8.4   Conclusions

In this work, we have presented a new methodology for applying the homotopy optimization technique to the parameter identification problem. We have considered the general problem of parameter identification for nonlinear parameters with partial state measurement corrupted by measurement noise. The proposed homotopy method can successfully find global minima given a wide range of initial parameter guesses. The effectiveness of the proposed technique for parameter identification has been demonstrated by several example problems. The authors are currently investigating the use of the homotopy method for recursive parameter identification in on-line applications, and for systems governed by differential-algebraic equations.

Acknowledgements. The financial support provided to the first author by the Ontario Centres of Excellence, to the second author by the Natural Sciences and Engineering Research Council of Canada (NSERC), to the third author by the MITACS Globalink Program, and to the fourth author by the NSERC/Toyota/Maplesoft Industrial Research Chair program is gratefully acknowledged. The first author also wishes to acknowledge Professor Kingshook Bhattacharyya, IIT Kharagpur and Professor Baris Fidan, University of Waterloo for several interesting discussions during the course of this study.

References

[1] Abarbanel, H.D.I., Creveling, D.R., Farsian, R., Kostuk, M.: Dynamical state and parameter estimation. SIAM J. Appl. Dyn. Syst. 8, 1341–1381 (2009)

[2] Brewer, D., Barenco, M., Callard, R., Hubank, M., Stark, J.: Fitting ordinary differential equations to short time course data. Philos. Trans. R. Soc. Lond., Ser. A 366, 519–544 (2008)

[3] Chatterjee, A., Cusumano, J.P.: Asymptotic parameter estimation via implicit averaging on a nonlinear extended system. J. Dyn. Syst. Meas. Control 125, 11–18 (2003)

[4] Chow, S.N., Mallet-Paret, J., Yorke, J.A.: A homotopy method for locating all zeros of a system of polynomials. Funct. Differ. Equs. Approx. Fixed Points 730, 77–88 (1979)

[5] Dunlavy, D.M., O'Leary, D.P.: Homotopy optimization methods for global optimization. Sandia National Laboratories, Report SAND2005-7495 (2005)

[6] Dunlavy, D.M., O'Leary, D.P., Klimov, D., Thirumalai, D.: HOPE: A homotopy optimization method for protein structure prediction. J. Comput. Biol. 12, 1275–1288 (2005)

[7] Eberhard, P., Schiehlen, W., Bestle, D.: Some advantages of stochastic methods in multicriteria optimization of multibody systems. Arch. Appl. Mech. 69, 543–554 (1999)

[8] Gershenfeld, N.: The nature of mathematical modeling. Cambridge University Press, Cambridge (1999)

[9] Goldberg, D.E.: Genetic Algorithms in Search, Optimization and Machine Learning. Addison–Wesley, Reading (1989)

[10] Gordon, T.J., Hopkins, R.: Parametric identification of multibody models for crash victim simulation. Multibody Syst. Dyn. 1, 85–112 (1997)

[11] He, Y., McPhee, J.: Application of optimisation algorithms and multibody dynamics to ground vehicle suspension design. Int. J. Heavy Veh. Syst. 14, 158–192 (2007)

[12] Hillermeier, C.: Generalized homotopy approach to multiobjective optimization. J. Optim. Theory Appl. 110, 557–583 (2001)

[13] Hu, J., Hirasawa, K.: A homotopy approach to identification of ARMAX systems. Trans. Inst. Electr. Eng. Jpn. 119, 206–211 (1999)

[14] Hu, J., Hirasawa, K., Kumamaru, K.: A homotopy approach to improving PEM identification of ARMAX models. Automatica 37, 1323–1334 (2001)

[15] Ioannou, P., Fidan, B.: Adaptive control tutorial. Society for Industrial and Applied Mathematics, Philadelphia (2006)

[16] Khalil, H.K.: High-gain observers in nonlinear feedback control. In: New Directions in Nonlinear Observer Design. Lecture Notes in Control and Information Sciences, vol. 244, pp. 249–268. Springer, Heidelberg (1999)

[17] Ljung, L.: System identification: Theory for the user. Prentice-Hall, Englewood Cliffs (1987)

[18] Ljung, L., Söderström, T.: Theory and practice of recursive identification. MIT Press, Cambridge (1983)

[19] Metz, L.D.: Dynamics of four-wheel-steer off-highway vehicles. SAE Paper No. 930765 (1960)

[20] Moré, J.J.: The Levenberg-Marquardt algorithm: Implementation and theory. Numer. Analysis 630, 105–116 (1977)

[21] Pacejka, H.B.: Tire and Vehicle Dynamics. SAE International (2002)

[22] Rao, S.S.: Optimization: Theory and Applications. John Wiley & Sons, Chichester (1983)

[23] Romeijn, H.E., Smith, R.L.: Simulated annealing for constrained global optimization. J. Global Optim. 5, 101–126 (1994)

[24] Sayers, M.W., Han, D.: Generic multibody vehicle model for simulating handling and braking. Veh. Syst. Dyn. 25, 599–613 (1996)

[25] Schmitke, C., Morency, K., McPhee, J.: Using graph theory and symbolic computing to generate efficient models for multibody vehicle dynamics. IMechE J. Multi-body Dyn. 222, 339–352 (2008)

[26] Serban, R., Freeman, J.S.: Identification and identifiability of unknown parameters in multibody dynamic systems. Multibody Syst. Dyn. 5, 335–350 (2001)

[27] Watson, L.T., Haftka, R.T.: Modern homotopy methods in optimization. Comput. Methods Appl. Mech. Eng. 74, 289–305 (1989)

[28] Watson, L.T., Wang, C.Y.: A homotopy method applied to elastica problems. Int. J. Solids Struct. 17, 29–37 (1981)

Part III
The Importance of Data

Chapter 9
A Tutorial on Applications-Oriented Optimal Experiment Design

Cristian R. Rojas, Jonas Mårtensson, and Håkan Hjalmarsson

Abstract. Presented in this chapter is a tutorial on the design of input signals for system identification. The chapter concerns the details of basic optimal experiment design, computational issues, implementation and conditions that lead to an optimal input signal. It is also addressed how the two fundamental issues associated with system identification, model structure identification and model validation relate to experiment design. Throughout the chapter, a simple FIR example is employed to illuminate the concepts presented.

9.1 Introduction

The problem of how to determine an adequate model from experimental data – system identification – is a multi faceted topic, imposing quite high demands on the user. Two of the core issues are model structure selection and model validation. The key to simplify these problems and others lies in ensuring that suitable data is available for the modeling. What this means depends very much on what the model is going to be used for. We illustrate this with a simple example.

Example 1. Consider the finite impulse response (FIR) system

$$y_t = \sum_{k=1}^{n} \theta_k u_{t-k} + e_t \qquad (9.1)$$

Cristian Rojas · Jonas Mårtensson · Håkan Hjalmarsson
ACCESS Linnaeus Center, School of Electrical Engineering,
KTH – Royal Institute of Technology, S-100 44 Stockholm, Sweden
e-mail: {cristian.rojas,jonas.martensson,hakan.hjalmarsson}@ee.kth.se

D. Alberer et al. (Eds.): Identification for Automotive Systems, LNCIS 418, pp. 149–164.
springerlink.com

where y_t is the output, u_t is the input and where $\{e_t\}$ is zero mean white noise with variance[1] $\lambda = \mathrm{E}[e_t^2]$, and where $\theta_k \in \mathbb{R}$, $k = 1, \ldots, n$ are the impulse response coefficients.

Let $\theta = \begin{bmatrix} \theta_1 \ldots \theta_n \end{bmatrix}^T \in \mathbb{R}^n$ and suppose that the objective is to estimate the frequency response $G(e^{j\omega}, \theta) = \sum_{k=1}^{n} \theta_k e^{-j\omega k}$ of the system. In particular let us assume that the frequency of interest is $\omega = 0$,, i.e. we are interested in the static gain $\sum_{k=1}^{n} \theta_k$ of the system. With a constant input $u_t = 1$, $t = 1, \ldots$, the system output will be

$$y_t = \left(\sum_{k=1}^{n} \theta_k \right) + e_t \tag{9.2}$$

i.e. the output will fluctuate around the sought after quantity making the estimation problem very simple regardless of the order of the system. However, with this input the individual impulse response coefficients $\{\theta_k\}_{k=1}^{n}$ cannot be identified and thus, e.g., the frequency response at other frequencies than 0 can not be recovered from the data. ■

Thus, a given data set may be suitable for one application but not for another. In this chapter we will outline how one can formulate optimal experiment design problems whereby the application is taken into account.

The bibliography on optimal experiment design is extensive and include the by now classical texts [14, 37, 2, 29]. The survey papers [1, 28] provide good overviews of the area. Recently there have been contributions to design in the frequency domain [15], set-membership identification [6], applications-oriented experiment design [18, 23, 19, 4], plant-friendly design [9], convexification of the associated optimization problems [18, 23], robust experiment design [34], least-costly identification where the cheapest experiment that fulfills given performance specifications is designed [8], and closed loop experiment design [20]. Also the robustness properties of optimal experiment designs have been highlighted [26]. The four PhD theses [22, 25, 3, 32] provide good coverage of recent developments.

In this tutorial we will follow the framework outlined in [19]. We will start in the next section by discussing how to formulate such problems as optimization problems. Next we will consider how to "translate" such optimization problems into a computationally tractable format. This is rooted in recent developments of convex optimization. It is well known that typically solutions to this type of problem depend on properties of the unknown system. Practical ways to handle this are discussed in Section 9.4. For applications where the amount of properties that have to be extracted from the system by way of system identification is limited it turns out that applications-oriented optimal experiment designs provide "added value". The way the estimated model affects the application becomes less sensitive to design choices in the identification step. This is discussed in Section 9.5.

---

[1] $\mathrm{E}[x]$ denotes expectation of a random variable x.

In order to not become submerged in technical details, we will keep the exposition simple by restricting ourselves to models of the type (9.1), estimated by way of standard least-squares.

9.2   Applications-Oriented Experiment Design

9.2.1   The Set of Acceptable Models

A simplified, but conceptually attractive way, to think of the objective of any modeling activity is that a model should be delivered to the user such that when it is used in the application, the resulting performance is acceptable. In fact, we could split the universe of all possible models into two sets: 1) $\mathcal{E}_{app}$: the set of models for which the performance of the application is acceptable, and 2) the set complementary to $\mathcal{E}_{app}$, consisting of all unacceptable models. The optimal experiment design problem thus becomes that of designing an experiment such that it is ensured that the resulting model belongs to the set of acceptable models $\mathcal{E}_{app}$. How to achieve this depends on the underlying assumptions about the data generating mechanism, i.e. the system, the used model and the estimation method. Here, for simplicity, we will consider standard least-squares estimation.

How to quantify what acceptable performance means is of course application dependent. Here we will for simplicity assume that this is captured by that a non-negative function V_{app} is less than some given value $1/\gamma$, we refer to Section 9.3.2 for details.

9.2.2   A Stochastic Framework

We will assume that data is generated by a system of the form (9.1). To simplify the discussion we will assume that the model structure is known, i.e. only the parameters $\theta = \begin{bmatrix} \theta_1 \ \dots \ \theta_n \end{bmatrix}^T \in \mathbb{R}^n$ are unknown. We will return to the situation when the structure is unknown in Section 9.5. With $\hat{\theta}_N$ denoting the parameter estimate based on N input-output samples $\{y_t, u_t\}_{t=1}^N$, the source of the discrepancy between $\hat{\theta}_N$ and θ will then, neglecting unknown initial conditions, be due solely to the noise $\{e_t\}$ in (9.1). Assuming this quantity to be zero mean white noise implies that the error in the parameter estimate will be random. More specifically, the least-squares estimate of θ is given by [24]

$$\hat{\theta}_N := R_N^{-1} f_N \quad \text{where} \quad R_N := \sum_{t=1}^{N} \varphi_t \varphi_t^T, \quad f_N := \sum_{t=1}^{N} \varphi_t y_t$$

where $\varphi_t = \begin{bmatrix} u_{t-1} & \ldots & u_{t-n} \end{bmatrix}^T$. Above, we have assumed that R_N is full rank which corresponds to that data is persistently exciting [24]. Using (9.1), simple algebra gives

$$\hat{\theta}_N - \theta = R_N^{-1} \sum_{t=1}^{N} \varphi_t e_t$$

Assuming the noise to be normal distributed, this implies that also the parameter estimate is normal distributed[2]

$$\hat{\theta}_N - \theta \in N(0, \lambda R_N^{-1}) \tag{9.3}$$

where λ is the variance of the noise $\lambda = \mathrm{E}[e_t^2]$. From (9.3) it follows that $(\hat{\theta}_N - \theta)^T \frac{R_N}{\lambda} (\hat{\theta}_N - \theta)$ is χ^2-distributed with n degrees of freedom. With α denoting the 99.5% percentile of this distribution, this in-turn implies that the parameter estimate will end up in the confidence ellipsoid

$$\mathcal{E}_{id} := \left\{ \bar{\theta} : (\bar{\theta} - \theta)^T \frac{R_N}{\lambda} (\bar{\theta} - \theta) \leq \alpha \right\} \tag{9.4}$$

with 99.5% probability [24]. Notice that the ellipsoid $\mathcal{E}_{id}$ depends on the experimental conditions since R_N depends on the used input sequence.

9.2.3 Stochastic Applications-Oriented Experiment Design

With the set of acceptable models $\mathcal{E}_{app}$ given, one way of realizing the experiment design paradigm outlined in Section 9.2.1 is to design the input sequence such that

$$\mathcal{E}_{id} \subseteq \mathcal{E}_{app} \tag{9.5}$$

since then with at least 99.5% probability, the model parameter estimate will belong to $\mathcal{E}_{app}$.

The condition (9.5) can be achieved by many different types of experiments and in order to make the problem well posed some criterion has to be introduced. In [8] it was proposed that experiment design should try to minimize

---

[2] $x \in N(m, P)$ denotes that the random variable x is normal distributed with mean m and covariance matrix P.

the experimental cost. Following this, one should try to achieve (9.5) by as "cheap" an experiment as possible. This leads to the following input signal design problem

$$\min_{u^N} V(u^N)$$

$$\text{s.t. } \mathcal{E}_{id} \subseteq \mathcal{E}_{app}$$

(9.6)

where $u^N = \begin{bmatrix} u_1, \ldots, u_N \end{bmatrix}^T$ and where the function $V(u^N)$ should measure the cost of the experiment.

To illustrate the approach we return to the frequency response estimation problem again.

Example 2 (Example 1 cont'd). The frequency response can be expressed as $G(e^{j\omega}, \theta) = \theta^T \Gamma(e^{j\omega})$ where $\Gamma(e^{j\omega}) = \begin{bmatrix} e^{-j\omega} \ldots e^{-j\omega n} \end{bmatrix}^T$, $\omega \in \mathbb{R}$. Suppose that we want the squared error

$$|G(e^{j\omega}, \hat{\theta}_N) - G(e^{j\omega}, \theta)|^2 = (\hat{\theta}_N - \theta)^T \Gamma(e^{j\omega}) \Gamma^*(e^{j\omega})(\hat{\theta}_N - \theta)$$

(9.7)

(x^* denotes the complex conjugate transpose of x) to be less than $1/\gamma$ where the positive constant γ represents the desired accuracy. This means that

$$\mathcal{E}_{app} = \left\{ \bar{\theta} : (\bar{\theta} - \theta)^T \Gamma(e^{j\omega}) \Gamma^*(e^{j\omega})(\bar{\theta} - \theta) \leq \frac{1}{\gamma} \right\}$$

(9.8)

■

9.2.4 *Alternative Formulations*

The formulation (9.6) was introduced in [19] but there exist other problem formulations that aim at achieving a similar objective. Here we briefly outline an approach that is closely connected to so called L-optimal design. We continue to use the frequency response estimation problem as example. The frequency response estimate is given by $G(e^{j\omega}, \hat{\theta}_N) = \hat{\theta}_N^T \Gamma(e^{j\omega})$, which, being linear in $\hat{\theta}_N$, has variance

$$\mathrm{E} \left[|G(e^{j\omega}, \hat{\theta}_N) - G(e^{j\omega}, \theta)|^2 \right] = \Gamma^*(e^{j\omega}) R_N^{-1} \Gamma(e^{j\omega})$$

(9.9)

due to (9.3). Thus the constraint

$$\Gamma^*(e^{j\omega}) R_N^{-1} \Gamma(e^{j\omega}) \leq \beta$$

(9.10)

for some suitably chosen constant β, captures that a certain quality of the estimate of $G(e^{j\omega}, \theta)$ is required. Replacing the set constraint in (9.6) by (9.10) gives

$$\min_{u^N} V(u^N)$$
$$\text{s.t. } \Gamma^*(e^{j\omega}) R_N^{-1} \Gamma(e^{j\omega}) \le \beta \tag{9.11}$$

We refer to [23] for further details on this approach.

9.3 Computational Aspects

When solving the input signal design problems (9.6) and (9.11) it is often more convenient to do it in the frequency domain. Below we will show that we can cast both these problems as convex programs in this domain.

9.3.1 Choice of Decision Variables

It is the matrix R_N in (9.6) and (9.11) that depends on the input. However, using the input sequence directly as decision variable leads to a non-convex problem. Turning to the frequency domain, the discrete Fourier transform (DFT) [27] of the input sequence is given by

$$U(e^{j\mu_k}) := \sum_{t=1}^{N} u_t e^{-j\mu_k t}, \quad \mu_k = 2\pi k/N, \quad k = 1, \ldots, N \tag{9.12}$$

and with a periodic extension of the input, the DFT of the regressor sequence $\{\varphi_t\}$ is given by $\sum_{t=1}^{N} \varphi_t e^{-j\mu_k t} = \Gamma(e^{j\mu_k}) U(e^{j\mu_k})$. By Parseval's theorem, [27], the matrix R_N can be expressed as

$$R_N = \sum_{t=1}^{N} \varphi_t \varphi_t^T = \frac{1}{N} \sum_{k=1}^{N} \Gamma(e^{j\mu_k}) |U(e^{j\mu_k})|^2 \Gamma^*(e^{j\mu_k}) \tag{9.13}$$

Now we let the input be parameterized by the coefficients $\{c_t\}_{t=0}^{N-1}$ as

$$|U(e^{j\mu_k})|^2 = \sum_{t=0}^{N-1} c_t e^{-j\mu_k t} \tag{9.14}$$

Then the (p,q)-element of R_N is equal to $c_{|p-q|}$. Thus R_N is linearly parametrized in terms of these cofficients. As we will see below this is instrumental in order to convexify both (9.6) and (9.11).

Since $|U(e^{j\mu_k})|^2 \geq 0$, it has to hold that

$$\sum_{t=0}^{N-1} c_t e^{-j\mu_k t} \geq 0, \quad k = 1, \dots, N \tag{9.15}$$

for $\{c_t\}_{t=0}^{N-1}$ to be a valid parametrization. One implication of this is that $c_{N-j} = c_j$. These constraints thus have to be included.

9.3.2   Using the Set of Acceptable Models

Let us now turn our attention to the set of acceptable models $\mathcal{E}_{app}$. Consider first the frequency response estimation example.

Example 3 (Example 2 cont'd). As we saw in (9.4) the confidence set $\mathcal{E}_{id}$ is an ellipsoid.

Furthermore, since (9.8) is quadratic in $\bar{\theta}$, $\mathcal{E}_{app}$ for the frequency response estimation problem also is an ellipsoid[3] It is easy to verify, see [19], that the set constraint $\mathcal{E}_{id} \subseteq \mathcal{E}_{app}$ is equivalent to

$$R_N - \gamma \lambda \alpha \Gamma(e^{j\omega}) \Gamma^*(e^{j\omega}) \geq 0 \tag{9.16}$$

Now, since R_N is linear in the decision variables $\{c_t\}_{t=0}^{N-1}$, this inequality is a linear matrix inequality (LMI) which is a convex constraint. If $V(u^N)$ is the input energy, i.e. $V(u^N) = \sum_{t=1}^{N} u_t^2$ which in turn, using Parseval's theorem, can be written

$$V(u^N) = c_0 \tag{9.17}$$

then the optimal input signal design problem (9.6) can be written

$$\min_{c_0, \dots, c_{N-1}} c_o \tag{9.18}$$

$$\text{s.t.} \quad \begin{bmatrix} c_o & c_1 & \cdots & c_{N-1} \\ c_1 & c_0 & c_1 & \cdots & c_{N-2} \\ \vdots & \ddots & \ddots & \ddots & \vdots \\ c_{N-2} & c_{N-1} & \ddots & \ddots & c_1 \\ c_{N-1} & c_{N-2} & \cdots & c_1 & c_0 \end{bmatrix} - \gamma \lambda \alpha \Gamma(e^{j\omega}) \Gamma^*(e^{j\omega}) \geq 0 \tag{9.19}$$

$$\sum_{t=0}^{N-1} c_t e^{-j\mu_k t} \geq 0, \quad k = 1, \dots, N \tag{9.20}$$

$$c_{N-j} = c_j, \quad j = 1, \dots, N-1 \tag{9.21}$$

which is a semi-definite program and hence convex.

---

[3] To be more precise it is a degenerate ellipsoid since $\Gamma(e^{j\omega})\Gamma^*(e^{j\omega})$ is singular.

For example for $\omega = 0$, i.e. the static gain estimation problem discussed in Example 1, we have

$$\min_{c_0,\ldots,c_{N-1}} c_o \tag{9.22}$$

$$\text{s.t.} \quad \begin{bmatrix} c_o & c_1 & \cdots & c_{N-1} \\ c_1 & c_0 & c_1 & \cdots & c_{N-2} \\ \vdots & \ddots & \ddots & \ddots & \vdots \\ c_{N-2} & c_{N-1} & \ddots & \ddots & c_1 \\ c_{N-1} & c_{N-2} & \cdots & c_1 & c_0 \end{bmatrix} - \gamma\lambda\alpha \begin{bmatrix} 1 & 1 & \cdots & 1 \\ 1 & 1 & \cdots & & 1 \\ \vdots & \vdots & & & \vdots \\ 1 & 1 & \cdots & & 1 \\ 1 & 1 & \cdots & & 1 \end{bmatrix} \geq 0 \tag{9.23}$$

$$\sum_{t=0}^{N-1} c_t e^{-j\mu_k t} \geq 0, \quad k = 1,\ldots,N \tag{9.24}$$

$$c_{N-j} = c_j, \quad j = 1,\ldots,N-1 \tag{9.25}$$

The diagonal elements in the matrix inequality give $c_0 \geq \gamma\lambda\alpha$. However, then it is straightforward to see that the choice $c_0 = c_1 = \ldots = c_{N-1} = \gamma\lambda\alpha$ both satisfies the matrix inequality and the positivity constraints (9.24). This solution corresponds to

$$|U(e^{j\mu_k})|^2 = \begin{cases} N\gamma\lambda\alpha & k = N \\ 0 & k = 1,\ldots,N-1 \end{cases}$$

which in turn (taking the inverse DFT) corresponds to

$$u_t = \sqrt{\gamma\lambda\alpha}, \quad t = 1,\ldots,N$$

Thus we have proven that a constant input signal is optimal when the static gain is the quantity of interest. ∎

In general, $\mathcal{E}_{app}$ may not be convex. One can then use the following approximation. Consider a set of acceptable models given by

$$\mathcal{E}_{app} = \left\{ \bar{\theta} : V_{app}(\bar{\theta}) \leq \frac{1}{2\gamma} \right\} \tag{9.26}$$

for some function $V_{app}(\bar{\theta})$ that has a global minimum equal to zero at the true parameter value θ. A second order approximation is then given by

$$V_{app}(\bar{\theta}) \approx \frac{1}{2}(\bar{\theta} - \theta)^T V_{app}''(\theta)(\bar{\theta} - \theta) \tag{9.27}$$

since the gradient $V_{app}'(\theta)$ is zero by construction. Also, by construction, the second derivative $V_{app}''(\theta)$ is positive definite at the global optimum and when the approximation (9.27) is used in (9.26) the set of acceptable models is an ellipsoid given by

$$\mathcal{E}_{app} \approx \left\{ \bar{\theta} : (\bar{\theta} - \theta)^T V''_{app}(\theta)(\bar{\theta} - \theta) \leq \frac{1}{\gamma} \right\} \tag{9.28}$$

Following the reasoning in Example (3) gives that the condition $\mathcal{E}_{id} \subseteq \mathcal{E}_{app}$ (9.5) can be approximated by

$$R_N - \gamma \lambda \alpha V''_{app}(\theta) \geq 0 \tag{9.29}$$

9.3.3 Using Variance Constraints

Now we turn to the alternative constraint formulation (9.10) which involves the inverse of R_N. By making use of Schur complements, see e.g. [38], it can be translated to a constraint that is linear in R_N. Consider the partitioned hermitian matrix $M = \begin{pmatrix} A & B \\ B^* & C \end{pmatrix}$ where A is square and non-singular. The Schur complement of A in M is defined as $S_A := C - B^* A^{-1} B$ and it holds that $M \geq 0$ if and only if $A > 0$ and $S_A \geq 0$. The constraint (9.11) can be written as $\beta - \Gamma^*(e^{j\omega}) R_N^{-1} \Gamma(e^{j\omega}) \geq 0$ where the left-hand side is the Schur complement of R_N in the partitioned matrix $M_R := \begin{pmatrix} R_N & \Gamma(e^{j\omega}) \\ \Gamma(e^{j\omega})^* & \beta \end{pmatrix}$. Since $R_N > 0$ by assumption it follows that (9.11) is equivalent to $M_R \geq 0$. Using the same arguments again we find that $M_R \geq 0$ if and only if the Schur complement of β in M_R is positive semi-definite (β is positive by assumption):

$$R_N - \frac{1}{\beta} \Gamma(e^{j\omega}) \Gamma^*(e^{j\omega}) \geq 0 \tag{9.30}$$

Note that, with $\beta = \frac{1}{\lambda \alpha \gamma}$, (9.30) is also equivalent to the constraint (9.29) with $V_{app}(\hat{\theta}_N)$ given by (9.7). Thus we end up with exactly the same optimization problem (9.18)–(9.21) as before, but with β replacing $\frac{1}{\lambda \alpha \gamma}$. We refer to [23] for generalizations.

9.4 How to Handle System Dependency of the Optimal Solution

In the formulation (9.6), typically either $\mathcal{E}_{id}$ or $\mathcal{E}_{app}$ (or both) depend on the true system, i.e., we have $\mathcal{E}_{id} = \mathcal{E}_{id}(\theta)$ and $\mathcal{E}_{app} = \mathcal{E}_{app}(\theta)$. This usually implies that the optimal input signal will be a function of something which is unknown prior to the experiment.

In our FIR example, $\mathcal{E}_{id}$ is independent of θ, since it is a linearly parameterized model structure (unlike, e.g., ARX or ARMAX structures, where the covariance does depend on θ). For the frequency estimation problem, $\mathcal{E}_{app}$ is also independent of θ. However, if we were interested instead in estimating the squared magnitude of the frequency response at a given frequency, $|G(e^{j\omega})|^2$, then a Taylor expansion shows that $\mathcal{E}_{app}$ is approximately

$$\mathcal{E}_{app}(\theta) = \left\{ \bar{\theta} : (|G(e^{j\omega}, \bar{\theta})|^2 - |G(e^{j\omega}, \theta)|^2)^2 \leq \frac{1}{\gamma} \right\}$$
$$\approx \left\{ \bar{\theta} : (\bar{\theta} - \theta)^T \Gamma(e^{j\omega}) |G(e^{j\omega}, \theta)|^2 \Gamma^*(e^{j\omega})(\bar{\theta} - \theta) \leq \frac{1}{4\gamma} \right\}$$

where the dependence on θ through $|G(e^{j\omega}, \theta)|^2$ is clear.

The fact that the optimal input signal is a function of the true plant is not necessarily a problem if we already had some prior information, say from a previous experiment, and our goal is to design a new experiment to improve an existing model. In this case, we can simply replace in the formulation (9.6) $\mathcal{E}_{id}$ and $\mathcal{E}_{app}$ by $\mathcal{E}_{id}(\hat{\theta})$ and $\mathcal{E}_{app}(\hat{\theta})$, respectively, where $\hat{\theta}$ is a basic estimate of the plant; the input signal obtained in this way is called a *locally optimal input signal design* [13].

However, there are situations where we do not have enough reliable prior information to design our input signal based on a previous estimate of the plant. In the following subsections we describe two approaches to overcome this difficulty.

9.4.1 Robust Designs

Even in the absence of a preliminary estimate of the plant, we might have some basic knowledge about its parameters. For example, from physical considerations it is sometimes possible to infer the order of magnitude of the dominant time constant, the static gain, etc. If this is the case, we can assume that we know a priori a set Θ such that $\theta \in \Theta$, and reformulate the optimal input signal design problem (9.6) to account for this information. One possibility is to consider the following robust version of (9.6):

$$\min_{u^N} V(u^N)$$
$$\text{s.t. } \mathcal{E}_{id}(\theta) \subseteq \mathcal{E}_{app}(\theta), \text{ for all } \theta \in \Theta \tag{9.31}$$

This problem corresponds to a *robust convex program*, which, save for a few exceptions, cannot be solved exactly in a computationally tractable way [7]. Here we present a simple approximate method to solve (9.31), based on a probabilistic relaxation technique known as the *scenario approach* [10]. For details, the reader is referred to [35]. The basic idea is to replace (9.31) by

$$\min_{u^N} V(u^N)$$
$$\text{s.t. } \mathcal{E}_{id}(\theta_i) \subseteq \mathcal{E}_{app}(\theta_i), \quad i = 1, \ldots, m \tag{9.32}$$

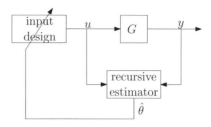

Fig. 9.1 Adaptive input signal design.

where $\theta_1, \ldots, \theta_m$ are independent random samples from a distribution P_θ in Θ. In [11] it was shown that if (9.32) is formulated in terms of d decision variables and m is chosen such that

$$\binom{m}{d}(1 - \epsilon)^{m-d} < \delta$$

for some constants $\delta, \epsilon > 0$, then the solution of (9.32) will, with probability greater than $1 - \delta$, satisfy 'most' of the constraints in (9.31). More precisely, $P_\theta(\{\theta \in \Theta : \mathcal{E}_{id}(\theta) \not\subseteq \mathcal{E}_{app}(\theta)\}) < \epsilon$ holds with probability $1 - \delta$.

In Corollary 1 in [10], the following more explicit (but conservative) expression for m was proposed: $m \geq (2/\epsilon) \ln(1/\delta) + 2d + (2d/\epsilon) \ln(2/\epsilon)$. For example, for $d = 30$, $\epsilon = 0.01$ and $\delta = 10^{-10}$, m should be at least 7864, which gives a tractable semidefinite program (solvable within 10 minutes on a computer with Intel Core 2 Duo CPU of 2.53 GHZ and a standard SDP solver)."

9.4.2 Adaptive and Sequential Designs

The idea of designing an experiment based on a previous estimate, and then redesigning the experiment from the previous one can be carried out even further, to arrive at the concept of adaptive input signal design. In this approach, a recursive estimator is employed to obtain a new estimate of the plant at each time instant, and the recursive estimate is then used to simultaneously redesign the input sequence. Figure 9.1 presents a block diagram of the adaptive technique.

As explained in [16], adaptive input signal design can achieve optimal performance (i.e., the same performance as if the experiment were designed based on the knowledge of the true system) under mild conditions, even though the convergence proofs are difficult and beyond the scope of the present tutorial. The interested reader is referred to [16, 17] for details, in particular the latter reference contains a convergence proof for adaptive signal design for ARX models.

9.5   Added Value of Optimal Input Signal Design

In the previous sections we have focused on the formulation of applications-oriented optimal input signal design problems and how to solve the associated optimization problems. An obvious question is how much that can be gained by optimal input signal design. The answer to this question is clearly problem dependent, and also depends on what one compares with. The reduction in experiment time in the two case studies (a process control application and a mechanical system) presented in [5] was roughly a factor 4 as compared to a Pseudo-Random Binary Sequence. The paper [12] presents results from re-tuning of an MPC controller of a process having 34 inputs and 90 outputs at one of the world's largest refineries, the Hovensa refinery in Virgin Islands, United States. Partially through the use of optimal experiment design, the total modeling time was reduced by 90%.

By solving the optimal applications-oriented input signal design problem for different model complexities and different performance specifications it is possible to characterize how the minimal experimental cost depends on these quantities. This cost of complexity quantification, as it has become known, may provide valuable information for trading off performance in the application versus experimental effort in the identification problem. In Section 9.5.1 we discuss this for our standing example.

It is fairly obvious that optimal input signal design helps make system properties of importance easy to detect in the measured signals. However, a less obvious mechanism is that the optimal input signal tries to avoid to excite system behaviours that are inessential for the application. Thus it *hides* properties which are not important to the user. This dual aspect of the input signal has some very pleasant consequences, e.g.

1. It keeps the cost of the experiment at a reasonable level for models of high order.
2. It sometimes allows for consistent estimation of the properties of interest, even in the presence of unmodelled dynamics.

We will discuss this further in Section 9.5.2.

9.5.1   Cost of Complexity

By focusing on specific properties of a given system, it is possible to reduce the identification effort, measured in terms of the input signal energy required to estimate the desired property within a given accuracy.

Example 4 (Example 3 cont'd). Suppose that (9.8) is modified to

$$\mathcal{E}_{app} = \left\{ \bar{\theta} : (\bar{\theta} - \theta)^T \Gamma(e^{j\omega}) \Gamma^*(e^{j\omega})(\bar{\theta} - \theta) \leq \frac{1}{\gamma}, \ \forall \omega \in [0, \omega_B] \right\}, \quad (9.33)$$

i.e. it is required that the accuracy of the frequency response is γ over the frequency range $[0, \omega_B]$, where $0 < \omega_B \leq \pi$. Then, following [33], it can be shown that the minimum input signal energy required (for large model orders n) is approximately

$$n\lambda\gamma\omega_B \tag{9.34}$$

where λ is the variance of the noise e_t and $1/\gamma$ is the maximum allowed variance of the frequency function estimate in the range $[0, \omega_B]$.

According to (9.34), the required input signal energy increases in proportion to the model order n. However, this cost can be kept low by focusing on a more specific frequency region, i.e., by reducing ω_B. This observation reenforces the statement that input signal design can hide those properties which are unimportant, thus providing an energy efficient experiment.

As a comparison, it can be shown, see [33] for details, that using a PRBS (Pseudo Random Binary Sequence) as input signal requires the energy $n\lambda\gamma\pi$ regardless of the bandwidth ω_B to ensure that the parameter estimate ends up in (9.33). Thus, for large model orders n, the reduction in required input energy through the use of optimal input signal design can be significant for small to moderate bandwidths ω_B, as compared to PRBS excitation.   ■

9.5.2   Consistent Estimation with Unmodelled Dynamics

A simple example of the phenomenon that sometimes optimal input signal design allows the use of restricted models is given by the static gain estimation problem studied in the examples of this chapter.

Example 5 (Example 4 cont'd). In Example 3 we saw that a constant input signal is optimal for the problem of estimating the static gain. However, this particular input signal has the additional property that it still allows for consistent estimation of the static gain if we use the following simple static model: $y_t = Ku_t$. The reason lies in the fact that a constant input signal completely hides all the other properties of the system by making them unidentifiable, This since, after the transient has died out, the system behaves as

$$y_t = \left(\sum_{k=1}^{n} \theta_k \right) u_t + e_t$$

when $u_t = u$ (constant).   ■

A perhaps more interesting example is given in [25], where it has been shown that a nonminimum phase zero of a system G (of arbitrary order) at $z = z_o$ can be consistently estimated with a model of the form $y_t = b_1 u_{t-1} + b_2 u_{t-2}$ by using an input signal

$$u_t = \frac{1}{1 - z_o^{-1}} r_t$$

where r_t is zero mean white noise. This input signal depends on the property of interest, but it can be implemented using the adaptive scheme of Section 9.4.2 [31].

9.6 Concluding Remarks

We have in this chapter provided the basics of applications-oriented experiment design. There exists a wide range of extensions. By viewing the input signal as a stationary signal with a continuous spectral density $\Phi_u(e^{j\mu})$, the spectral density takes the place of $|U(e^{j\mu_k})|^2$ in the formulation above and the discrete-time Fourier transform (DTFT) is used instead of the DFT. An input signal with the desired spectral density can be generated by filtering white noise through the spectral factor of $\Phi_u(z)$. The Kalman-Yakubovich-Popov lemma [36, 30] can be used to replace the positivity condition on the input spectrum by a linear matrix inequality. This allows a wide range of model structures to be used, e.g. ARX, ARMAX, output-error and Box-Jenkins models. We refer to [22] for details. Also closed loop experiment design can be handled in this way [20] and there is emerging work on applications-oriented experiment design for non-linear systems [21].

References

[1] Atkinson, A., Bailey, R.: One hundred years of the design of experiments on and off the pages of Biometrika. Biometrika 88, 53–97 (2001)
[2] Atkinson, A., Doner, A.: Optimum experiment design. Clarendon Press, Oxford (1992)
[3] Barenthin, M.: Complexity issues, validation and input design for control in system identification. Doctoral thesis, KTH, Stockholm, Sweden (2008)
[4] Barenthin, M., Bombois, X., Hjalmarsson, H., Scorletti, G.: Identification for control of multivariable systems: Controller validation and experiment design via LMIs. Automatica 44(12), 3070–3078 (2008)
[5] Barenthin, M., Jansson, H., Hjalmarsson, H.: Applications of mixed $\mathcal{H}_\infty$ and $\mathcal{H}_2$ input design in identification. In: 16th World Congress on Automatic Control, IFAC, Prague, Czech Republik (2005) Paper Tu-A13-TO/1
[6] Belforte, G., Gay, P.: Optimal input design for set-membership identification of Hammerstein models. International Journal of Control 76(3), 217–225 (2003)
[7] Ben-Tal, A., Nemirovski, A.: Robust convex optimization. Mathematics of Operations Research 23(4), 769–805 (1998)
[8] Bombois, X., Scorletti, G., Gevers, M., Van den Hof, P.M.J., Hildebrand, R.: Least costly identification experiment for control. Automatica 42(10), 1651–1662 (2006)

[9] Braun, M., Ortiz-Mojica, R., Rivera, D.: Application of minimum crest factor multisinusoidal signals for "plant-friendly" identification of nonlinear process systems. Control Engineering Practice 10(3), 301–313 (2002)

[10] Calafiore, G.C., Campi, M.C.: The scenario approach to robust control design. IEEE Transactions on Automatic Control 51(5), 742–753 (2006)

[11] Campi, M.C., Garatti, S.: The exact feasibility of randomized solutions of uncertain convex programs. SIAM Journal on Optimization 19(3), 1211–1230 (2008)

[12] Celaya, P., Tkatch, R., Zhu, Y., Patwardhan, R.: Closed-loop identification at the Hovensa refinery. In: NPRA Plant Automation & Decision Support Conference, San Antonio, Texas (2004)

[13] Chernoff, H.: Locally optimal designs for estimating parameters. Annals of Mathematical Statistics 24, 586–602 (1953)

[14] Fedorov, V.V.: Theory of Optimal Experiments, Probability and Mathematical Statistics. Academic Press, London (1972)

[15] Forssell, U., Ljung, L.: Some results on optimal experiment design. Automatica 36(5), 749–756 (2000)

[16] Gerencsér, L., Hjalmarsson, H.: Adaptive input design in system identification. In: Proc. of the 44th IEEE Conference on Decision and Control and European Control Conference, Seville, Spain, pp. 4988–4993 (2005)

[17] Gerencsér, L., Hjalmarsson, H.: Identification of ARX systems with non-stationary inputs – asymptotic analysis with application to adaptive input design. Automatica 45(3), 623–633 (2009)

[18] Hildebrand, R., Gevers, M.: Identification for control: Optimal input design with respect to a worst-case ν-gap cost function. SIAM J. Control Optim. 41(5), 1586–1608 (2003)

[19] Hjalmarsson, H.: System identification of complex and structured systems. European Journal of Control 15(4), 275–310 (2009); Plenary address. European Control Conference

[20] Hjalmarsson, H., Jansson, H.: Closed loop experiment design for linear time invariant dynamical systems via LMIs. Automatica 44(3), 623–636 (2008)

[21] Hjalmarsson, H., Mårtensson, J.: Optimal input design for identification of non-linear systems: Learning from the linear case. In: American Control Conference, New York City, USA (2007)

[22] Jansson, H.: Experiment design with applications in identification for control, Doctoral thesis, KTH, Stockholm, Sweden (2004)

[23] Jansson, H., Hjalmarsson, H.: Input design via LMIs admitting frequency-wise model specifications in confidence regions. IEEE Transactions on Automatic Control 50(10), 1534–1549 (2005)

[24] Ljung, L.: System Identification: Theory for the User, 2nd edn. Prentice-Hall, Englewood Cliffs (1999)

[25] Mårtensson, J.: Geometric analysis of stochastic model errors in system identification, Doctoral thesis, KTH, Stockholm, Sweden (2007)

[26] Mårtensson, J., Hjalmarsson, H.: How to make bias and variance errors insensitive to system and model complexity in identification. IEEE Transactions on Automatic Control 56(1), 100–112 (2011)

[27] Oppenheim, A., Schafer, R.: Discrete-Time Signal Processing. Prentice-Hall, Englewood Cliffs (1989)

[28] Pronzato, L.: Optimal experimental design and some related control problems. Automatica 44(2), 303–325 (2008)

[29] Pukelsheim, F.: Optimal design of experiments. John Wiley, Chichester (1993)

[30] Rantzer, A.: On the Kalman-Yakubovich-Popov lemma. Systems and Control Letters 28(1), 7–10 (1996)

[31] Rojas, C., Hjalmarsson, H., Gerencsér, L., Mårtensson, J.: Consistent estimation of real NMP zeros in stable LTI systems of arbitrary complexity. In: 15th IFAC Symposium on System Identification, Saint-Malo, France, pp. 922–927 (2009)

[32] Rojas, C.R.: Robust experiment design, Ph.D. thesis, The University of Newcastle (2008)

[33] Rojas, C.R., Barenthin, M., Welsh, J.S., Hjalmarsson, H.: The cost of complexity in identification of FIR systems. In: 17th IFAC World Congress, Seoul, South Korea, pp. 11,451–11,456 (2008)

[34] Rojas, C.R., Welsh, J.S., Goodwin, G.C., Feuer, A.: Robust optimal experiment design for system identification. Automatica 43(6), 993–1008 (2007)

[35] Welsh, J.S., Rojas, C.R.: A scenario based approach to robust experiment design. In: Proceedings of the 15th IFAC Symposium on System Identification (SYSID 2009), Saint-Malo, France (2009)

[36] Yakubovich, V.A.: Solution of certain matrix inequalities occurring in the theory of automatic control. Docl. Acad. Nauk. SSSR, 1304–1307 (1962)

[37] Zarrop, M.: Optimal Experiment Design for Dynamic System Identification. Lecture Notes in Control and Information Sciences, vol. 21. Springer, Berlin (1979)

[38] Zhang, F. (ed.): The Schur Complement and Its Applications. Numerical Methods and Algorithms, vol. 4. Springer, Heidelberg (2005)

Chapter 10
Engine Calibration Using Nonlinear Dynamic Modeling

Karsten Röpke, Wolf Baumann, Bert-Uwe Köhler, Steffen Schaum,
Richard Lange, and Mirko Knaak

Abstract. In recent years, engine calibration efforts have increased dramatically in order to fulfill legislation requirements, in particular the reduction of undesired emissions and fuel consumption while maintaining drivability and meeting comfort demands. Dynamic modeling has been shown to be a useful tool in this context. In this paper it is shown how nonlinear dynamic modeling is applied to the virtual calibration of engines. Two different kinds of excitation signals for the training of the models are compared and their usability for engine calibration is analysed. It is shown that virtual engine calibration can lead to a significant reduction of test bench costs during engine and vehicle development while maintaining a high level of calibration accuracy.

10.1 Introduction to Engine Calibration

Today's combustion engines not only have to meet customer requirements, but also have to comply with strict legal regulations. These regulations concern particularly the combination of the emissions (e.g. NOx, CO, HC) and the fuel consumption. Engine efficiency, emissions and fuel consumption in this regard, form a tradeoff which highly increases the complexity of the task. To achieve these target criteria, modern engine control systems provide a large number of parameters that can be adjusted during operation. For example, when operating a modern direct injection (DI) gasoline engine in homogeneous mode, the parameters spark timing, intake camshaft, exhaust camshaft, fuel injection timing and air-fuel ratio are set depending on the engine's current operating condition (see Fig. 10.1). For modern Diesel engines the number of parameters may be even larger. Here, start of injection,

Karsten Röpke
IAV GmbH Berlin, Carnotstr. 1, 10587 Berlin, Germany
e-mail: `Karsten.Roepke@iav.de`

D. Alberer et al. (Eds.): Identification for Automotive Systems, LNCIS 418, pp. 165–182.

air mass flow, boost pressure, rail pressure, exhaust gas recirculation (EGR) and the timing and quantity of pilot injections can be adjusted.

Dynamic Engine Model

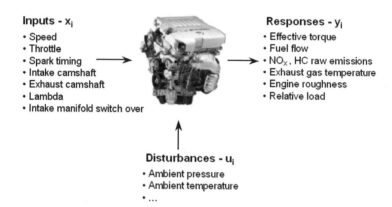

Inputs - x_i
- Speed
- Throttle
- Spark timing
- Intake camshaft
- Exhaust camshaft
- Lambda
- Intake manifold switch over

Responses - y_i
- Effective torque
- Fuel flow
- NO_x, HC raw emissions
- Exhaust gas temperature
- Engine roughness
- Relative load

Disturbances - u_i
- Ambient pressure
- Ambient temperature
- ...

Fig. 10.1 Engine inputs and responses

The control of all those actuators is carried out by the engine's control unit (ECU), which basically is a computer with digital and analog inputs and outputs. For example, the ECU determines the optimal control settings from the current engine status defined by engine speed, environmental temperature, intake air mass and the demand of the driver. The basis for the calculation is the physical relationship between all involved factors, controllers and/or stored multi-dimensional data tables. The goal of the engine calibration process is to fill all values (up to 30,000), data tables and maps in the ECU, so that the engine meets all requirements.

Basic calibration is a major constituent component in calibrating gasoline and Diesel engines. It provides the starting point for a whole host of stages in the calibration process. This is why basic calibration needs to be done with meticulous care since modifications later on come with far-reaching consequences. Basic calibration of the control unit is understood to mean determining optimum steady-state maps on the engine test bench with the aim of ensuring safe and reliable engine operation. Optimization criteria encompass emissions, fuel economy and running smoothness on the gasoline engine side, and, additionally, combustion noise in the case of Diesel engines. With gasoline engines, this involves optimizing the maps for charge sensing, valve timing and ignition angle as well as calibrating any torque interface that may exist. In the case of Diesel engines, the focus is on air mass, boost pressure, rail pressure and the various injection events.

Many engine developers employ models for the basic calibration process, i.e. just a few measurement data are taken as the basis for the training of a - mostly statistical - model which is then used for optimizing the basic maps. The Design of Experiments (DoE) method is often employed for this purpose as it can help to slash measurement input. However, as frequent changes must be expected on the hardware side - particularly in the early stages of developing an engine - maps from a previous construction stage or a comparable engine are often used. This can lead to incorrect component assessments since the maps used are not ideal for the data record in hand. For the early phase of development, therefore, methods must be found that permit high-speed basic calibration. At this point in time, it is admissible to accept minor compromises on the accuracy side. In the late development phase, when it comes to generating production-level data, it is calibration accuracy and robustness that are of paramount importance. Here too, of course, processes and methods must be used that save time and resources while still ensuring quality.

10.2   Nonlinear Statistical Dynamic Modeling

Based on measurements, nonlinear statistical dynamic modeling determines the mathematical relation between inputs and outputs without trying to adopt the physical system structure. The relation is not unique and the modeling assumptions made will affect the model behavior and the necessary effort to obtain it. Generally, the model terms cannot be associated with a physical meaning (black box modeling).

Linear system identification is well established and a sound theoretical foundation has been reached [8]. The extension to nonlinear system identification has only recently found a broader attention, where a main focus is put on single input/single output (SISO) applications. State of the art combustion engines, however, are influenced by several parameters and are therefore consequently multiple input/single output (MISO) systems.

System identification is closely related to response surface modeling in statistics [2]. The main idea is to use a sequence of designed experiments to obtain an optimal response, i.e. a statistical model. DoE methodology assures a sufficiently accurate description with a minimum number of experiments. Since the latter can be extremely costly, their number is the limiting factor in many disciplines, such as modeling combustion engines.

Statistical modeling is originally defined for static conditions and is well established for internal combustion engines [10]. Its extension to non-linear system identification has been described e.g. in [5]. The formal method of DoE can consequently be used for planning the measurements and minimizing the experimental effort.

Starting from a general system with only little a-priori information is equivalent to a highly redundant representation of information. This means that the general model will have much more parameters than the system itself. Through efficient parameter identification, the relevant parameters need to be selected.

For the modeling of internal combustion engines, the state of the art consists of data-based steady-state models with an abstract, non-linear approximator for reproducing or modeling of steady-state operating states. This could be polynomials, neural networks, local linear neuro-fuzzy models (Lolimot), statistical interpolation methods or probabilistic models [9, 6]. Any possible combination of these approaches with physical models that inherently allow for time dependencies will entail complementing the statistical approximators with dynamic components. This is usually based on the approach of external dynamics [10], resulting in the currently most common model types:

- Volterra series
- time-delay neural networks (TDNN)
- dynamic local linear neuro-fuzzy models

The structure of parametric Volterra series is made up of a nonlinear transformation of the input quantities by polynomials with a subsequent finite response filter (FIR) stage. Additionally an infinite response filter (IIR) stage is used to cope with dynamic systems with large time constants. In general the model equation is given by

$$y(k) = g[x(k), x(k-1), \ldots] + a_1 y(k-1) + a_2 y(k-2) + \ldots + a_N y(k-N), \quad (10.1)$$

with $g(\cdot)$ as the nonlinear transformation of the inputs x(k), x(k-1), ... The noteworthy properties of Volterra series are:

- the linearity of parameters
- the high flexibility (ability to approximate Wiener and Hammerstein systems)
- and the easy stability check (by methods of linear system theory)

The linearity of parameters allows for the estimation of a global solution on the basis of least squares or related methods, like weighted least squares (WLS) or total least squares (TLS). In case of outliers in the data there are also appropriate methods available, like e.g. robust regression [7]. In case of an autoregressive (AR) stage, the use of instrumental variables usually gives better parameter estimation results since the noise spectrum is not required to be shaped like the AR spectrum.

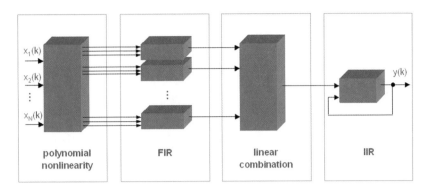

Fig. 10.2 Schematic structure of parametric Volterra series

10.3   Basic Engine Calibration Using Nonlinear Dynamic Modeling

10.3.1   Test Design

Common to all data based modeling techniques is the great importance of the data quality, i.e. the distribution of measurement points. In many cases the user has the possibility to adjust the controlled variables and to design the test as necessary.

A sound framework for the design of system identification tests is given by the method "Design of Experiment" (DoE), which considers the influence of measurement points to parameter estimation. By means of DoE it is possible to optimize the distribution of measurement points in a statistical sense and to achieve an optimal ratio of the number of measurements to parameter accuracy. It is no surprise that DoE methods have become very popular in the field of engine calibration as they allow for the high-dimensional modeling of modern combustion engines with 8 and more parameters. But the application of DoE to nonlinear dynamic modeling is not always a straightforward task. The basic DoE principle, however, remains the same: The appropriate excitation depends on the system to be identified or the underlying model. That means, the more prior information is available the better the test can be adapted to the system under investigation.

A peculiar difficulty for nonlinear dynamic systems in combustion engine modeling is given by the fact that there is only little prior knowledge available about suited model structures for transient emission modeling. Especially the unknown number of system states makes it practically impossible to design statistically optimal excitation sequences. Besides adaptive approaches based on online modeling and test design [4] there are only a few alternative

excitation signal types, which are used in engine modeling. The basic wave-
forms are

- ramp
- step
- sinus

Ramp-like sequences are often used for validation purposes, e.g. as a speed-
ramp to simulate an acceleration phase. For identification tasks we use either step-like or sinusoidal signals, designed as Amplitude Modulated Pseudo
Random Binary Sequences (APRBS) or chirp, respectively.

10.3.1.1   APRBS

A pseudo-random binary sequence (PRBS) is a white stochastic sequence
with good applicability to linear system identification. It can take on levels
of plus and minus one, and has a constant power density as it is a white
sequence. In case of good prior knowledge of the system it may also be useful
to apply non-white PRBS for parameter training.

In addition to the variation of hold times, the APRBS is generated by
a variation of the amplitude levels. An example of an APRBS is shown in
figure 10.3. There, the amplitude modulation is pseudo-random and has the
purpose to cover the a priori unknown nonlinear characteristics. In case of
polynomial nonlinearities in a Hammerstein setup [9] it is also useful to design the amplitude levels in a statistical optimal way, e.g. as D-optimal points
within the legal hull. This approach helps to improve the safety of the engine
measurements at the test bench and results in a higher accuracy of the estimated parameters. The major drawback of APRBS signals is their step-like
nature which may not be appropriate for all engine inputs, like e.g. relative
load of gasoline engines. It may also cause an unstable operation at the test
bench.

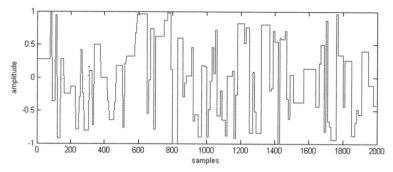

Fig. 10.3 Example of an APRBS

10.3.1.2   Sinusoidal Excitation

An alternative possibility of practical engine excitation is given by sinusoidal sequences with variable instantaneous frequency, like e.g.

$$x(t) = \sin\left(2 \cdot \pi \cdot f \cdot t^2\right). \tag{10.2}$$

These so called chirp sequences are easy to create and have proven to be very well suited for combustion engine modeling [1]. Within IAV, chirp sequences were successfully applied to Diesel as well as gasoline engines. The direct design parameters are the frequency range of interest, the distribution of frequencies, i.e. linear or logarithmic, the chirp sweep duration and also the phase shifts between different inputs. These excitation signal parameters affect the quantities of interest: coverage of the test space, the condition number of the design and the inter-input correlation. A nonlinear optimization algorithm, like e.g. Particle Swarm Optimization [3] may be used to determine a suited parameter set.

The big advantage of using sinusoidal signals for combustion engine excitation is their continuous slope, but there are also advantages from a theoretical point of view [1]:

- excitation of important frequency regions
- high D-optimality value
- easy control of the gradient of adjustment
- polynomial transformations are orthogonal for sinusoids

Having designed a suited excitation sequence, it is still necessary to scale the inputs for safe engine operation, i.e. to avoid violations of the engine hull. This is done by amplitude modulation on basis of the valid hull, as shown in figure 10.4.

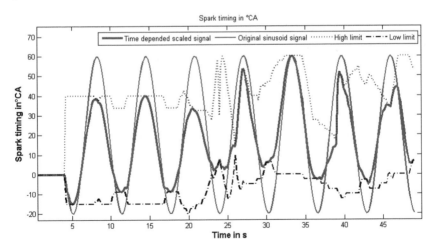

Fig. 10.4 Amplitude modulation of a sinusoidal sequence

The disadvantage of sinusoidal excitation signals can be seen as the lack of steady state excitation phases and the related hold times which are required to identify systems with larger time constants. However, by adding steady state excitation sequences this disadvantage can be overcome.

10.3.1.3 Practical Realization

Depending on the calibration task different frequencies and adjustment ranges need to be considered. E.g. in case of hardware in the loop (HIL)/ software in the loop (SIL) environments or the calibration of ECU models for torque or temperature, the excitation signals have to cover broader model parameter input ranges than in case of modeling near a known optimum. Furthermore it plays a crucial role whether the engine model is used for controller tuning or for prediction of steady state conditions.

In the subsequent example, the task was to create an engine model which is able to predict the transient and steady state behaviour of a gasoline V6 engine under warm operating conditions. This model is mainly used for complete SIL testing of a complex test bench functionality, where good transient and steady state prediction is required. The steady state behavior and the boundaries of the system are well known from previous steady state DoE and grid measurements. The engine excitation boundaries are described by a convex hull. This convex hull information is required for the design of excitation signals within legal engine limits. The following six quantities are used for engine excitation: Engine speed, relative engine load, spark timing, intake cam phase, exhaust cam phase, air-fuel ratio (lambda). Figure 10.5 shows the designed excitation sequence with chirp signals before the scaling into the convex hull.

We used two different types of excitation: a pure sinusoidal approach and a mixed design with three channels as APRBS. Similar designs were used as validation sequences. Each design fulfills the demanded constraints regarding ranges of amplitudes and maximum allowed gradient of adjustment. However, during the design process of APRBS it was difficult to consider limited step sizes and to ensure full modulation at the same time. This finally resulted in a slightly less modulation of APRBS amplitudes, as can be observed by comparison of Fig. 10.6 and 10.7.

10.3.2 Modeling

For the virtual optimization of engine parameters it is necessary to establish models for a number of engine quantities such as emissions, torque, knocking, IMEP-COV (Indicated Mean Effective Pressure - Coefficient of Variance), pre-catalyst exhaust gas temperature, fuel consumption and load. As model

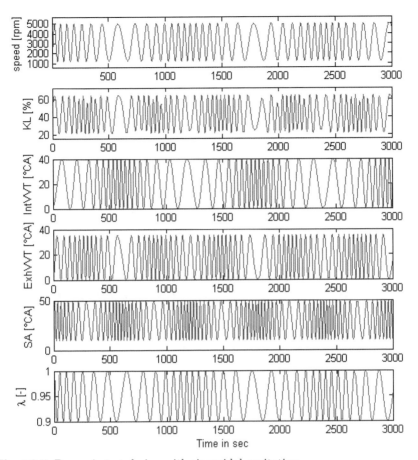

Fig. 10.5 Dynamic test design with sinusoidal excitation

type parametric Volterra series are applied. Depending on the target quantity polynomial orders up to 4 and interaction orders up to 3 are used. Input quantities are engine speed, throttle angle, intake valve timing, exhaust valve timing, spark timing and air-fuel ratio. Also depending on the target quantity, delayed versions of the inputs are required to account for system dynamics with short time constants. The pre-catalyst exhaust gas temperature model uses one additional feed back term. The significant terms of the polynomial are selected using the orthogonal least squares (OLS) term selection procedure as described in [9]. For modeling it is necessary to synchronize the data of the different sensors (time alignment).

Dynamic measurements usually provide a large amount of data within a short time. The number of training data for the sinusoidal excitation is about 33000 samples, for APRBS about 18000 samples. The sampling rate is

10 samples/second. That means the overall measurement time is 55 min for sinusoidal and 30 min for APRBS signals. The quality of the data is ensured by the test design as described in the previous subsection and by a visual inspection of the data to check the plausibility, the modulation amplitude and the noise level.

To evaluate the modeling results, the following measures are used:

- RMSE (root mean square error):

$$RMSE = \sqrt{\frac{1}{N-1}\sum_{i=1}^{N}(y_i - \hat{y}_i)^2} \qquad (10.3)$$

- nRMSE (normalized RMSE):

$$nRMSE = \frac{100 \cdot RMSE}{\max_i(y_i) - \min_i(y_i)}, \qquad (10.4)$$

where N is the number of measured samples, y_i is the modeling target and $\hat{y}_i$ is the estimated target signal. The $RMSE$ is measured in the same unit as the target quantity wheras the $nRMSE$ is measured in %.

Table 10.1 and table 10.2 show the modeling results for sinusoidal excitation and for APRBS excitation, respectively.

Table 10.1 Modeling results for sinusoidal excitation signals

modeling target	training		validation	
	RMSE	nRMSE [%]	RMSE	nRMSE [%]
CO[a]	0.2 %	4.07	0.2 %	7.94
HC[b]	187.0 ppm	5.50	349.0 ppm	6.61
NOx[c]	227.0 ppm	5.69	212.0 ppm	5.52
relative load	2.2 %	3.27	3.4 %	4.52
torque	3.7 Nm	1.64	4.8 Nm	2.13
temperature[d]	7.2 °C	2.32	18.8 °C	5.65
knocking	10.7 %	6.09	11.2 %	5.48
fuel mass flow	0.6 kg/h	2.19	0.8 kg/h	2.94
COV[e]	0.6 %	7.43	1.0 %	8.64

[a]carbon monoxide, [b]hydrocarbon, [c]nitrogen oxide, [d]exhaust gas temperature before catalyst, [e]coefficient of variance (IMEP).

For a visual impression of the modeling results, Fig. 10.6 and Fig. 10.7 show *modeled vs simulated plots* for the torque, modeled on the basis of sinusoidal and on the basis of APRBS signals, respectively. Between sinusoidal training tests and sinusoidal validation tests the engine was dismounted from the test bench and installed again later. Due to minor setup changes the differences between training and validation results are larger than for APRBS.

Table 10.2 Modeling results for APRBS excitation signals

modeling target	training RMSE	nRMSE [%]	validation RMSE	nRMSE [%]
CO[a]	0.2 %	4.26	0.2 %	6.07
HC[b]	144.0 ppm	5.33	179.0 ppm	6.07
NOx[c]	260.0 ppm	6.89	226.0 ppm	5.81
relative load	2.2 %	4.38	2.2 %	5.35
torque	3.9 Nm	2.03	4.4 Nm	2.99
temperature[d]	6.2 °C	2.33	11.2 °C	4.14
knocking	5.8 %	8.27	7.0 %	9.08
fuel mass flow	0.6 kg/h	2.27	0.7 kg/h	2.57
COV[e]	0.5 %	5.40	0.4 %	9.71

[a]carbon monoxide, [b]hydrocarbon, [c]nitrogen oxide, [d]exhaust gas temperature before catalyst, [e]coefficient of variance (IMEP).

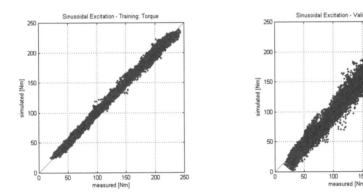

Fig. 10.6 Torque modeling using sinusoidal excitation: training (left) and validation (right) results (measured-vs.-simulated plots)

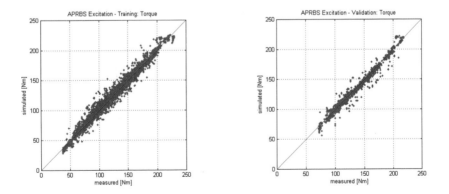

Fig. 10.7 Torque modeling using APRBS excitation: training (left) and validation (right) results (measured-vs.-simulated plots)

10.3.3 Virtual Engine Calibration

The models from subsection 12.2.2 are used for calibration of optimum valve timings in various operating points. This comprises the calibration of spark timing to produce maximum torque and eventually using the optimized spark in further valve timing calibration.

As a first step, optimal spark timing values for each extended operating point given by the quadruple {engine speed, load, intake cam phase, exhaust cam phase} are calculated. The optimal spark timing for a given quadruple is assumed to be the spark timing with a maximum related torque. The torque is determined using the dynamic torque model. As a constraint the modeled knocking and exhaust gas temperature values must not exceed user defined thresholds. So, for each extended operating point a spark sweep diagram as depicted in Fig. 10.8 is calculated. If knocking and exhaust gas temperature are below the given thresholds, the spark timing of the maximum torque is considered as the optimum. Since dynamic models are used, it is required to calculate the static final torque value, i.e. the torque value after completion of the settling phase.

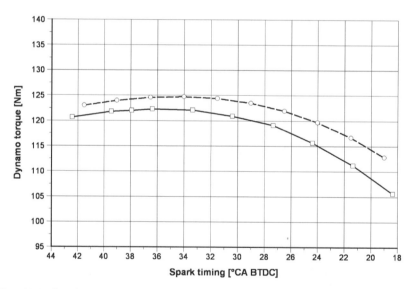

Fig. 10.8 Spark sweep vs. torque: measured (square) and simulated (circle)

Using this optimum spark timing, within the next step the optimal combinations of valve timings are determined for each extended operating point. The optimum valve timing is selected according to the maximum torque subject to the constraint of running smoothness (COV).

The described steps are accomplished for the models based on sinusoidal excitation and the models based on APRBS excitation separately. The results are subsequently compared to the optimization results of the grid measurement reference. The comparison for the operating point {engine speed = 3000 rpm, relative load = 50 % } is depicted in Fig. 10.9 to 10.10. For the shown example, the grid measurement for torque in Fig. 10.9 has its best value at the combination of[1] IVVT=40 °/ EVVT=20 °. The dynamic model with sinuoidal excitation shows its optimum at the combination of IVVT=40 °/ EVVT=22 °; for APRBS excitation models the optimum is at IVVT = 40 °/ EVVT = 15 °(interpolated).

The deviations between the grid measurement reference and the models for brake specific fuel consumption (BSFC) are depicted in Fig. 10.11 and 10.12 for the sinusoidal and the APRBS excitation, respectively.

---

[1] IVVT: intake cam phase, EVVT: exhaust cam phase.

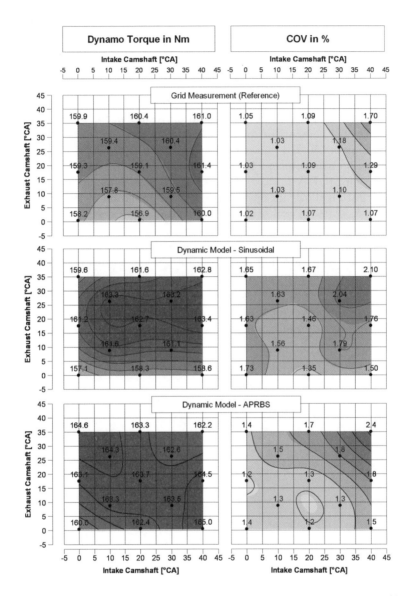

Fig. 10.9 Optimization results for engine speed 3000 rpm and load 50 %: torque (left) and COV (right)

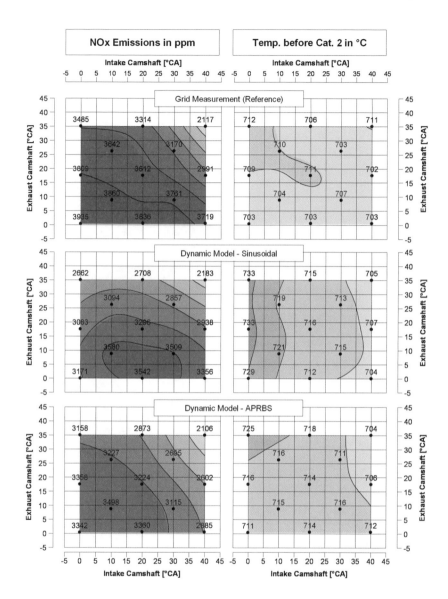

Fig. 10.10 Optimization results for engine speed 3000 rpm and load 50 %: NOx (left) and temperature before catalyst (right)

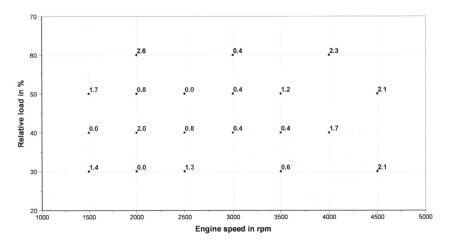

Fig. 10.11 Deviation of BSFC as calculated with the dynamic model based on sinusoidal excitation signals from grid measurement reference in [%]

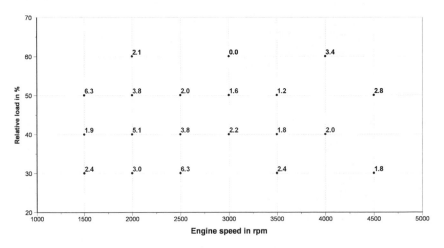

Fig. 10.12 Deviation of BSFC as calculated with the dynamic model based on APRBS excitation signals from grid measurement reference in [%]

10.4   Discussion

The results of the virtual engine calibration show that grid measurement, sinusoidal excitation based calibration and APBRS excitation based calibration arrive approximately within the same range of the parameter settings. This can be considered as a proof of concept of the virtual engine calibration methodology. The error plots in Fig. 10.11 and 10.12 depict the error of the virtual calibration with respect to the calibration based on grid measurements. The errors are in general very small. It depends on the development stage whether this error is acceptable or not. The essential benefit of the virtual calibration technique is the required measurement time of only about one hour. So, early development stages, where a fast calibration is more important than accuracy, can take benefit from this virtual calibration process.

Beside the virtual engine calibration it becomes apparent that both excitation techniques, i.e. sinusoidal excitation and APRBS excitation, can be successfully used for modeling and for the subsequent optimization. The result tables show relative mean errors within an acceptable range of around 5 % for training and validation for both excitation techniques. Since the number of training samples and also the amplitude range of the excitation signals are not the same it is not advisable to compare the numbers directly. This applies particularly to the error plots in Fig. 10.11 and 10.12.

As Fig. 10.8 shows, the optimum of the spark sweep is rather flat. That is, a relatively large variation of the spark timing causes only small changes of the torque. It is hence important to have a good model accuracy in order to have a high certainty about the position of the spark optimum. Furthermore, for the determination of the optimum, a sufficient amplitude modulation of the excitation signals is essential. Otherwise, the optimum of the spark may be outside the model range which limits the usability of the model and the optimization. This underlines that a careful test design is crucial for a successful application of the proposed method.

10.5   Conclusion

In this paper, nonlinear statistical dynamic modeling is applied to the virtual calibration of engines. It is shown that there is a tradeoff between calibration accuracy and calibration time. While the loss of accuracy is usually very small and below a few per cent, the savings in calibration and measurement time are significant. Consequently, the integration of this method into the development cycle may lead to great cost savings. Furthermore, it has been shown that there is no preference for APRBS or sinusoidal excitation within the modeling phase. The integration of the nonlinear dynamic modeling approach into different, already existing, toolchains is feasible without great adaptation efforts.

References

[1] Baumann, W., Schaum, S., Röpke, K., Knaak, M.: Excitation Signals for Non-linear Dynamic Modeling of Combustion Engines. In: IFAC World Conference 2008, Seoul, Korea (2008)

[2] Box, G.E.P., Wilson, K.B.: On the Experimental Attainment of Optimum Conditions (with discussion). J Royal Stat. Soc. B 13(1), 1–45 (1951)

[3] Clerc, M.: Particle Swarm Optimization. Wiley-ISTE, Chichester (2006) ISBN: 1905209045

[4] Deflorian, M., Klöpper, F.: Design of dynamic experiments. In: Röpke, K. (ed.) DoE in Engine Development IV, p. S. 31. expert verlag, Renningen (2009)

[5] Knaak, M., Schoop, U., Barzantny, B.: Dynamic Modelling and Optimisation - the Natural Extension to Classical DoE. In: Röpke, K. (ed.) DoE in Engine Development III. expert-Verlag, Renningen (2007)

[6] Köhler, B.-U.: Konzepte der statistischen Signalverarbeitung. Springer, Heidelberg (2005) ISBN 3-540-23491-8

[7] Kötter, H., Schneider, F., Fang, F., Güsner, T., Isermann, R.: Robust Regression and Outlier-Detection for Combustion Engine Measurements. In: Röpke, K. (ed.) DoE in Engine Development III. expert verlag (2007)

[8] Ljung, L.: System Identification - Theory for the user. Prentice Hall, Upper Saddle River (1999)

[9] Nelles, O.: Nonlinear System Identification. Springer, Berlin (2001)

[10] Röpke, K. (ed.): DoE - Design of Experiments. Verlag Moderne Industrie, Berlin (2005)

Part IV
Applications of Identification Methods in Automotive Systems

Chapter 11
Representation Limits of Mean Value Engine Models

Carlos Guardiola, Antonio Gil, Benjamín Pla, and Pedro Piqueras

Abstract. Mean Value Engine Models (MVEMs) have been widely used for internal combustion engine modelling with main application areas on the design and development of engine control systems. However, modellers must be aware of the limitations of these MVEMs which are associated to the simplification of the geometry and the time scale, and the partial consideration of the physical phenomena involved. This chapter analyses through several real-life examples the effects of some of the most important simplifications done in MVEMs.

11.1 Introduction

Modelling has become a key tool in many engineering applications. The automotive internal combustion engine field is not an exception, and along the last decades a huge variety of models have been developed and applied for different design and analysis problems. Models range from complex 3D Computational Fluid Dynamics (CFD) to simple linear transfer functions and they usually are selected for a given application according to different criteria, which include the accuracy and prediction capabilities of the model, the development and tuning cost, and the computational cost of the model.

Model accuracy, understood as the ability of the model to accurately represent the engine behaviour, is a critical issue. It must be also considered that it is possible for a model to be extremely accurate only when simulating the engine operation in the design conditions (or the those used for the model tuning), while failing to properly predict the engine behaviour on different conditions. Model prediction capability is also crucial for computer-aided engineering (CAE) applications, where design decisions are usually founded on

Carlos Guardiola · Antonio Gil · Benjamín Pla · Pedro Piqueras
CMT-Motores Térmicos, Universidad Politécnica de Valencia
e-mail: {carguaga,angime,benplamo,pedpicab}@mot.upv.es

D. Alberer et al. (Eds.): Identification for Automotive Systems, LNCIS 418, pp. 185–206.
springerlink.com   © Springer-Verlag London Limited 2012

simulation results while no experimental data is available for model tuning. In general, the more the model is based on the physical representation of the system, the better it is able to predict the off-design behaviour.

Model development and tuning is a time demanding task. Although many advances have been done for automated modelling (specially for *data driven* models), trial-and-error is still needed in many steps of the model development process and a slight redefinition of the model architecture can imply to completely restart the identification task. For model tuning, pre-existent data is needed. Available data will compromise model accuracy, and in the case of pure data driven models, the representation range of the model is restricted to the range of the available data. In the case of *physical-based* models, usually less data is needed due to their higher prediction capabilities, however model development process and tuning can become tedious because of the more complex mathematical structure.

Finally, the program complexity and execution time stands a limitation for the model selection. Despite the huge advancements in computing power, a CFD simulation can involve several days for a complex simulation in current workstations. Slower models are restricted for the assistance in design and analysis during the engine development phase while simple physical based models are used for extensive simulations for control system development. Only simpler models are envisaged for online tasks in control, like model-based control, since computation time must be kept several times below real time.

Plenty of models have been proposed for engine simulation along the last 5 decades [1, 2, 3]; models are usually classified according to the spatial and time resolution adopted for the problem representation and, at least, the following groups can be identified:

- *Mean value models.* Mean value models neglect the in-cycle variation that most engine physical quantities (flow, pressure, etc.) experience during the engine cycle. This in-cycle variation is related to the fact that the internal combustion engine is dominated by processes which occur periodically during the engine cycle (intake process, compression, combustion, expansion and exhaust). As its name suggest, mean value models only predict the average value of the different variables. On the other hand, the system representation in this kind of models is reduced to a (limited) number of states. Hence they are usually referred as 0D models or lumped parameters models. Within this group, two families of models can be differentiated, although differences may become clinal:

  - *Data driven models (DDMs).* Derived from pure mathematical identification process, data driven models are based on the direct fitting of a mathematical model structure. Different alternatives have been explored, being some of the most popular state space models, transfer functions, neural networks, fuzzy models or hybrid models. Prediction capability of pure data driven models can be compromised because of their nature, which is not directly based on the physical description of the system; however, this drawback can be shortcut through an adequate selection of the tuning dataset, or the application of adaptive techniques. DDMs can be very fast (several times faster than the real engine), thus allowing their implementation in optimisation solvers for model-based predictive control.

– Physical-based *mean-value engine models (MVEMs)*. Derived from simple mass and energy balances, MVEMs (also known as first-principle models) are quite popular for being a compromise between pure mathematical DDMs and slower physical models. MVEMs can benefit of a user-friendly component based approach, where interchangeable components are combined for modelling the full engine. On the other hand, moderate computing time have made them very popular for testing and developing software components. Generally DDMs and MVEMs are not grouped together, however they are quite similar in the simplification adopted, and hence in their limitations. Additionally, there are some models, usually referred as grey-box models, which combine the physical-based approach with a data driven final tuning.

- *Emptying-and-filling models (E&FMs)*. Emptying and filling models are first-principle models that, in opposition to the MVEMs, describe the in-cycle variation of the engine quantities. Although they still have a simple spatial representation of the engine (i.e. they are 0D or lumped parameter models), they are able to provide an insight in the internal evolution of the gas inside the cylinder and to represent pulsations in the manifolds (which are modelled as volumes). In some cases E&FMs are referred in the literature as *crank-angle solved* (CAS) models, because they are computed in the crank-angle domain instead of the time domain, or because of their ability of representing the in-cycle evolution of the different variables as a function of the crank-angle.

- *Wave action models (WAMs)*. One-dimensional geometric description of the engine manifolds permits the inclusion of the momentum conservation equation (which added to the mass and energy balance results in the Euler equation set), and thus the representation of wave effects. In many engines wave dynamics influence on the air management process cannot be neglected, since the pressure at the intake and exhaust ports is affected by strong pulsations. Wave action models usually combine the 1D description of the manifolds with first-principle E&FMs of cylinders and different engine subsystems.

- *Multi-dimensional models (CFD)*. On the basis of a detailed geometrical description of the system, Navier–Stokes equations can be formulated and solved. CFD simulations solution provides a description of the velocity and composition field for complex geometries, being the only way to

Table 11.1 Main characteristics of the different modelling techniques.

	DDM	MVEM	E&FM	WAM	CFD
Spatial resolution	0D	0D	0D	1D	3D
Time resolution	MV	MV	CAS	CAS	CAS
Physical description	no/very low	low	low	medium	complex
Prediction capabilities	no	low	medium	medium	high
Computational cost	very low	low	medium	high	very high
Main application	control	control develop.	CAE	CAE	CAE

simulate 3D mixing problems. At present understanding, Navier-Stokes equations are able to describe the fluid-dynamics accurately. However, the direct numerical simulation (DNS) is extremely time consuming, and different techniques are needed for lowering computing time, like the inclusion of turbulence models (e.g k–ε), Reynolds-averaged Navier-Stokes equations (RANS), or Large-eddy simulation (LES), where larger turbulent scales are explicitly solved. When considering combustion modelling, a combination of a spray model (describing fuel injection, jet break-up and droplet evaporation), a mixing model, an ignition model, a turbulent combustion model and a chemistry model (which includes pollutant formation) must be considered [4]. The complexity of the problem and the broad range of time and length scales do not allow a generalised approach, and it stand a hot topic in research [5].

Table 11.1 summarises the main general characteristics of the models presented in terms of spatial and time resolution, physical description of the system, prediction capabilities and computational cost; main application area is also identified. When analysing the model accuracy, it must be remarked that the models describing the system physics (right part in Table 11.1) are able to provide a better prediction of the system behaviour when no experimental data is available (on the basis of the physical properties of the system). However, after being tuned with available experimental data, DDMs and MVEMs can provide a good input-to-output mapping within the identification range and conditions. In this case, their accuracy can be even superior of that of the more complex physical approaches, and they can also exhibit limited extrapolation capabilities, specially in the case of MVEMs.

Although Table 11.1 could suggest that DDMs and MVEMs cannot compete with other modeling techniques as CFD or WAM, this is not by far the actual situation: because of the low computational power required and the existence of identification methods, and their accuracy within the identified range, DDMs and MVEMs are usually preferred (and sometimes the only modeling approaches available) for control and diagnosis in automotive applications.

In this context, significant advances have been done in improving DDM and MVEM accuracy, computation time and in the generation of automated model generation and identification tools. However, the modeller must be aware that, because of the simplistic physical representation assumed in these models, there are problems that need a more complex approach. The rest of the chapter is devoted to the identification of the main phenomena that cannot be modelled through the MVEM (and DDM) approach, and the presentation of some application examples in which these phenomena are important.

11.1.1   Main Limitations of the Mean Value Representation

In general, both MVEMs and DDMs are based on a simplistic representation of the engine system, with the following assumptions:

- The in-cycle variation of any variable in the engine is disregarded. Usual time resolution is in the order of the engine firing frequency, and hence in-cycle variations are neglected.
- A limited number of states are used for the system representation, and hence neither 3D nor 1D flow characteristics can be reproduced. Usually intake manifold is lumped into a single volume, and the same occurs with the exhaust manifold; ducts are usually represented by equivalent nozzles.

A summary of the equations usually used in MVEMs can be found in [6]. Today, MVEMs are a common tool: simplified versions of MVEMs are internally run in current electronic control units (ECU) for estimating unmeasured quantities, while more complex MVEMs are used for the design and development of engine control systems. For many applications, the improvement in precision obtained does not usually justify the computing time needed even for E&FMs[7].

However, the modeller must be aware of the limitations of these MVEMs, which are associated to the simplification of the geometry, the time scale, and the oversimplification of the physical phenomena involved. In order to cope with the simplified physical representation, the modeller usually must increase the number of tuning parameters. As far as the physical phenomena are not properly represented and hence most of the tuning parameters strongly depend on the operation conditions, a collection of maps are usually needed in order to provide a full representation of the engine behaviour.

Some of the most troubling issues in engine modelling through MVEMs are the following:

- *Engine volumetric efficiency.* Due to the reciprocating operation of the engine and the cyclic opening and closing of the valves, neglecting the in-cycle variation of the pressure at the intake and exhaust ports can result in significant deviations in the resulting breathing capacity of the engine. Furthermore, wave effects on the manifolds can improve (or worsen) the trapped air mass beyond 20%. MVEMs partially solve the problem through volumetric efficiency maps, which must be carefully tuned. Although these mapped corrections can be sufficient during steady state operation, pulsating flow effects or exhaust gas recirculation (EGR) transient cannot be mapped nor easily predicted.

- *Three-dimensionality and asymmetries in the intake system.* Three-dimensional nature and asymmetries of intake and exhaust subsystems of real engines may take significant effects on the cylinder-to-cylinder charge distribution, influencing the trapped air mass in each individual cylinder. In case of high pressure exhaust gas recirculation, significant differences in the cylinder-to-cylinder gas composition can exist.

- *Unsteady flow and distributed pressure losses.* In cases where pressure loss is distributed, as in the case of porous media, or when pulsating flow occurs, important differences can appear between the equivalent nozzle representation and the real engine behaviour.

- *Turbine swallowing capacity and efficiency.* Automotive engines use pulse turbocharging, that is, the turbine profits from the pressure pulses in the exhaust manifold. Using mean value assumption results in a significant bias in the estimate of the available isentropic power. Additionally, instantaneous efficiency is affected by the ratio between the gas and blade speed, thus resulting in important variations of the efficiency during the engine cycle. Directly using exhaust manifold mean values and the efficiency obtained in steady-state standard gas stands can be the cause of significant deviations. Hence, tuning coefficients are needed for adjusting the behaviour of the turbine.

- *Compressor surge occurrence.* Compressor performance is usually mapped with steady tests; as far as significant pulsations do not reach the compressor, this mapping can reproduce on-engine behaviour with fidelity. Although in some conditions pulsations can be significant, or unsteady effects associated with the engine transients can appear, the most complex compressor behaviour to predict is surge occurrence. In most MVEMs, surge effect is included through the surge line, which assumes that the surge occurrence is only dependent on the compressor corrected flow and the compressor speed. However surge is a complex dynamical process, which is related with the internal flow conditions in the compressor. Flow pulsating conditions or compressor inlet geometry can significantly mitigate (or increasing) the surge occurrence.

- *Detailed combustion and pollutant formation modelling.* Combustion and pollutant formation is strongly dependent on the in-cycle evolution of the gas and fuel mix in the cylinder; computing this evolution is out of the MVEM approach (but simple input-output models can be used for control purposes [8, 9]). Although E&FMs or 1D spray description can provide quite good results for some combustion modes and some pollutants [10], in other cases the three-dimensional mixing and reacting processes inside the geometrically complex combustion chamber cannot be neglected; thus

Table 11.2 Suitability of the different modelling techniques for solving different problems.

	DDM	MVEM	E&FM	WAM	CFD
Volumetric efficiency	-	-	Fair	Good	Good
Wave effects	-	-	-	Good	Good
Asymmetries	-	-	-	1D	3D
Unsteady flow	-	-	-	1D	3D
Turbine	Poor	Poor	Fair	Fair	Good
Surge	-	Poor	Poor	Fair	Good
Combustion	-	-	Poor	Poor	Good

CFD simulation are the only feasible approach. Even then, probabilistic assumptions must be usually done, and still many advances are needed for providing an accurate general solution to the problem.

Table 11.2 provides an overview of the suitability of the different modelling techniques for the detailed modelling of different subsystems and phenomena occurring in the internal combustion engine. Note that suitability does not ensure that the effect is properly modelled with present knowledge. Next sections provide several real-life examples that illustrate situations where the simplifications assumed in the MVEMs are not feasible. The fact that MVEMs (including DDMs) are not able to reproduce these effects with sufficient accuracy does not implies they are useless, but highlights the necessity of a right selection of the modelling technique depending on the model objective and the relevant phenomena involved.

11.2   In-Cycle Variation

Most quantities associated with the operation of the internal combustion engine exhibit an in-cycle variation due to the periodicity of the engine events (intake, combustion, exhaust, etc.), and the mean value assumption is in many cases hardly acceptable. For example, Figure 11.1 illustrates the pressure and the mass flow in-cycle variations for a heavy duty diesel engine; for the case flow in the EGR duct (which corresponds to the lateral entry of the venturi in the figure) peak-to-peak amplitude variation is higher than 200% the mean value.

Although in-cycle variations do not completely invalidate the mean-value modelling concept, as far as the system behaves in a non linear way, MVEM results will result on significant deviations from the actual engine behaviour. Unfortunately, many relevant phenomena occurring in the internal combustion engine are extremely non-linear, and neglecting the signal pulsation implies the inclusion of correction parameters, whose validity is restricted to the tuned conditions.

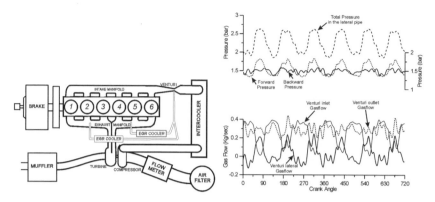

Fig. 11.1 Flow through a venturi used for improving EGR rate in conditions of boost pressure exceeding mean exhaust manifold pressure. Modelled values have been computed with OpenWAM™.

In the simplest cases, the pulsation of the different signals can be explained through the emptying-and-filling process; however, in other cases wave action is needed for accurately predicting the amplitude and phasing of the pulsation. Thus, E&FMs can suffice in some situations, while WAMs are needed for the more general case.

In some cases, including important processes like the turbine swallowing capacity and power, the flow through elements subject to important pressure pulsations (e.g. EGR circuit) or the cylinder intake process, the pulsating effect is not negligible.

In the case of the turbine swallowing capacity, the non-linearity on the flow characteristic of the turbine causes significant deviations if averaged variables are considered. According to Zinner [11], the predicted mean-value mass flow exceeds the actual flow in pulsating conditions. The difference between both quantities strongly depends on the inlet pressure profile; in cases with important peak-to-peak pulsation, the difference can exceed 20%.

Equivalently, pulsation also modifies the power provided by the turbine. Here two effects must be considered: pulsating flow has higher power availability than mean-value flow [11, 12, 13], and instantaneous turbine efficiency is affected by the ratio between blade and gas speeds, being the last significantly modified according to the pressure pulses. Significant variations in the expected turbine power exceeding 40% can thus appear [11].

Both effects occurring in automotive turbines are usually solved using turbine E&FMs, while 1D WAMs are usually used for the exhaust system modelling when wave effects are significant.

With respect to the flow through elements, significant divergences in the MVEM approach are obtained with highly pulsating conditions providing a low mean flow. For example, Chevalier et al. [14] report that using a MVEM approach in a throttle valve can cause up to 50% error (because the nonlinear pressure drop-flow characteristic), while a revised model including some corrections due to compressibility and pressure pulsation was used for lowering the error down to 11%. In the case of distributed pressure losses, like flow through an EGR duct for wide-open EGR valve positions, a WAM is needed for proper modelling because of the combination of pressure pulsations and wave effects. However, E&FMs behave quite well for punctual pressure losses.

Last interesting example is related with the cylinder intake process. Since the valve opening varies during the engine cycle the problem is highly non-linear. The existence of pressure waves in the cylinder ports, and the different timing of the pressure peaks, with regards to the valve closing, dramatically affect the resulting volumetric efficiency. MVEMs usually assume that the volumetric efficiency is known before-hand, and intake mass flow presents a proportional variation with the mean density at the intake plenum (known as the speed density equation); steady-state volumetric efficiency is mapped and used in a quasi-steady approach. In opposition, WAMs can be used for estimating the volumetric efficiency, as shown in Figure 11.2.

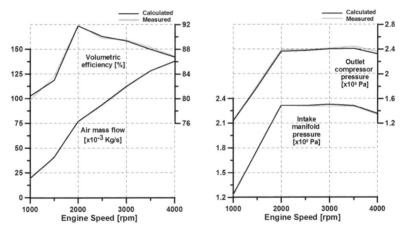

Fig. 11.2 Example of volumetric efficiency prediction through OpenWAM™[15].

MVEM approach for the volumetric efficiency has important limitations: on one hand, volumetric efficiency can differ from the nominal one when engine settings are modified (for example EGR valve opening, which permits the wave interaction between the intake and exhaust manifold); on the other hand, during engine transients (including speed acceleration, variable nozzle turbine closing, EGR valve position variation, etc.), instantaneous volumetric efficiency can differ from its steady state value. Although [14] reports divergences below 2%, this results are hardly extensible to all engines and every kind of transients (specially if turbocharged engines are considered).

11.3 Distributed Pressure Losses: Flow through Porous Media

Because of their simplification in the spatial representation of the system, both MVEMs and E&FMs concentrate the pressure losses along ducts into singular pressure drops, often assumed to be equivalent to orifices. However the actual behaviour of distributed pressure losses can be significantly different, specially if strong flow pulsations occur. Another specific problem that is hardly described by equivalent 0D models is the flow through porous walls, as in the case of diesel particulate filters (DPF).

The definition and complexity of the DPF models differ depending on the extent of the objectives. Basic analysis to evaluate the pressure drop taking place across the porous medium is traditionally raised with simple 0D models (MVEM or E&FM), which approximate on a lumped term the inertial effect at inlet and outlet monolith cross-sections [16]. If only the monolith is analysed, the pressure drop at clean conditions neglecting the Forchheimer's effect and assuming incompressible flow can be expressed as

$$\Delta p_{DPF} = \frac{\mu Q}{2V_{DPF}} (\alpha + w_w)^2 \left[\frac{w_w}{k_w \alpha} + \frac{8FL^2}{3\alpha^4} \right] +$$
$$+ \frac{\rho Q^2}{V_{DPF}^2 \alpha^2} (\alpha + w_w)^4 \left[\frac{\beta w_w}{4} + 2\varsigma_t \left(\frac{L}{\alpha} \right)^2 \right] \qquad (11.1)$$

where Δp_{DPF} is the stagnation pressure drop; μ is the dynamic viscosity; Q represents the volumetric flow rate; V_{DPF} and L are the monolith volume and lenght respectively; α and w_w are the honeycomb cell size and the porous wall thickness and define the cellular structure of the monolith. The non-inertial pressure drop is a function of the porous wall permeability k_w and the friction process. The pressure drop due to friction in square porous channels depends on the momentum transfer coefficient F. On the other hand, the inertial pressure drop depends on the porous wall contribution given by the Forchheimer's coefficient β_F and the pressure drop due to local contraction and expansion at the inlet and outlet of the monolith channels. These contributions are considered together in a solely pressure drop coefficient ς_t.

The potential of this sort of lumped-parameter approach lies on the possibility to provide the value of the porous wall permeability, which is a property determined by only the pore structure and then not dependent on fluid properties nor flow mechanisms. According to equation 11.1, the pressure drop to volumetric flow ratio is a linear function of the volumetric flow. The independent term is a function of the dynamic viscosity, the monolith geometry and the porous wall permeability and represents the non-inertial contribution to the pressure drop whereas the slope covers the inertial mechanisms affecting the pressure drop phenomena:

$$\frac{\Delta p_{DPF}}{Q} = \underbrace{b}_{\substack{Inertial \\ term}} Q + \underbrace{a}_{\substack{Non-inertial \\ term}} \tag{11.2}$$

The value of the non-inertial term can be determined from experimental monolith pressure drop measurement carried out under steady flow conditions. As a consequence, the wall permeability can be determined as:

$$k_w = \frac{w_w}{\alpha} \frac{1}{\frac{a}{\frac{\mu}{2V_{DPF}}(\alpha+w_w)^2} - \frac{8FL^2}{3\alpha^4}} \tag{11.3}$$

The model can be expanded to the calculation of the pressure drop under load filter conditions, which besides the estimation of the inertial pressure drop coefficient (ς_t) needs the additional estimation of the particulate layer properties, such as soot permeability and thickness.

Another similar model to deal with the calculation of the DPF pressure drop consists in the definition of a pressure drop coefficient K. This coefficient relates the stagnation pressure drop and the kinetic energy at the filter inlet:

$$\Delta p_{DPF} = K\frac{1}{2}\rho u^2 \tag{11.4}$$

This approach allows the solution of a quasi-steady boundary condition where the pressure drop is locally concentrated [17], which can be used indistinctly for MVEMs or for E&FMs (although final result could differ depending on the variables to be averaged because of the existence of quadratic terms). The same solution can be included as boundary condition in a WAM, as sketched at central plot in Figure 11.3.

There, the inlet and outlet cones are modelled as volumes whereas the monolith geometry is defined by two ducts separated by the boundary condition solving the pressure drop. The total length of the ducts coincides with the length of the monolith. All of them are half-length so that it is assumed that all the flow covers the half length of the inlet channels, where the pressure drop is locally imposed, and goes into the outlet channels. Thus, although the pressure drop is calculated with a quasi-steady approach, the model accounts for the influence of the monolith length on the flow dynamics. The pressure drop coefficient has been determined from experimental data and corrected taking into consideration the pressure drop that the model geometry involves. This treatment leads to very accurate predictions of the pressure drop as bottom plot in Figure 11.3 shows for a wide range of cold steady flow operating conditions.

Despite the interest of mean value models for engine modelling and wall-flow monolith pre-design [19], one-dimensional models (top plot in Figure 11.3) provide a more detailed description of the flow dynamics in a wall-flow DPF. Under steady flow conditions, 1D models yield a pressure drop prediction of high agreement with experimental data, but similar to that provided by lumped-parameter approaches as (a) plot in Figure 11.4 shows.

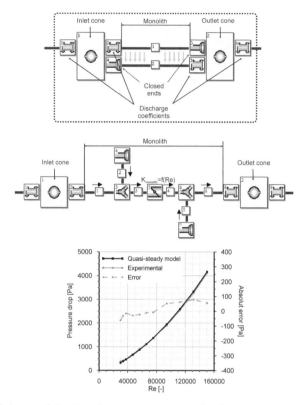

Fig. 11.3 Scheme of the distributed pressure loss (top) and of the lumped pressure loss (centre) for the DPF pressure drop prediction into the gas dynamics software OpenWAM^{TM}. Bottom: results of the model with steady flow.

However, main differences and improvements arise from the fact that 1D models account for the pressure drop through the porous medium as an axial distributed phenomenon along the channels. As a consequence, the effect of the plug end at the inlet channel and the flow regime on the porous media properties and flow distribution are successfully simulated. Graphs (b) and (c) of Figure 11.4 show the normalised channel and filtration velocity profiles along the length of a pair of inlet-outlet channels for low and high Reynolds number at the inlet of the DPF respectively.

The clearly different gas properties distribution shown along the porous wall depending on operating conditions underlines the importance of considering the axial dimension: the flow distribution controls the soot deposition both during deep bed and cake filtration regimes. The soot and ash distribution along the porous wall determines the properties of the particulate layer, such as permeability, thickness or soot packing density; these characteristics are also related to the regeneration process, due to their influence on heat

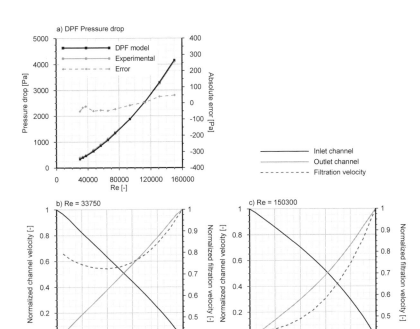

Fig. 11.4 Modelling of steady state wall-flow DPF response with a one-dimensional unsteady compressible non-homentropic flow model implemented in OpenWAM^{TM}software.

transfer phenomena, location of the regeneration start and the existence of hot spots under expected engine operation conditions.

As shown in Figure 11.4, the essentially one-dimensional nature of the porous wall pressure drop process and its dependence on flow regime do not invalidate lumped-parameter results under steady flow. However, under unsteady flow, the increase of spatial resolution results in deeper knowledge of DPF response, fundamental to explore the effects that the filter introduces on the engine response depending on the exhaust line architecture [20, 21]. The accuracy in the simulation of the filter dynamic response governs the reliability of further analysis regarding filter placement influence on engine performance or turbine efficiency. A current example is the evaluation of multi-functional exhaust manifold configurations containing the DPF aftertreatment device and EGR flow as thermal insulator. As example of 1D DPF model features, a comparison of the transmitted and reflected waves by a wall-flow DPF is performed in Figure 11.5 under impulsive flow. Graph (a) compares experimental data with the results of mean value model based on a quasi-steady treatment of the pressure drop presented in Figure 11.3. In spite of the inclusion of the monolith length, underprediction of the attenuation, as the transmitted wave results shown, and phase shift in the reflected wave appears leading to a poor description of the filter dynamics. On the contrary, the one-dimensional unsteady compressible non-homentropic flow model [17], whose results are plotted in graph (b) of Figure 11.5, provides a more accurate prediction. In engine modelling context, these results would mean reliable data for task demanding filter geometry optimization and the development of engine control strategies in several aspects, such as VGT or EGR valves actuators, consideration of DPF operating conditions regarding soot loading or regeneration state, etc.

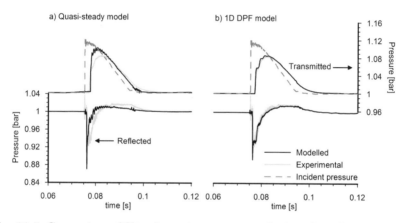

Fig. 11.5 Comparison of filter dynamics response under impulsive flow conditions.

11.4 Three-Dimensional Flow

While 0D codes cannot calculate the spatial distribution of the flow properties in engine components, 1D codes are able to compute the flow field in internal combustion engines resolving the flow equations only in the main flow direction, and assuming uniformity of the flow field in the perpendicular plane. Despite the advantages in terms of computing time of 0D and 1D models, the optimisation of engine components with complex geometries requires the use of 3D-CFD codes. The main reason for the use of 3D-CFD codes is the need of resolving three-dimensional flow effects that cannot be properly captured by the more simplistic 0D and 1D models. In contrast, both 0D and 1D codes allow modelling the whole engine, while computation requirements of CFD codes limit their application to the modelling of engine subsystems. In this sense, 0D-1D codes can be employed to provide the boundary conditions for the 3D simulation of particular components. Next sections are devoted to two examples of situations where 3D modelling is required, since the intrinsically three dimensional nature of the phenomenon must be considered. Selected examples are the EGR flow repartition amongst cylinders, and the simulation of the effects of the compressor inlet geometry on the surge phenomenon.

11.4.1 EGR Flow Repartition

Since current HSDI engines operate with EGR rates as high as 50% the effect of the unequal EGR distribution becomes important. While the high linearity between intake CO_2 concentration and NO_x emissions allows cylinders with higher EGR rates to compensate the NO_x emitted by cylinders with lower EGR rates, particulate emissions do not increase linearly with intake CO_2 concentration. Then, the uneven EGR distribution between cylinders may increase particulate emissions[22]. In addition, cylinder-to-cylinder charge dispersion becomes a critical aspect on the control of low temperature combustion systems [23]. For these reasons, some investigations have been carried out in recent years in order to improve the inlet manifold design or to develop devices that assure a suitable mixing between air and EGR [24]. Since the time scale governing the mixing process between species in the intake manifold of an engine is too low for most of the measuring devices, the use of CFD codes has gained importance in the process of engine manifold design [25]. In fact, engine models can calculate engine parameters such as instantaneous mass flows at different locations which cannot be obtained experimentally, but are necessary to understand gas mixing phenomena.

The engine intake manifold can be modelled with three different approaches, in increasing order of complexity they are:

- 0D models, where the intake manifold is represented by a single volume attached to the air mass flow duct coming from the intercooler, the EGR duct and the pipes which guide the intake charge to the cylinders.
- 1D models that simulate the intake manifold by means of a network of flow-splits, pipes and orifices.
- 3D models in which the spatially resolved flow distribution is obtained by dividing the actual manifold in cells with dimensions of tipically 1 to 10mm.

In figure 11.6 an scheme of the three different approaches is represented.

Despite of the valuable information that 0D and 1D models provide concerning the evolution of thermodynamic variables such as temperatures, pressures and concentrations, when a spatial resolution of those variables is required in an element with such a complex geometry as that presented in figure 11.6, 3D modelling becomes imperative. In fact, it is worth noting that the 0D approach for modelling intake manifolds, i.e. a volume attached to the cylinders, is not appropriate because flow would be distributed to every pipe

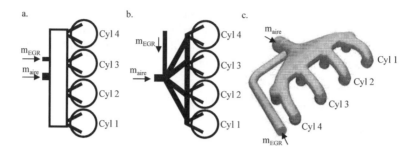

Fig. 11.6 Schemes of the intake manifold employed in the different simulation approaches. a: 0D model. b: 1D model. c: 3D model

in the same way, losing the manifold geometry effect on the gas dynamics. As far as 1D models are concerned, the grid of pipes shown in figure 11.6 only solves the problem partially. Nevertheless, it should be underlined that despite of the limitations of the 0D and 1D models, they are key tools in the study of the gas distribution amongst cylinders since they are required to provide the boundary conditions for the 3D model, serving as time dependent tabulated inputs for the 3D-CFD code. Figure 11.7 shows the methodology employed for analysing the EGR distribution amongst cylinders [26]. First, a set of experimental test is done in order to obtain the necessary information to fit 0D and 1D engine models. In spite of the fact that one-dimensional models provide some information about mass flow distribution between cylinders, one-dimensional hypothesis is too hard for obtaining suitable results. However, 0D or 1D models become essential for feeding the CFD models with the required boundary conditions. The results of the CFD model are employed to retrofit the 0D-1D model in order to take into account the EGR distribution effects on the engine performance. Finally, the CFD results are compared with experimental data to validate the model.

Figure 11.8 shows the comparison between the measured and calculated EGR rate at the four cylinders under various working conditions. A strong linear correlation between measured and calculated values can be observed. Small differences are attached to modelling and measuring uncertainties, however the combination of 0D-1D engine codes and 3D models of particular engine parts can be validated by means of this picture.

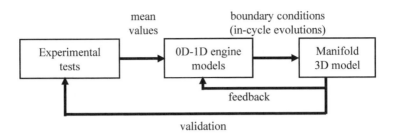

Fig. 11.7 Scheme of the methodology employed to obtain the EGR flow distribution amongst cylinders

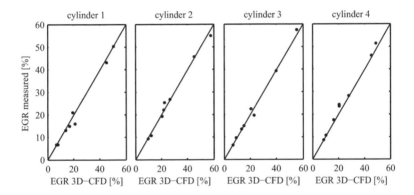

Fig. 11.8 Measured EGR versus values predicted by the 3D model with the proposed methodology.

11.4.2 Compressor Surge

Compressor operating ranges are limited by surge when running at low mass flows and high compression ratios. Downsizing and the high low-end torque requirements of current diesel engines, have make surge to become a major concern in the boosting system definition. In the design process of the engine air system, the turbocharger selection mainly consists of choosing a set turbine-compressor whose operating maps fit the engine air mass flow and pressure demands to reach a given engine performance. Those maps are obtained in particular facilities under steady conditions [27]. Nevertheless, in a real engine, both compressor and turbine work under pulsating conditions, and then important differences between the actual turbocharger behaviour and the steady state maps predictions can appear [28]. In addition, due to packaging limitations, the air system layout in a current engine differs from those used in the turbocharger tests benches. Particularly, the ideal straight compressor inlet pipe arrangement can seldom be used, and a 90° elbow is often employed for reducing packaging volume. Some studies in the literature have highlighted the important effect of the compressor inlet flow distribution on surge phenomenon [29]. Both the pulsating conditions and the particular air system layout give rise to the fact that the surge limit obtained in turbocharger tests benches hardly fits with surge occurrence on engines.

Concerning the compressor modelling, compressor maps have been traditionally employed in 0D and 1D models successfully [30]. Even there are 0D models able to manage with compressor surge [31]. Nevertheless, since those models are based on compressor maps and cannot take into account the compressor inlet flow pattern if high accuracy is required near the surge line, their application is limited to cases where the compressor inlet line coincide with that employed in the turbocharger characterisation. As far as 3D models are concerned their high accuracy make them the suitable option to carry out surge comprehension studies.

The modelling approach for gaining insight into the flow inside the turbocharger is the application of 3D calculation models, which are able to solve the governing flow equations. When applied to internal combustion engines, 3D models have to address specific problems linked to the flow unsteadiness, high Reynolds numbers involved, and the complex variable geometry of the solid boundaries. However, these problems have been partially solved in recent years as result of the significant improvement in power and speed of modern computers, and in the ability of the 3D models to solve complex flows.

A combined methodology that includes 3D calculations and experiments has been successfully applied in order to study the behaviour of centrifugal compressors working near the surge conditions. CFD calculations have been performed on a 3D domain reproducing the real geometry of the compressor an the inlet and outlet ducts for different flow conditions at 150000 rpm. The selection of this speed is justified because this point presents significant instabilities for a wide range of air mass flows before the deep surge occurs.

The steady state calculations allow to reproduce with accuracy the compressor behaviour in its stable operation zone, as sketched in Figure 11.9. Near the surge conditions, the calculation diverges below a critical mass flow rate, due to the transient nature of the surge phenomenon. Therefore, unsteady calculations are necessary to reproduce the instability processes, namely rotating stall and surge, that appear in the compressor when the mass flow rate is reduced to these critical values. The instability is detected in the CFD simulations map because higher oscillations in the pressure ratio are detected, as shown in Figure 11.9.

Also analysis of the internal 3D flow has been performed in order to get a better understanding of the compressor behaviour. Although the visualised steady point is inside the normal operating zone of the compressor, pressure and velocity contours show two instability zones (lower pressure levels) that

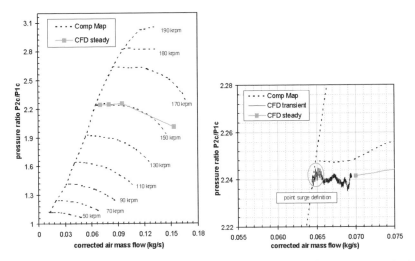

Fig. 11.9 Calculated and experimental compressor map. (left) Stable zone, (right) near surge zone.

lead to back flow in the periphery of sections immediately upstream of the rotor, see Figure 11.10 (top). These instability cells grow (low pressure zones fill the whole section) as the mass flow rate decreases to levels near to those producing surge, and recirculation is more evident, as shown in Figure 11.10 (bottom).

CFD techniques are thus able not only to predict surge occurrence, but provide a valuable tool for inferring the causes yielding to surge. On the other hand, CFD calculations have proved their value for analysing the effect of different geometries at the compressor inlet.

11.5   Summary

This chapter has reviewed several alternatives for internal combustion engine modelling, ranging from detailed 3D description of the system through CFD simulation, to simple lumped-parameter MVEMs and DDMs. Model structure selection is usually based on a trade-off between computing time and model performance, and also of development and tuning effort. MVEM and DDM approaches are able to provide cost-effective solutions with adequate accuracy, and are more and more used in engine control and diagnosis.

However, because mean value models are limited in the spatial (a discrete number of states are used) and temporal (only mean value is considered) resolution, they have difficulties for properly representing phenomena that

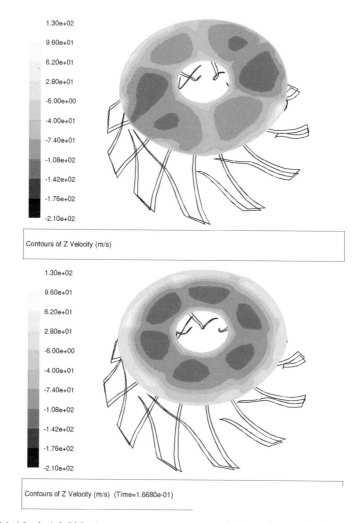

Fig. 11.10 Axial Velocity contours upstream of the inducer. (top) stable zone, (bottom) near surge zone.

depend of a complex spatial or temporal pattern. This includes cylinder filling and emptying process, compressor and turbine performance, unsteady flow in manifolds or three-dimensional flow mixes. In cycle-pulsations affecting turbine behaviour and emptying and filling process in the cylinders, distributed pressure losses, three dimensional EGR flow repartition and compressor surge are some characteristic examples involving physical phenomena hardly included in mean value models. However, even in these cases mean value models can usually map the input-to-output behavior but are not able to predict the behaviour when operation conditions widely differ from the reference situation.

References

[1] Heywood, J.B.: Internal combustion engine fundamentals. Mcgraw-Hill, New York (1989)

[2] Rakopoulos, C.D., Giakoumis, E.G.: Diesel engine transient operation: principles of operation and simulation analysis. Springer, Berlin (2009)

[3] Rakopoulos, C.D., Giakoumis, E.G.: Review of thermodynamic diesel engine simulations under transient operating conditions. SAE Paper 2006-01-0884 (2006)

[4] Tap, F.A., Angel, B.N.: Including detailed chemistry effects in industrial 3D engine simulations. In: International Conference on Diesel Engine, Lyon (2006)

[5] Haworth, D.C.: A review of turbulent combustion modeling for multidimensional in-cylinder CFD. SAE Paper 2005-01-0993 (2005)

[6] Eriksson, L., Wahlström, J., Klein, M.: Physical modeling of turbocharged engines and parameter identification. In: del Re, L., et al. (eds.) Automotive Model Predictive Control: Models, Methods and Applications, pp. 53–71. Springer, Berlin (2010)

[7] Karlsson, J., Fredriksson, J.: Cylinder-by-cylinder engine models Vs mean value engine models for use in powertrain control applications. SAE Paper 1999-01-0906 (1999)

[8] Stobart, R.: Control oriented models for exhaust gas aftertreatment; a review and prospects. SAE paper 2003-01-1004 (2003)

[9] Hirsch, M., Oppenauer, K., del Re, L.: Dynamic engines emission models. In: Automotive Model Predictive Control: Models, Methods and Applications, pp. 73–88. Springer, Berlin (2010)

[10] Arregle, J., Lopez, J.J., Guardiola, C., Monin, C.: On board NOx prediction in diesel engines: a physical approach. In: del Re, L., et al. (eds.) Automotive Model Predictive Control: Models, Methods and Applications, pp. 25–36. Springer, Berlin (2010)

[11] Zinner, K.: Supercharging of internal combustion engines. Springer, Berlin (1978)

[12] Winterbone, D.E., Pearson, R.J.: Turbocharger turbine performance unsteady flow—a review of experimental results and proposed models. Inst. Mech. Eng. paper C554/031/98 (1998)

[13] Macián, V., Luján, J.M., Bermúdez, V., Guardiola, C.: Exhaust pressure pulsation observation from turbocharger instantaneous speed measurement. Measurement Science and Technology 15, 1185–1194 (2004)

[14] Chevalier, A., Müller, M., Hendricks, E.: On the validity of mean value engine models during transient operation. SAE paper 2000-01-1261 (2000)

[15] Torregrosa, A.J., Galindo, J., Guardiola, C., Varnier, O.: A combined experimental and modelling methodology for intake line evaluation in turbocharged diesel engines. International Journal of Automotive Technology (2011)

[16] Masoudi, M.: Hydrodynamics of diesel particulate filters. SAE Technical Paper 2002-01-1016 (2002)

[17] Piqueras, P.: Contribución al modelado termofluidodinámico de filtros de partículas diesel de flujo de pared, PhD. Thesis (text in spanish), Universidad Politécnica de Valencia (2010)

[18] OpenWAM website, CMT-Motores Térmicos (Universidad Politécnica de Valencia), www.openwam.org (Cited June 1, 2010)

[19] Konstandopoulos, A.G., Skaperdas, E., Warren, J., Allansson, R.: Optimized filter design and selection criteria for continuously regenerating diesel particulate traps. SAE Technical Paper 1999-01-0468 (1999)

[20] Payri, F., Desantes, J.M., Galindo, J., Serrano, J.R.: Exhaust manifold of a supercharged reciprocating internal combustion engine (text in spanish), Patent application P200900482. Priority date 13/02/2009. Oficina Española de Patentes y Marcas (2009)

[21] Windsor, R.E., Baumgard, K.J.: Internal combustion engine with dual particulate traps ahead of turbocharger. Patent Application Publication, US 2009/0151328 A1, United States (2009)

[22] Payri, F., Luján, J.M., Climent, H., Pla, B.: Effects of the intake charge distribution in HSDI engines. SAE Paper 2010-01-1119 (2010)

[23] Beatrice, C., Avolio, G., Bertoli, C., Del Giacomo, N., Guido, C., Migliaccio, M.: Critical aspects on the control in the low temperature combustion systems for high performance DI diesel engines. Oil & Gas Science and Technology 62, 471–482 (2007)

[24] Siewert, R.M., Krieger, R.B., Huebler, M.S., Baruah, P.C., Khalighi, B., Wesslau, M.: Modifying an Intake Manifold to Improve Cylinder-to-Cylinder EGR Distribution in a DI diesel Engine Using Combined CFD and Engine Experiments. SAE Paper 2001-01-3685 (2001)

[25] Wehr, D., Huurdeman, B., Spennemann, A.: EGR- A challenge for modern plastic intake manifolds. SAE Paper 2002-01-0902 (2002)

[26] Luján, J.M., Galindo, J., Serrano, J.R., Pla, B.: A methodology to identify the intake charge cylinder-to-cylinder distribution in turbocharged direct injection diesel engines. Meas. Sci. Technol. 19, 1–11 (2008)

[27] Galindo, J., Serrano, J.R., Guardiola, C., Cervelló, C.: Surge limit definition in a specific test bench for the characterization of automotive turbochargers. Exp. Thermal Fluid Sci. 30, 485–496 (2006)

[28] Galindo, J., Serrano, J.R., Climent, H., Tiseira, A.: Experiments and modelling of surge in small centrifugal compressor for automotive engines. Exp. Thermal Fluid Sci. 32, 818–826 (2007)

[29] Kim, Y., Engeda, A., Aungier, R., Derinzi, G.: The influence of inlet flow distortion on the performance of the centrifugal compressor and development of an improved inlet using numerical simulations. Proc. of the Inst. Mech. Eng. Part A: J. of Power and Energy 215, 323–338 (2001)

[30] Galindo, J., Serrano, J.R., Arnau, F., Piqueras, P.: Description and analysis of a onedimensional gas-dynamic model with independent time discretization. J. Eng. Gas Turb. Power - Trans. ASME 131, 34504 (2009)

[31] Galindo, J., Climent, H., Guardiola, C., Tiseira, A.: On the effect of pulsating flow on surge margin of small centrifugal compressors for automotive engines. Exp. Thermal Fluid Sci. 33, 1163–1171 (2009)

Chapter 12
Identification Methods for Reliable and High Dynamic Off-Road Engines

Christian Benatzky and Gerd Schlager

Abstract. With the introduction of the emission level EU Stage 3B / US Tier 4i in 2011 for Off-Road engines, new engine concepts are mandatory, due to the necessity to introduce exhaust aftertreatment equipment for all kind of applications with more than 130kW power. To utilize these technologies and their components a function based approach for the whole system is necessary. Therefore, modeling and identification techniques are the key technologies to achieve the control performance which is necessary to fulfill customer and legislative requirements. In order to discuss these points, the paper presents two examples from the fuel path. First, a systematic way to derive a high pressure pump model, which is able to run in real time, is shown. The second example discusses an approach to use identification methods to compensate injector remanence effects. In the outlook different future identification topics and challenges are outlined as starting points for future investigations.

12.1 Introduction

With the introduction of the emission level EU Stage 3B / US Tier 4i in 2011 a completely new situation for all Diesel Off-road engine manufacturers is given. To reach the emission limits, exhaust aftertreatment systems are necessary for all kind of applications with more than 130kW rated engine power. Therefore, the engine concept hast to be changed from a simple layout as used for EU Stage 3A (low rate, internal or external exhaust gas recirculation (EGR)) to engine concepts which can provide optimal conditions for the exhaust

Christian Benatzky
Liebherr Machines Bulle SA, Rue de l'Industrie 45, 1630 Bulle, Switzerland
e-mail: `christian.benatzky@liebherr.com`

Gerd Schlager
Liebherr Machines Bulle SA, Rue de l'Industrie 45, 1630 Bulle, Switzerland
e-mail: `gerd.schlager@liebherr.com`

D. Alberer et al. (Eds.): Identification for Automotive Systems, LNCIS 418, pp. 207–221.

aftertreatment system. According to the application (excavator, wheel loader, mobile crane, et cetera) the following concepts are feasible:

- High EGR rate engine with Diesel Particulate Filter (DPF)
- Stage 3A (low or no EGR) engine with high performance Selective Catalytic Reaction (SCR) catalytic converter
- Medium rate EGR Engine with DPF and SCR

Due to the missing SCR infrastructure on construction sites, Liebherr decided to use the engine concept with high EGR rate and DPF for earth moving equipments. Thus, the following technologies, which are new within the Liebherr group, had to be introduced:

- 2 stage turbo charging system to realize the high pressure ratio which is necessary for the EGR rates (30% at rated engine speed)
- Common Rail System (CRS) with the possibility to realize multiple injections within one cycle (post injection to achieve low soot values, due to after oxidation effects)
- DPF with HC doser to realize a reliable active regeneration also for low load applications
- Actuators and sensors necessary for engine thermal management

The comprehensive introduction of exhaust aftertreatment systems also significantly influences the whole vehicle behavior. To compensate negative effects (e.g. dynamics) due to high EGR rates and to realize a better (or at least the same) behavior as those of the predecessor Stage 3A engines, advanced control strategies are necessary. To achieve these goals a function based approach was introduced on the engine sub-system level. The new strategy also required a new ECU platform and a new tool chain to cover the model based approach. The main challenges with respect to the function development are:

- Common Rail System (CRS):
  - High pressure fuel path: Injector and high pressure pump (HPP) control
  - Low pressure fuel path with the functional integration of the HC doser. Due to the large number of applications different hardware designs are necessary. Each of the applications has its own engine and temperature characteristics and each design its own boundary conditions. Therefore, the design of the functions should be done in such a way that a high calibration effort is avoided.

- Air path:
  - EGR control, especially dynamic issues. Due to the high EGR rates the engine dynamic is reduced in a way which is not acceptable for the customer. Thus, new methods like predictive control functions are necessary. Liebherr Machines Bulle SA develops and produces the whole power train for a construction machine, consisting of the

engine, the power split box and also the hydraulic pumps and motors. Thus, the optimization of the dynamic behavior can be done on the basis of the sub-system power train. This leads to an additional benefit, because the signals of the servo control valve from the hydraulic pump can be utilized in a feed forward control structure to reduce the dynamic lag between torque demand (operator) and torque generation (diesel engine).

– Thermal management to fulfill the exhaust aftertreatment requirements. To achieve optimal conditions for the HC doser, the Diesel Oxidation Catalyst (DOC) and also the DPF should have optimal temperatures. Therefore, a thermal management is mandatory, which must work in all operating conditions without influencing the machine dynamics.

In 2014 the emission level Tier 4, which includes also On Board Diagnostics (OBD) will be introduced. To fulfill the requirements for OBD in a cost effective and reliable way, that is without a high number of sensors, new approaches and methods in identification and control are necessary.

12.1.1 The Fuel Path as System Example

In order to point out some identification needs, the fuel path is a suitable example of a system built from different components governed by multiple physical domains. The system is shown schematically in Fig.12.1 where the main components are

- fuel pre- (LPP) and high pressure pumps (HPP) both driven at pump speed n
- actuators (volume control valve VCV, pressure control valve PCV, HC doser HCD driven by currents i_{VCV}, i_{PCV}, and i_{HCD})

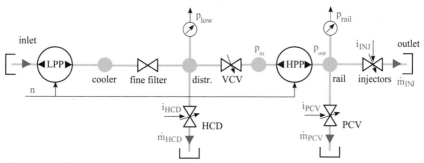

Fig. 12.1 Fuel path system

- sensors (low and high pressure sensor yielding p_{low} and p_{rail})
- injectors driven by the current profile i_{INJ}
- low and high pressure circuit elements (cooler, fine filter, distributor, daisy chain system, et cetera).

The first step in developing the overall fuel path system is to design, improve and optimize each of the components themselves. This task relies on detailed physical models in order to understand the underlying physical principles and to improve the component itself by extensive utilization of simulation and optimization. With the optimized components, one is able to put together the overall system, where the main goal is to achieve some prescribed, desired functions. These may include sophisticated control strategies, means for fault detection and fault diagnosis or some compensating functionality. For each of the functions a model of the system or subsystem can be used in the ECU to compute state variables, error signals or state dependent corrections.

In order to realize models which are able to run on the ECU and to achieve an overall desired system function the complexity of the simulation models has to be reduced. This consequently leads to the necessity of applying identification methods to estimate function and system parameters as well as to minimize the output error between complex and simplified models.

12.2   Fuel Path Subsystem Examples

In the following two examples are discussed to demonstrate the needs for identification in the field of Off-Road applications. The first example shows a simplification procedure for a high pressure pump (HPP) model which is able to satisfy both requirements, accuracy of the predicted flows (an thus the related pressures) as well as fast computation. As a second example the compensation of the remanence effects of a fuel injector solenoid on multiple injections is discussed.

12.2.1   Efficient High Pressure Pump Model

Generally, during the hardware development cycle a simulation model of the considered component is built in order to support the engineers with insight into the underlying physical phenomena as well as to reduce development time and costs by shifting the workload from the testbed to the simulation. Therefore, such a model is usually of high complexity and thus infeasible for real time application purposes. Considering an HPP model one such example is the small mass and high contact stiffness of the inlet and outlet valves, respectively. Clearly, such a model can not run in real time on an ECU and furthermore the memory footprint would be too large. In order to meet the calculation run time requirements a different approach has to be chosen to achieve a fast and accurate fuel path model.

In the case at hand a highly sophisticated model of a two piston in-line HPP [4] was available (compare Fig.12.2). Therefore, the basic idea is to replace the full dynamic HPP model by a functional description of the flows at the pump in- and outlets augmented by the mass balance of the pump.

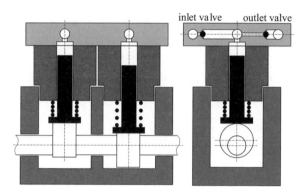

Fig. 12.2 HPP functional scheme

12.2.1.1   Functional Flow Description and Data Generation

The relevant inputs for a functional description of the pump flows are:

- pump speed n,
- pump inlet pressure p_{in} after VCV,
- and pump outlet pressure p_{out}.

In order to be able to fit functional models for the pump flows Q_{in} (inlet) and Q_{out} (outlet) steady state data are generated utilizing the validated fully dynamical model of the HPP implemented in an overall fuel path model according to Fig.12.1. Considering the system model, these flows as well as the pressures at the in- p_{in} and the outlets p_{out} are crank angle dependent system states and have to be averaged over one pump revolution. Thus, for each triple (n, p_{in}, p_{out}) values for Q_{in} and Q_{out} result. In Fig.12.3 the resulting, already averaged, data for n=1000rpm are shown. Since the input space is of dimension 3, one possibility would be to fit a map $\chi(p_{in}, p_{out})$ over p_{in} and p_{out} for some pump speeds n_j and then interpolate between the different maps, as utilized in [6]. The disadvantage is the amount of storage capacity required, especially when the input dimension increases. To reduce this problem χ, at every engine speed n_j, could also be fitted with an appropriate function and the individual functions could then be combined via activation functions $\tau(n_j)$ (radial basis function approach [3])

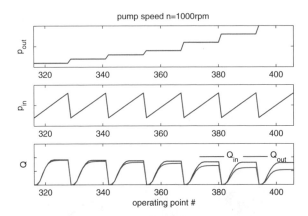

Fig. 12.3 $n=1000$rpm: influence of p_{in} and p_{out} on the flows Q_{in} and Q_{out}

$$Q = \sum_{j=1}^{N} \chi(p_{in}(n_j), p_{out}(n_j)) \cdot \tau(n_j). \tag{12.1}$$

Another possibility can be derived for the special shapes of Q_{in} and Q_{out} (Fig.12.3):

1. The pressure at the pump outlet p_{out} determines the maximal flow that is possible for an optimally filled HPP cylinder (provided enough pressure p_{in} is available at the pump inlet → VCV fully open).
2. The pressure at the pump inlet p_{in} determines only how well the HPP cylinder is filled.

Therefore, a more simple model is given by

$$Q = f(p_{in}, \Theta_1(n)) \cdot g(p_{out}, \Theta_{2i}(n)); \quad i = a, b, c. \tag{12.2}$$

Such a model has to be fitted for both flows, Q_{in} and Q_{out}.

12.2.1.2 Model Structure and Parameter Identification

Since the inlet pressure p_{in} determines the filling state of the HPP cylinders (0→not filled, 1→completely filled) $f(p_{in}, \Theta_1(n))$ is chosen to be a tanh-function with a center at 0.78bar and an exponent of 3.5 to provide for anti-symmetric behavior

$$f = \left\{ \frac{1}{2} \left[1 + \tanh \left(\frac{p_{in} - 7.8 \cdot 10^4}{\Theta_1} \right) \right] \right\}^{3.5}, \tag{12.3}$$

whereas for $g(p_{out}, \Theta_{2i}(n))$ a parabola is sufficient

$$g = \Theta_{2a} + \Theta_{2b} \left(\frac{p_{out}}{2.5 \cdot 10^8} \right) + \Theta_{2c} \left(\frac{p_{out}}{2.5 \cdot 10^8} \right)^2. \tag{12.4}$$

It has to be remarked, that for each pump speed n the parameters Θ_1, Θ_{2a}, Θ_{2b}, and Θ_{2c} for the functions f and g have to be fitted. Therefore, each of the parameters $\Theta_1 - \Theta_{2c}$ itself (8 parameters; 4 for Q_{in} as well as Q_{out}) is modeled as a parabola

$$\Theta_k = \alpha_k + \beta_k \left(\frac{n}{4000} \right) + \gamma_k \left(\frac{n}{4000} \right)^2 = h(\alpha_k, \beta_k, \gamma_k). \tag{12.5}$$

For the above given functions, f reflects the (normalized) influence of p_{in} on Q and g adjusts the flow to the real physical values. Additionally, in (12.4) and (12.5) g and h have been normalized to a maximum outlet pressure of 2500bar and a maximum pump speed of 4000rpm. Finally, it has to be pointed out that for f also the center and the exponent could have been chosen as (nonlinear) parameters to be optimized, which has been omitted for reasons of simplicity. Thus, the overall identification procedure is:

1. First identify Θ_1 as well as Θ_{2a}, Θ_{2b}, Θ_{2c} of (12.3) and (12.4) by nonlinear and linear identification methods [1], respectively.
2. Then fit the parameters α_k, β_k, γ_k of (12.5) by linear identification.

The results are given in Fig.12.4 and Fig.12.5 for the flow at the pump inlet Q_{in}.

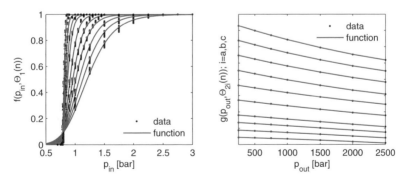

Fig. 12.4 Fitting of the functions f and g for Q_{in}; each curve represents a constant pump speed n

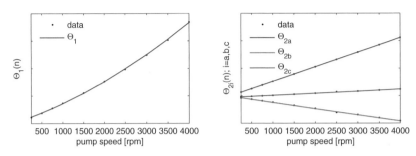

Fig. 12.5 Fitting of the parameters Θ_1, Θ_{2a}, Θ_{2b}, and Θ_{2c} as functions of the pump speed n for Q_{in}

12.2.1.3   Simplified HPP Model

In Fig.12.6 a comparison between the flows computed with the full HPP model and the calculation results of the functional approach is given. One can observe, that the simplified model fits the data quite well. In order to replace the full HPP description in the overall fuel path model (compare Fig.12.1) by a simpler version, the functional flow description is combined with the mass balance of the pump as shown in Fig.12.7. Since the simplified HPP model is to be coupled into the overall fuel path system model, the pressures at the pump in- and outlet are states of the overall system as well as inputs to the functional flow description $f*g$, which is a part of the simplified HPP model. Applying the mass balance to the flow source one obtains the leakage of the pump

$$Q_{leak} = Q_{in} \frac{\rho_{in}}{\rho_{out}} - Q_{out}. \tag{12.6}$$

Furthermore, utilizing the standard flow equation for a valve with flow coefficient α, cross section A, according diameter d_{valve}, and density ρ

$$Q_{leak} \cong \alpha A \sqrt{\frac{2p_{out}}{\rho_{out}}} = \alpha \frac{d_{valve}^2 \pi}{4} \sqrt{\frac{2p_{out}}{\rho_{out}}}, \tag{12.7}$$

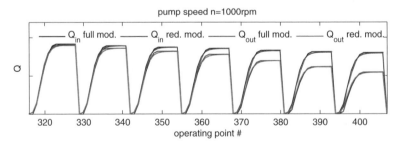

Fig. 12.6 Comparison of pump flows for fully dynamical and simplified models

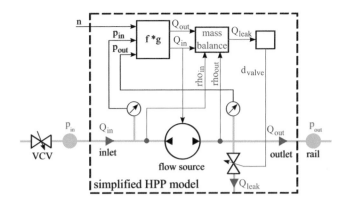

Fig. 12.7 Model structure of the simplified HPP model including the functional flow description f^*g

the state dependent diameter of the valve can be calculated from

$$d_{valve} = \sqrt{\frac{4Q_{leak}}{\alpha\pi}} \sqrt{\frac{\rho_{out}}{2p_{out}}}. \tag{12.8}$$

In (12.7) the counter pressure downstream of the valve has been omitted, since in general the pressure at the pump outlet p_{out} (rail pressure) is much larger. Additionally, the valve diameter d_{valve} represents the overall, state dependent cross section, through which the leakage is flowing from the cylinders via the pistons to the pump crank shaft.

Equations (12.2), (12.3), (12.4), (12.5), (12.6), and (12.8) are implemented in SimulationX in order to achieve a simple high pressure pump model which then is implemented in the overall fuel path model. Thus, a system model capable of predicting steady state and dynamic pressure changes is obtained. The simulation results are shown in Fig.12.8, where i_{VCV} is the current applied to the VCV and d_B is the diameter of a valve representing the flow through the injectors. Consequently, the rail pressure p_{rail} results from the injector flows as well as Q_{out} and Q_{leak}. This simple model reflects the system dynamics in a suitable way. For $n = 2000$rpm a slight disturbance at the beginning of the simulation can be observed. This results from the fact, that in the case of the full model the cylinders of the pump are completely filled at $t = 0$s, whereas the simple model directly yields the mean value of the flow.

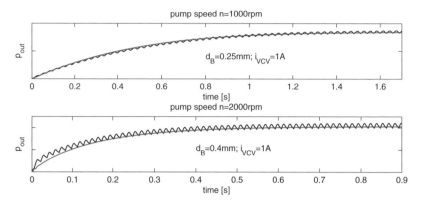

Fig. 12.8 Comparison of simulation results for the full dynamic model (black) and the simplified one (red)

12.2.2 Injector Remanence Compensation

In order to remain inside the boundaries of exhaust emission imposed by legislation an efficient fuel injection equipment (FIE) calibration is necessary. To reach this goal the behavior of the injector has to be well-known and controlled for a wide range of operating conditions. One effect distorting the injection functionality is the remanence behavior of the injector solenoid.

12.2.2.1 Solenoid Remanence Influence on Multiple Injection

On the left hand side of Fig.12.9 the effect of the solenoid remanence can be observed from the increased mass of the second injection, although the electrical injection time t_{INJ} is the same for both. On the right hand side both the mass for primary and secondary injection are given over the distance t_D, which is the duration between the end of the first and the beginning of the second injection (a definition of the current profile parameters is shown in Fig.12.10). With increasing t_D the influence of the primary injection on the secondary one decreases.

The reason for this effect is the magnetic remanence of the injector solenoid which results in a time delay between the electric current i_{INJ} and the magnetic force. Thus, the hydraulic opening delay $t_{SOI,h}$ is not only dependent on the parameters of the current profile, but also on t_D. The main difficulty now is, that the mass flow profile $\dot{m}$ and thus the injected mass m_{INJ} as well as the hydraulic start of injection $t_{SOI,h}$ depend on a large number of parameters, which are:

- time between two injections t_D
- duration of the first injection $t_{INJ,1}$
- rail pressure p_{rail}

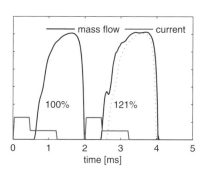
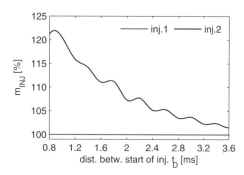

Fig. 12.9 Left: mass flow $\dot{m}$ and current profiles i_{INJ} for t_{INJ}=1200μs, p_{rail}=1800bar, $i_{H1} = 25A$, and t_D=800μs; right: injected masses m_{INJ} for first and second injection over t_D

- fluid/injector temperature T
- Hold 1 current i_{H1} (boost current level to open the injector)
- position of injector in the daisy chain.

Therefore, a perfect compensation of all these influences on m_{INJ} and $t_{SOI,h}$ is nearly impossible with standard map-based corrections. In the underlying case a more powerful strategy is necessary to efficiently reflect and thus reduce the remanence effects on the mass flow profile $\dot{m}$.

12.2.2.2   Compensation Function Structure

Since the magnetic force depends on the parameters of the current profile, the secondary injection can be influenced either by the value of the Hold 1 current i_{H1} or the electric starting point of the injection $t_{SOI,e}$. A possible structure for a compensation function is given in Fig.12.10 on the left, where additionally on the right hand side the current profile parameters are defined. The given structure suggests, that from the desired mass m_{INJ} and the desired hydraulic start of injection $t_{SOI,h}$ the electric start of injection $t_{SOI,e}$ and the injection duration t_{INJ} are calculated (**model 1**; reflects primary and secondary injection nominal and thus remanence free operation). Then, in a second step (**model 2**), a t_D-dependent update either for the hold 1 current (*method 1*) or the electrical start $t_{SOI,e}$ (*method 2*) of the secondary injection are calculated and applied to the injector. A suitable way in dealing with the input space dimensions of **models 1** and **2** as well as the strong coupling between input space directions is to apply Neural Network (NN) methods [3] or polynomial models [2].

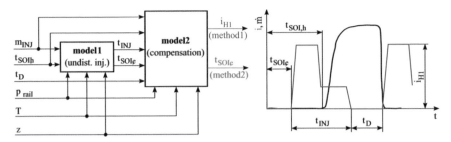

Fig. 12.10 Left: structure of compensation scheme; right: current and mass flow parameters

12.2.2.3 Modeling

To be able to derive **model1** and **model2**, according data of the variables to be modeled have to be generated. This can be done by either utilizing a simulation model of the injector or directly measuring the injected mass m_{INJ} and hydraulic start of injection $t_{SOI,h}$ on the hydraulic test bed. In this study, simulation data are applied in the modeling procedure and as structures for models **1** and **2** fully coupled polynomial models of third order are chosen and fitted utilizing the Matlab Model-Based Calibration Toolbox.

Generally the fitting of models **1** and **2** is straight forward: after the necessary data have been obtained the polynomial coefficients can be calculated/optimized. Additionally, for **model2** an intermediate step is necessary, since the quantity to be modeled, the correction value, has to be obtained iteratively at every operating point. The according work flow is given in Fig.12.11 for *method 2* (correction of electrical start of injection $t_{SOI,e,2}$; analogous work flow for *method 1*). In a first step the reference case (single injection, remanence free) is simulated, which provides as output the injected

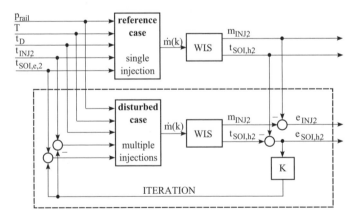

Fig. 12.11 Structure of data generation scheme

mass quantity $m_{INJ,2}$ and the hydraulic starting point $t_{SOI,h,2}$ of the second injection. Regard that these data can directly be utilized for the fitting of **model1**. Then, in the second step the disturbed case (multiple injections, remanence effects) is simulated and compared to the reference case. As a consequence the errors of the injected mass $e_{INJ,2}$ and the hydraulic start of injection $e_{SOI,h,2}$ are obtained. Normally, the remanence effects lead to an early start of injection and thus the correction is calculated as $K \cdot e_{SOI,h,2}$. Since a correction of the start of injection also adjusts the injected mass quantity, a correction based solely on $e_{SOI,h,2}$ suffices. Here the gain K determines the speed as well as the stability of the iteration/adaptation. Regard, that the calculated correction is applied to the electrical start of injection $t_{SOI,e,2}$ and the electrical duration of injection $t_{INJ,2}$.

One difficulty in this scheme is to obtain in an automated and robust way the hydraulic start of injection $t_{SOI,h,2}$ from the mass flow profile $\dot{m}(k)$, recorded with the sampling time ΔT. This is achieved via the WLS blocks indicated in Fig.12.11. There, the rising (or falling) edge of $\dot{m}(k)$ is fitted with a line and the data range is restricted via the weighting function W in a Weighted Least Squares (WLS) procedure [1]. In the current case the WLS algorithm is given by

$$W\dot{m}(k) = [t(k)\ 1] \cdot \begin{bmatrix} a \\ b \end{bmatrix} = X\theta, \tag{12.9}$$

where $t(k)$ is the time vector corresponding to $\dot{m}(k)$ and $\theta^T = [a\ b]^T$ is a vector of the line parameters. Then the parameter vector θ is derived from

$$\theta = \left(X^T W^T W X\right)^{-1} X^T W^T W \dot{m}(k), \tag{12.10}$$

where the diagonal $diag(W)$ of the weighting matrix W is given by

$$diag(W) = \frac{\psi}{\max \psi} \tag{12.11}$$

and the function ψ by

$$rising\ edge\ :\ \psi = \begin{cases} t \leq t_{max} & \left(\frac{\dot{m}(k)-\dot{m}(k-1)}{\Delta T}\right)^2 \\ t > t_{max} & 0 \end{cases}$$

$$falling\ edge : \psi = \begin{cases} t < t_{max} & 0 \\ t \geq t_{max} & \left(\frac{\dot{m}(k)-\dot{m}(k-1)}{\Delta T}\right)^2 \end{cases} \tag{12.12}$$

In (12.12) t_{max} results from

$$t_{max} = t\big|_{\max(\dot{m}(k))}. \tag{12.13}$$

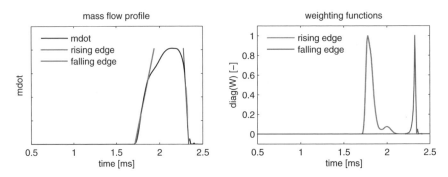

Fig. 12.12 Left: $\dot{m}$ and slopes; right: ψ for rising and falling edge

The off-diagonal elements of W are set to zero. In Fig.12.12 an example is shown demonstrating the chosen WLS procedure. Clearly, the mass flow profile is only fitted at the rising and the falling edge, respectively.

12.2.2.4 Simulation Results

In Figure 12.13 simulation results for the proposed *methods 1* and *2* are given for mass flow and current profiles of the second injection. Compared to the reference mass flow profile without remanence effects (no pre-energization of the injector solenoid) both methods are basically capable of correcting the mass flow profile. The choice of the method to be implemented depends on the ECU software architecture as well as the underlying compensating philosophy. Where *method 1* is more difficult to implement in the ECU (shot-to-shot change of the Hold 1 current controller set point), it also leaves t_{INJ} and $t_{SOI,h}$ the same for both injections. Therefore, *method 1* may be more intuitive for practical implementation.

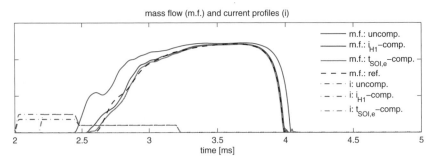

Fig. 12.13 Mass flow and current profiles of secondary injection for $t_{INJ}=1200\mu s$, $p_{rail}=1800$bar, and $t_D=800\mu s$: uncompensated, reference (without first injection) and compensated mass flows (*method 1*: i_{H1} compensated; *method 2*: $t_{SOI,e}$ compensated)

Finally, it has to be pointed out that with the adaptation/modeling framework presented above a validated simulation model can be utilized to derive a pre-calibration of the models to be used on the ECU before going on the test bench, thus minimizing test bed costs.

12.3   Summary and Outlook

In this paper two examples have been utilized to demonstrate the necessity of identification methods in order to fulfill future requirements for Off-Road engines. Both examples have been taken from the function development domain for the FIE system in order to highlight their potential in this industrial field. Another area, not addressed here but also of high interest, is the topic of virtual sensors e.g. the replacement/enhancement of the mass flow sensor by a mass flow estimation/prediction based on a measurement of the turbo charger speed. The replacement of complex hardware with equivalent models increases the reliability of the engine and leads to a reduction of the calibration costs. One of the challenges for this kind of model is to additionally take into account aging effects.

From all these examples one concludes, that following the automotive industry, the Off-Road sector exhibits an increasing amount of topics in which identification methods are not only helpful or desirable but strongly needed.

References

[1] Isermann, R.: Identifikation dynamischer Systeme. Band I. Springer, Heidelberg (1988)
[2] Montgomery, D.C.: Design and Analysis of Experiments. John Wiley & Sons, Chichester (2001)
[3] Nelles, O.: Nonlinear System Identification. Springer, Berlin (2000)
[4] Ivantysyn, J., Ivantysynova, M.: Hydrostatische Pumpen und Motoren. Vogel Verlag (1993)
[5] Bosch GmbH, R.: Diesel-Engine Management. John Wiley & Sons Ltd., Chichester (2005)
[6] Clever, S., Isermann, R.: Modellgestützte Fehlererkennung und Diagnose für Common-Rail-Einspritzsysteme. Motortechnische Zeitschrift, 114–121 (February 2010)

Chapter 13
Dynamic Mapping of Diesel Engine through System Identification*

Maria Henningsson**, Kent Ekholm, Petter Strandh,
Per Tunestål, and Rolf Johansson

Abstract. From a control design point of view, modern diesel engines are dynamic, nonlinear, MIMO systems. This paper presents a method to find low-complexity black-box dynamic models suitable for model predictive control (MPC) of NO_x and soot emissions based on on-line emissions measurements.

A four-input-five-output representation of the engine is considered, with fuel injection timing, fuel injection duration, exhaust gas recirculation (EGR) and variable geometry turbo (VGT) valve positions as inputs, and indicated mean effective pressure, combustion phasing, peak pressure derivative, NO_x emissions, and soot emissions as outputs. Experimental data were collected on a six-cylinder heavy-duty engine at 30 operating points. The identification procedure starts by identifying local linear models at each operating point. To reduce the number of dynamic models necessary to describe the engine dynamics, Wiener models are introduced and a clustering algorithm is proposed. A resulting set of two to five dynamic models is shown to be able to predict all outputs at all operating points with good accuracy.

13.1 Introduction

The heavy-duty engine market is dominated by compression-ignition diesel engines, due to their high energy conversion efficiency. Traditionally, this high

Maria Henningsson · Rolf Johansson
Department of Automatic Control, Lund University

Kent Ekholm · Per Tunestål
Department of Energy Sciences, Lund University

Petter Strandh
Volvo Powertrain Corp.
e-mail: `maria.henningsson@control.lth.se`

* This work was supported by Vinnova and Volvo Powertrain Corp.

** Corresponding author.

D. Alberer et al. (Eds.): Identification for Automotive Systems, LNCIS 418, pp. 223–239.
springerlink.com   © Springer-Verlag London Limited 2012

efficiency has come at the price of high levels of nitrogen oxides (NO_x) and soot particle emissions.

In the last decade, the emissions legislation has been dramatically tightened which has pushed for a shift in technology, either through introducing an aftertreatment system for NO_x reduction or through in-cylinder techniques. In the latter case, the combustion process is cooled down using a lower compression ratio, high levels of exhaust gas recirculation (EGR), and suitably chosen fuel injection timings resulting in so called Low Temperature Combustion (LTC) [7].

Control is a key factor for successful LTC diesel engines. One reason is that slow dynamics involved in e.g. EGR flow and cylinder wall temperature play a more important role when ignition delays are prolonged. Another reason is that tight requirements both in terms of emissions legislation and fuel economy make error margins tight. Diesel engine control design is normally based on a mix of physical mean-value models and experimental maps specifying optimal operating points in terms of emissions and fuel efficiency [1]. Feedback control uses indirect measurements such as manifold pressures and gas flows.

New sensor technology makes additional information from the combustion process available. Recently, NO_x sensors have been introduced in production engines giving the opportunity to adjust actuator settings according to measured emissions instead of precalibrated maps. These developments open up for e.g. model predictive control (MPC) that explicitly handles emission trade-off optimization based on on-line measurements of emissions. To that purpose, low-complexity dynamic models of the engine that also include emission formation are needed, this objective being the purpose of the work presented in this paper. The approach has been to use black-box system identification.

The focus has been to use direct measured information on the interesting engine outputs instead of relying on indirect measurements combined with models and maps. Measurements from a NO_x sensor, an opacimeter, and in-cylinder pressure sensors are therefore used. Though not all of these sensors are used in today's production engines, one can expect that if the benefits are large enough, the technology will eventually be available.

Different actuators can be used to optimize emissions, e.g. EGR and variable geometry turbine (VGT) valve positions [9, 8], fuel injection parameters [2], variable valve actuation [15], or some combination of these [5]. In this paper, the combined effect of the most significant diesel engine actuators (EGR, VGT, injection timing, and injection duration) is examined. These actuators are standard in heavy-duty diesel engines today and are all essential for optimization of engine operation. It would be desirable also to include engine speed as an input to the model, but for practical constraints in the laboratory setup that was not possible in this work.

The paper is organized as follows. Section 13.2 presents the experimental equipment and Section 13.3 briefly describes the thoughts behind the choice of engine outputs and the experiment design. Section 13.4 describes the identification procedure, starting with local linear models that are expanded to Wiener models. A clustering algorithm is proposed to reduce the number of dynamic models required to describe the engine dynamics. Section 13.5 evaluates the identified models. Finally, conclusions and directions for future work are presented in Section 13.7.

13.2   Experimental Equipment

The experiments were conducted on a six-cylinder turbo-charged heavy-duty diesel engine with a displaced volume of $V_D = 12\text{e-}3[m^3]$. The compression ratio of the engine was reduced to $r_c = 14.1$ to facilitate low-temperature combustion. The engine was equipped with unit injectors for diesel where fuel injection timings could be set individually for each cylinder, with a low-pressure exhaust gas recirculation (EGR) loop where the EGR rate could be adjusted by a valve in the exhaust pipe, and with a variable geometry turbo (VGT) where the turbocharging also could be adjusted by a valve.

All cylinders were equipped with piezo-electrical, water-cooled pressure transducers of type Kistler 7061B [12], with cylinder pressure data sampled every 0.2 crank angle degrees using a Microstar DAP 5400a/627 data acquisition board [13]. The control system was based on a standard PC running Linux enabling cycle-to-cycle control. Fuel injection timings were updated every engine cycle, and the setpoints for the valve positions were updated with a frequency of 10 Hz.

The pressure measurements p as a function of crank angle θ from the in-cylinder pressure sensors were used to compute indicated mean effective pressure y_{IMEP}, combustion phasing α_{50}, and maximum pressure derivative d_p. The indicated mean effective pressure is defined as

$$y_{\text{IMEP}} = \frac{1}{V_D} \int p dV, \tag{13.1}$$

where the integral is taken over an engine cycle, and the maximum pressure derivative as

$$d_p = \max_{\theta} \frac{dp}{d\theta} \tag{13.2}$$

From the cylinder pressure p, the heat release rate dQ is computed using the relation

$$\frac{dQ}{d\theta} = \frac{\gamma}{\gamma - 1} p(\theta) \frac{dV}{d\theta} + \frac{1}{\gamma - 1} V(\theta) \frac{dp}{d\theta} \tag{13.3}$$

for the apparent heat release rate based on a fixed ratio of specific heats [3]. From the heat release rate, α_{50} is defined as the crank angle degree where 50 % of the heat has been released,

$$\frac{Q(\alpha_{50})}{\max_\theta Q(\theta)} = 0.5 \qquad (13.4)$$

Emissions of NO_x were measured using a Siemens VDO / NGK Smart NOx Sensor [14]. Soot emissions were measured using an opacimeter from SwRI measuring the percentage of light absorbed by the exhausts in the exhaust pipe.

13.3 Experiment Design

Four control variables act as inputs to the model:

- the crank angle of start of fuel injection u_{SOI}
- the fuel injection duration measured in crank angle degrees u_{FD}
- the position of the EGR valve u_{EGR}
- the position of the VGT vanes u_{VGT}

such that

$$u = \begin{pmatrix} u_{SOI} & u_{FD} & u_{EGR} & u_{VGT} \end{pmatrix}^T. \qquad (13.5)$$

For the main part of the work, five model outputs were considered; the net indicated mean effective pressure y_{IMEP}, the crank angle degree of 50 % fuel burnt α_{50}, the peak cylinder pressure derivative over the cycle d_p, the nitrogen oxide concentration y_{NOx} and the opacity of the exhausts y_{op} giving a measure of soot concentration

$$y = \begin{pmatrix} y_{IMEP} & \alpha_{50} & d_p & y_{NOx} & y_{op} \end{pmatrix}^T. \qquad (13.6)$$

The first three output variables were cylinder-individual. The last two outputs, y_{NOx} and y_{op} were common to all cylinders. These five outputs were chosen because they are all required in the model predictive control setup for which the model is intended to be used. The planned control design setup is to minimize y_{op} and y_{NOx}, to let y_{IMEP} follow a reference, to keep α_{50} at a setpoint corresponding to maximum brake torque, and to limit d_p to avoid combustion modes that may damage the engine.

Many alternative measured outputs could be considered. By including more measured outputs in the model, prediction of the original outputs to be used for control could be expected to improve when more information is available for prediction. An extended set of outputs was thus also considered, where intake manifold pressure p_{in} and ignition delay α_{ID} were added to the output vector,

$$y_{ext} = \begin{pmatrix} y_{IMEP} & \alpha_{ID} & d_p & y_{NOx} & y_{op} & p_{in} & \alpha_{50} \end{pmatrix}^T. \qquad (13.7)$$

The ignition delay was defined as $\alpha_{ID} = \alpha_{10} - u_{SOI}$, the time measured in crank angle degrees from start of injection until 10 % of fuel was burnt.

The study is limited to low and medium load operating points, where the possible impact of low temperature combustion is the greatest and where the associated slow dynamics are most dominant. A fixed engine speed of 1200 rpm was chosen, the study could easily be expanded to multiple engine speeds using the same methodology. Three different loads were chosen, corresponding to $y_{IMEP} = \{4\ \text{bar}, 7\ \text{bar}, 10\ \text{bar}\}$. At each load three different values of start of injection u_{SOI} were chosen, and at each such value four points in the u_{EGR}-u_{VGT} plane were determined. A few initial steady-state maps were used to define suitable operating points in terms of emissions and brake efficiency. At each operating point suitable amplitudes for pseudo-random binary sequence (PRBS) signals were determined. The switching frequency of the PRBS signal was adjusted to the dynamic characteristics of each of the inputs (more frequent switching for the fuel injection variables u_{SOI} and u_{FD}, and less frequent for the gas flow variables u_{SOI} and u_{VGT}). In total, 30 operating points were defined: 6 at load $y_{IMEP} = 4$ bar, 12 at load $y_{IMEP} = 7$ bar, and 12 and $y_{IMEP} = 10$ bar. At $y_{IMEP} = 4$ bar, it was concluded that keeping the VGT valve fully open at $u_{VGT} = 100$ was optimal in terms of emissions and fuel economy, so there was no need for excitation of that input. An overview of the operating points is given in Table 13.1.

13.4   Identification Procedure

Data of length 2800 engine cycles were collected at each operating point. The sample period was one engine cycle. Constant offsets were removed. From each data set, two separate data sets each of length 1300 engine cycles were taken for identification and validation. To avoid any cross-correlation between the identification and validation data, they were separated by 200 engine cycles. It was concluded that offsets differed between the cylinders whereas the dynamics were similar, so models were only identified for cylinder 5 which was determined to be representative for all six cylinders. The outputs were scaled to obtain the same order of magnitude.

The identification procedure is divided into two parts. First, linear models are identified separately for each operating point. Then, the number of models needed are reduced by introducing Wiener models and performing a clustering algorithm.

13.4.1   Identification of Local Linear Models

Linear state space models for each operating point of the form

$$x_{k+1} = Ax_k + Bu_k + Kw_k$$
$$y_k = Cx_k + Du_k + w_k$$

$$(13.8)$$

Table 13.1 Operating points

Op. point	y_{IMEP} (bar)	u_{SOI} (CAD)	u_{FD} (CAD)	u_{EGR} (%)	u_{VGT} (%)	speed (rpm)
1	4	−5	5.8	10	100	1200
2	4	−12	5.8	10	100	1200
3	4	−20	5.8	10	100	1200
4	4	−5	5.8	13	100	1200
5	4	−12	5.8	13	100	1200
6	4	−20	5.8	13	100	1200
7	7	−5	7.1	53	33.5	1200
8	7	−10	7.1	53	33.5	1200
9	7	−15	7.5	53	33.5	1200
10	7	−5	7.1	56	40	1200
11	7	−10	7.3	56	40	1200
12	7	−15	7.5	56	40	1200
13	7	−5	7.2	50	23	1200
14	7	−10	7.4	50	23	1200
15	7	−15	7.6	50	23	1200
16	7	−5	7.3	56	28	1200
17	7	−10	7.4	56	28	1200
18	7	−15	7.5	56	28	1200
19	10	−9	10.4	76	29	1200
20	10	−13	10.8	76	29	1200
21	10	−17	10.7	76	29	1200
22	10	−5	10.5	82	29	1200
23	10	−10	10.6	82	29	1200
24	10	−15	10.8	82	29	1200
25	10	−5	10.6	82	40	1200
26	10	−10	10.6	82	40	1200
27	10	−15	10.8	82	40	1200
28	10	−5	10.5	90	40	1200
29	10	−10	10.6	90	40	1200
30	10	−15	10.9	90	40	1200

where $x_k \in \mathbb{R}^n$, $u_k \in \mathbb{R}^r$, and $y_k \in \mathbb{R}^m$ were identifed using the n4sid algorithm of the System Identification Toolbox in Matlab. As validation criterion, variance accounted for (VAF) was used [4]. For output i, VAF is defined as

$$VAF(i) = 100 \left(1 - \frac{\text{var}(y_i - \hat{y}_i)}{\text{var}(y_i)} \right) \tag{13.9}$$

where y_i is the measured output in the validation data, and $\hat{y}_i$ is the predicted output of the model. The VAF gives the percentage of the variance of the output that is described by the model.

Models of orders $n = 2$ to $n = 12$ were identified at each operating point. The mean VAF over the five outputs as a function of model order for 20-step prediction is shown in Figure 13.1.

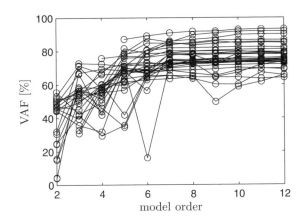

Fig. 13.1 VAF as a function of model order for the 30 operating points.

Low model orders, $n < 5$, lead to poor prediction. At some operating points, low-order models even lead to negative VAF (not included in Figure 13.1 for the sake of clarity) corresponding to a higher variance in the prediction error than in the measured outputs. Increasing the model order beyond $n = 7$ can give a small improvement in VAF, but not sufficient to motivate the greater computational burden of such large models for control design. From Figure 13.1, it can be concluded that model order $n = 7$ is suitable to describe the data. There is a large spread in VAF between the operating points for a fixed model order. This spread is mainly caused by different excitation of the outputs compared to the noise level at different operating points.

Figure 13.2 shows how VAF varies with prediction horizon for the five outputs for operating point 1.

13.4.2 *Identification of Models for Extended Output Set*

Models were also identified for the extended set of outputs in Eq. (13.7) and validated in terms of prediction of the original five outputs. Figure 13.3 shows the result for operating point 1. For low orders, the VAF for these models are significantly lower than for the original output set. This outcome could

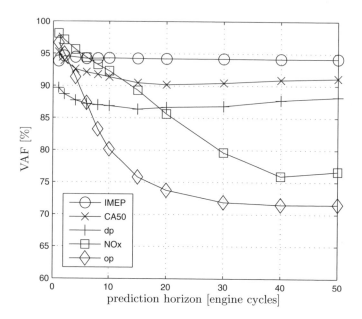

Fig. 13.2 VAF for different prediction horizons for the five outputs.

be expected, since with additional outputs the models need to incorporate additional dynamics particulary the slow intake manifold pressure evolution. At higher model orders, the VAF is similar for the original and the extended output set. It was thus concluded that there was not much to be gained from including the extra two outputs in the models and Kalman filter. In the following, only the original output set is considered.

13.4.3 Reducing the Number of Models

The purpose of the identification scheme is to find appropriate models for model-based control design. Since the process is clearly nonlinear, the controller must be based on more than one linear model leading to additional complexities of gain-scheduling and model-switching. With the computational resources available today, a feasible controller with cycle-to-cycle control in real-time cannot be based on 30 local linear models. Moreover, the models identified here are for a single engine speed and low-to-medium loads only. Expanding the mapping to cover the entire load-speed range of the engine, more models would have to be added.

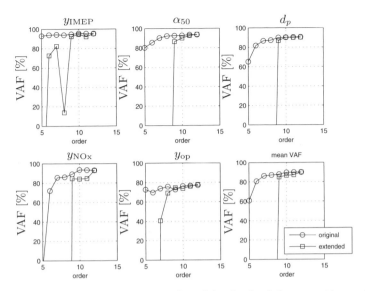

Fig. 13.3 VAF for 20-step prediction of models obtained for operating point 1 with the original and extended sets of outputs.

It is therefore of interest to see how close in behavior the models are to one another, i.e., if the same model could be used for more than one operating point. To that purpose, cross validation of all identified models to validation data at all operating points was performed. Figure 13.4 shows the resulting VAF for 20-step predictions.

It can be seen that, in general, the best model for data from an operating point is the model identified at that operating point, as would be expected. Some combinations of models and operating points give good predictions, others do not. Note that models identified at low load (operating points 1 to 6) cannot be used at higher loads since these models do not contain the u_{VGT} input, as described in Section 13.3.

Here, two steps are taken to find a limited number of dynamic models describing the data at all operating points. First, Wiener models are introduced to compensate for different gains at different operating points. A clustering algorithm is then applied to group the operating points and find a suitable model for each group.

13.4.3.1   Wiener Models

A powerful extension of linear models are the Wiener and Hammerstein model classes [4]. A Wiener model is a linear dynamic model in series with a static nonlinearity on the output, see Figure 13.5. A Hammerstein model is the

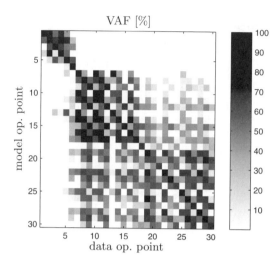

Fig. 13.4 Mean VAF for cross validation of the models identified at the 30 operating points with validation data at all operating points. Values on the diagonal are high, corresponding to good fit between model and data at the same operating point.

equivalent with the nonlinearity on the input. The static nonlinearity $f(\cdot)$ adds a large flexibility to the linear dynamic model at a moderate cost. If the nonlinearity can be inverted, control design can be based entirely on linear techniques. Here, we choose to use Wiener models rather than Hammerstein models because we have a 4-input-5-output system, which gives one more degree of freedom when transforming the output rather than the input. Previously, models of the form

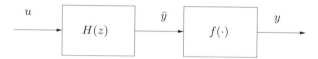

Fig. 13.5 A Wiener model consisting of a linear model $H(z)$ with a static nonlinearity on the output $f(\cdot)$.

$$x_{k+1} = A^i x_k + B^i u_k + K^i w_k$$
$$y_k = C^i x_k + D^i u_k + w_k,$$

where $x_k \in \mathbb{R}^n$, $u_k \in \mathbb{R}^r$, $y_k \in \mathbb{R}^m$, were identified for operating points $i = 1, \cdots, 30$. A Wiener model extends this representation to

$$x_{k+1} = A^i x_k + B^i u_k + K^i w_k$$

$$\bar{y}_k = C^i x_k + D^i u_k + w_k$$

$$y_k = f^i(\bar{y}_k)$$

Consider the dynamic model $\mathcal{M}_i(A^i, B^i, C^i, D^i, K^i)$ to be given from the previous identification procedure. We now wish to find a representation of the nonlinearities $f^i(\cdot)$ to let model i better fit data at another operating point j. This can be done through a local linearization of $f^i(\cdot)$ at operating point j,

$$M^{ij} = \left.\frac{\partial f^i}{\partial \bar{y}}\right|_{\text{op point}=j} \tag{13.10}$$

From the set of local linearizations M^{ij} and the offsets removed at each operating point, approximations of the nonlinearities $f^i(\cdot)$ can be reconstructed as e.g. piecewise linear functions.

We now wish to find the matrices M^{ij} to optimize the fit between model i and identification data j. To make the representation simpler and to avoid over-parameterization of the model, the matrices M^{ij} are constrained to be diagonal, $M^{ij} = \text{diag}(m_1^{ij}, \cdots m_m^{ij})$, such that the nonlinearity $f^i(\cdot)$ is decoupled into m scalar nonlinearities, one for each output.

Denote by $\hat{\mathcal{Y}}_{ij} \in \mathbb{R}^{N \times m}$ the data series of predicted outputs at operating point j using model i with the nominal gain matrix $M^{ij} = I$. It then holds for the output data series of Wiener model $\hat{\mathcal{Y}}_{ij}$

$$\hat{\mathcal{Y}}_{ij} = \hat{\bar{\mathcal{Y}}}_{ij} M^{ij^T} \tag{13.11}$$

The optimal gains m_k^{ij}, $k = 1, \cdots, m$, can now be found by solving the convex optimization problems

$$\min_{M^{ij} \text{ diagonal}} ||\mathcal{Y}_j - \hat{\bar{\mathcal{Y}}}_{ij} M^{ij^T}||_F^2 \tag{13.12}$$

where $\mathcal{Y}_j$ is the measured output data series at operating point j.

The cross validation was redone with an optimized scaling matrix M^{ij} for each pair of model and data operating points i and j. The resulting VAF can be seen in Figure 13.6. Notice the large improvement in VAF when matching models to different operating points compared to Figure 13.4.

13.4.3.2   Model Clustering

From Figure 13.6, we can see that some of the linear models with adapted gains well describes data at many operating points. We define by V the VAF matrix illustrated in Figure 13.6, where element v_{ij} corresponds to the mean VAF over the five outputs when using Wiener model i for a 20-step prediction of validation data at operating point j. The matrix V can be used to group

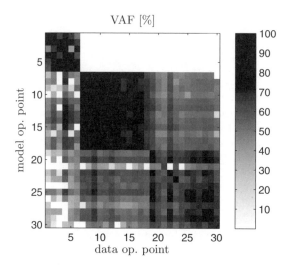

Fig. 13.6 Mean VAF for cross validation of the models identified at the 30 operating points with validation data at all operating points with gains adapted to form Wiener models. Note the large improvement compared to Figure 13.4.

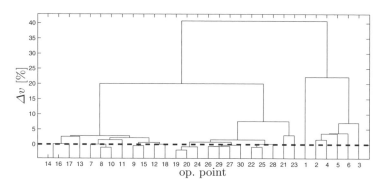

Fig. 13.7 Dendrogram of operating point clustering.

operating points to a limited number of models. A hierarchical agglomerative clustering algorithm is used to that respect [6].

The algorithm is based on the relative VAF loss matrix ΔV, where the elements are given by

$$\Delta v_{ij} = v_{jj} - v_{ij}, \tag{13.13}$$

and a clustering matrix C initialized as

$$C = \Delta V \tag{13.14}$$

The element Δv_{ij} is a measure of how much VAF that is lost by removing model j and using model i instead for that operating point. The hierarchical clustering algorithm successively merges two of the existing clusters at each step until only one cluster remains. At each step, the algorithm first finds the minimum element of C, corresponding to the smallest VAF loss, and merges models in cluster j into the cluster for model i. A new column is added to C representing the new cluster. The distance to other models for the newly formed cluster is defined as the worst-case distance over the included operating points to the model. The details of the algorithm are described in Algorithm 2.

Algorithm 2. Clustering algorithm based on the relative VAF loss matrix ΔV and the clustering matrix C.

1: Let K = number of operating points.
2: Initialize $C = \Delta V$.
3: Set $\mathrm{diag}(C) = \infty$ to mark that a cluster cannot merge with itself.
4: **for** $k = K + 1$ to $2K - 1$ **do**
5:     Find indices i, j of minimum element of C
6:     Merge cluster i and j into new cluster k, assign cluster i model to cluster k.
7:     Add a column to matrix C corresponding to cluster k, let $c_{\star k} = \max(\Delta v_{\star s_1}, \cdots, \Delta v_{\star s_L})$, where s_1 to s_L are the operating points previously assigned to clusters i and j, now to cluster k.
8:     Set $c_{\star k}$ to ∞ for models that have previously been removed.
9:     Remove clusters i and j by letting $c_{\star i} = \infty$, $c_{\star j} = \infty$.
10:    Remove cluster j model by letting $c_{model(j)\star} = \infty$.
11: **end for**

13.5   Identification Results

The result of the clustering procedure can be illustrated in a dendrogram, see Figure 13.7. The dendrogram shows the merges of operating points into clusters as a tree structure, starting at the bottom where every operating point is in its own cluster. On the y-axis, the loss of VAF is shown for each merge. It can be seen that a few models may be removed at no loss at all, which means that a model identified at a different operating point proved to be marginally better than the corresponding model in some cases. This result is likely due to the choice of prediction horizon in the computation of VAF. A large number of models may be removed at a moderate cost; if a loss of 5 % is accepted only six clusters remain, if 10 % can be accepted only four clusters are needed.

It can be noted that the clusters are not randomly assembled. With only two clusters remaining, one consists of the low load operating points 1 to 6, and the other to the medium and high load operating points 7 to 30. The medium-to-high load cluster is further divided at a lower level into one cluster for medium load operating points 7 to 18 and one for high load operating points 19 to 30.

If we settle for five clusters, corresponding to a VAF loss of 7.5%, the worst-case fit between model and operating point will occur for operating point 3 which has been assigned to model 5. Figure 13.8 shows the measured output validation data at operating point 3, the 20 step prediction using model 3, and the 20 step prediction using model 5 with the adapted Wiener gain for operating point 3. It can be concluded that prediction of all outputs is very good using model 3, and only slightly worse using the adapted model 5. It can be expected that model 5 with the adapted Wiener gain is accurate enough to be used for control design.

13.6 Discussion

In general, it can be concluded that prediction based on the identified local linear models was very good for all five outputs. Standard system identification algorithms were used to fit dynamical black-box model of moderate order to four-input-five-output data sets. Considering the complexity of the task, the results are surprisingly good. Furthermore, the Wiener modeling combined with the clustering algorithm reduces the number of dynamic models needed to an acceptable level while keeping the predictive ability of the remaining models high. The results are a promising start for model based control design that uses direct emissions measurements.

Alternative approaches to the Wiener gain approximation and clustering procedures could be taken. For larger freedom in fitting data and models at different operating points, non-diagonal Wiener gain matrices could be considered. Whereas no detailed analysis of non-diagonal structure vs. cluster complexity was done, we would expect that the increase of coefficients to estimate would deteriorate statistical properties and thus require more data. Other clustering techniques might also be considered, such as the Tagaki-Sugeno model structure [11].

The choice of measured variables to be included as model outputs is not evident. From the purpose of the study, it is clear that y_{IMEP}, $y_{\mathrm{NO_x}}$, and y_{op} need to be included. Besides reduced emissions, an important goal of diesel engine control is to minimize fuel consumption. In the test engine, fuel consumption measurements were slow and inexact, so they were not a viable choice of output. Since combustion phasing α_{50} is one important factor in determining the efficiency of the engine, it was included as an output to allow for some influence over brake efficiency in the control design. To limit audible noise and avoid damage to the engine, it is necessary to limit the peak pressure derivatives over the cycle d_p, which was thus included as an output.

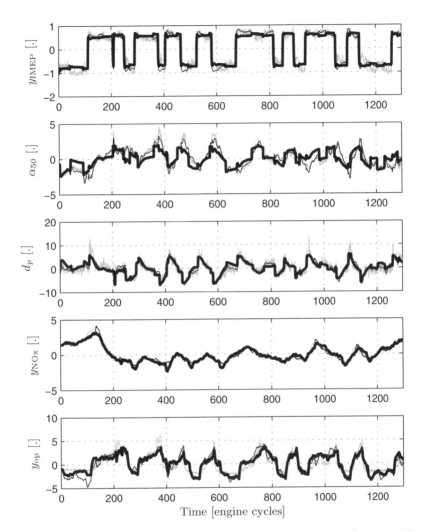

Fig. 13.8 Measured output of validation data at operating point 3 (gray solid line), 20-step ahead predicted output of local linear model identified at the same operating point (dashed line), 20-step ahead predicted output using model assigned through clustering algorithm (black solid line). The clustering algorithm was specified to set the level of accepted VAF loss at 7.5 %, yielding 5 dynamic models for the 30 operating points. Data are provided in scaled, arbitrary units.

Identification using the extended output vectors did not show any benefits by including intake manifold pressure and ignition delay as outputs to the models. These quantities are common in research on diesel engine combustion and intuitively one would expect that using information from these additional measurements would result in better prediction. It is likely that the information could be of use if some other model structure were considered.

A common issue in system identification of control-oriented models of diesel engines is to handle the MIMO and nonlinear nature of the process. As more actuators are added to provide new degrees of freedom for optimizing emissions and fuel consumption, the experiment design and control design tasks suffer from the curse of dimensionality. The same is true for the approach suggested here. Expanding the study to cover the entire speed-load range of the engine and possibly also other inputs such as a second fuel injection would substantially increase the number of operating points required for identification.

Thus, new developments in actuator technology make the control design task harder. However, we could expect that this will to a certain extent be counteracted by new developments in sensor technology. With added on-line measurements of the important outputs, such as NO_x, the requirements on off-line calibration could be substantially reduced. A fine grid of pre-calibrated static maps could be replaced by a coarse grid of dynamic models. If done right, the total experimental effort required for developing the engine controller need not increase.

Physical modeling could be advocated in order to reduce experimental efforts. The results presented here could give some insights into the necessary structure of a physical model to be used for the purpose of on-line control of emissions. The authors are not aware of a practical physical model of moderate order with the chosen set of inputs and outputs. It would be interesting to pursue work in that direction, and to see a discussion of what physical quantities that best represent the 7 states needed for prediction.

A number of approaches to model MIMO, nonlinear dynamics of the diesel engine presented in the literature use neural networks [2, 10, 8]. In contrast to these works, the procedure presented here produces models on state-space form including noise models that fits directly into the framework of common control design methodologies such as MPC and LQG. State-space models are convenient for MIMO processes since they provide a compact representation of the response in all inputs and outputs and only one order needs to be chosen for all inputs and outputs.

13.7 Conclusions

A method to model diesel engine dynamics over a range of operating points using system identification was presented. The model captures the influence of the most significant actuators on a set of important engine outputs. Notably, the model can predict measured emissions very well. For future work, it will be of interest to see what can be achieved using direct closed-loop control of measured emissions instead of indirect control through other measured variables which has been the dominant approach hitherto.

Acknowledgement

The authors would like to thank T. Henningsson for helpful suggestions on optimization of Wiener gain matrices.

References

[1] Guzzella, L., Onder, C.H.: Introduction to Modeling and Control of Internal Combustion Engine Systems. Springer, Berlin (2004)

[2] Hafner, M., Schüler, M., Nelles, O., Isermann, R.: Fast neural networks for diesel engine control design. Control Engineering Practice 8(11), 1211–1221 (2000)

[3] Heywood, J.B.: Internal Combustion Engine Fundamentals. McGraw-Hill, New York (1988)

[4] Johansson, R.: System Modeling and Identification. Prentice-Hall, Englewood Cliffs (1993)

[5] Karlsson, M., Ekholm, K., Strandh, P., Johansson, R., Tunestål, P.: LQG control for minimization of emissions in a diesel engine. In: Proc. of the IEEE Multi-Conference on Systems and Control, San Antonio, TX, USA, pp. 245–250 (2008)

[6] Manning, C.D., Raghavan, P., Schütze, H.: Introduction to Information Retrieval. Cambridge University Press, Cambridge (2008)

[7] Musculus, M.P.B.: Multiple simultaneous optical diagnostic imaging of early-injection low-temperature combustion in a heavy-duty diesel engine. SAE Technical Paper 2006-01-0079 (2006)

[8] Omran, R., Younes, R., Champoussin, J.C.: Optimal control of a variable geometry turbocharged diesel engine using neural networks: Applications on the ETC test cycle. IEEE Transactions on Control Systems Technology 17(2), 380–393 (2009)

[9] Ortner, P., del Re, L.: Predictive control of a diesel engine air path. IEEE Transactions on Control Systems Technology 15(3), 449–456 (2007)

[10] Schreiber, A., Isermann, R.: Identification methods for experimental nonlinear modeling of combustion engines. In: Proc. for the Fifth IFAC Symposium on Advances in Automotive Control, Aptos, CA, pp. 351–357 (2007)

[11] Takagi, T., Sugeno, M.: Fuzzy identification of systems and its application to modeling and control. IEEE Transactions on System Man and Cybernetics 15(1), 116–132 (1985)

[12] http://www.kistler.com

[13] http://www.mstarlabs.com

[14] http://www.vdo.com

[15] Yilmaz, H., Stefanopoulou, A.: Control of charge dilution in turbocharged diesel engines via exhaust valve timing. In: Proc. American Control Conference, pp. 761–766 (2003)

Chapter 14
Off- and On-Line Identification of Maps Applied to the Gas Path in Diesel Engines

Erik Höckerdal, Lars Eriksson, and Erik Frisk

Abstract. Maps or look-up tables are frequently used in engine control systems, and can be of dimension one or higher. Their use is often to describe stationary phenomena such as sensor characteristics or engine performance parameters like volumetric efficiency. Aging can slowly change the behavior, which can be manifested as a bias, and it can be necessary to adapt the maps. Methods for bias compensation and on-line map adaptation using extended Kalman filters are investigated and discussed. Key properties of the approach are ways of handling component aging, varying measurement quality, as well as operating point dependent model quality. Handling covariance growth on locally unobservable modes, which is an inherent property of the map adaptation problem, is also important and this is solved for the Kalman filter. The method is applicable to off-line calibration of maps where the only requirement of the data is that the entire operating region of the system is covered, i.e. no special calibration cycles are required. Two truck engine applications are evaluated, one where a 1-D air mass-flow sensor adaptation map is estimated, and one where a 2-D volumetric efficiency map is adapted, both during a European transient cycle. An evaluation on experimental data shows that the method estimates a map, describing the sensor error, on a measurement sequence not specially designed for adaptation.

14.1 Introduction

There are high demands on the control and diagnosis functionality in engine control units and to achieve the desired performance they need accurate information about the system state. State information is acquired using

Erik Höckerdal · Lars Eriksson · Erik Frisk
Vehicular Systems, Department of Electrical Engineering,
Linköping University, Sweden
e-mail: {hockerdal,larer,frisk}@isy.liu.se

D. Alberer et al. (Eds.): Identification for Automotive Systems, LNCIS 418, pp. 241–256.
springerlink.com

measurements with sensors as well as through estimation algorithms that utilize models of the system. The latter is important since measurements are not always possible because sensors are either; unavailable, impossible to install, or too expensive for the intended application.

In control and diagnosis systems, models play an important role. These models can be represented in many ways either as equations or, as is very common, by utilizing maps (or look-up tables) that represent a function [1, 10]. Examples of maps can be found in sensor calibration data (1-D) or the volumetric efficiency of the engine (often 2-D). In particular, maps are frequently used to describe relations where physical models are unavailable or too complex for on-line implementation, e.g. sensor and actuator characteristics, cooler efficiency, injector characteristics, and aftertreatment systems.

A common situation is that the dynamics is well captured by the model but there are stationary errors (bias), which will be referred to as a biased model. This has been observed in for example truck engines [2]. Another important issue is that the system ages and there is need for adjusting the models accordingly to capture and account for such effects. There is thus a need for calibrating maps off-line at development time, and adapt the maps on-line while the system is running to capture aging.

For the off-line case it is worth noting that it is straight forward to include the map parameters in the total parameter vector θ and apply standard identification methods. Routines for on-line map adaptation have been considered in [10, 12, 4] and the approach discussed below builds upon the latter. In particular it includes a systematic way for handling aging and other slowly varying uncertainties. Simultaneous bias compensation and on-line map adaption is a key property which is of industrial value since the method integrates well with existing map-based solutions.

14.2   Method Outline

The sections to come presents a systematic approach for designing an observer that adapt maps on-line at the same time as it reduces stationary errors in a model. The starting point is a default model, in discrete time state space form,

$$x_{t+1} = f_{\mathrm{def}}(x_t, u_t)$$
$$y_t = h(x_t), \tag{14.1}$$

with states $x \in \mathbb{R}^{n_x}$, inputs $u \in \mathbb{R}^{n_u}$, and outputs $y \in \mathbb{R}^{n_y}$, that suffers from stationary errors, see Figure 14.1 for an example. These model errors can exhibit both fast and slow dynamics, arising from for example operating point dependent bias and aging respectively.

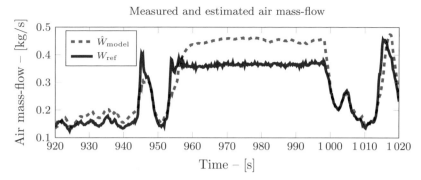

Fig. 14.1 Typical example of model output from a biased model [2], where W_{ref} is the air mass-flow measured by a reference sensor. As often is the case the model captures the dynamics well but suffers from operating point dependent stationary errors.

The objective is to handle these model errors in a systematic way by: using an on-line recursive algorithm, identifying model errors, and adapting the model. A direct way to achieve this is by introducing a parameterized function,

$$q_{\mathrm{fcn}}(x_t, u_t, \theta_t), \tag{14.2}$$

in (14.1) and augmenting the state vector with the parameter vector θ

$$\theta_{t+1} = \theta_t \tag{14.3a}$$
$$x_{t+1} = f(x_t, u_t, q_{\mathrm{fcn}}(x_t, u_t, \theta_t)) \tag{14.3b}$$
$$y_t = h(x_t). \tag{14.3c}$$

The idea with a construction like this is to let the parametrization (14.2) capture the operating point dependence, and use the parameters, θ, introduced as new states, to track the aging. With (14.3) as basis an observer that estimates the augmented state vector, (x, θ), can be designed. Note that it is possible to incorporate prior information into the map identification process, such as limitations, smoothness, trends, etc., by altering the right hand side of (14.3a).

The following example of a parameterized function q_{fcn}, from an automotive application [2], has served as motivation for the work and will later be used in the evaluation.

Example 1. In a heavy duty diesel engine with exhaust gas recirculation (EGR) and variable geometry turbocharger (VGT), the air mass-flow through the compressor is vital information for safe and clean engine control. Therefore the engine is typically equipped with a mass-flow sensor. However the sensor signal is subjected to an operating point dependent error, due to sensor installation and local flow fields, and this measurement error has to

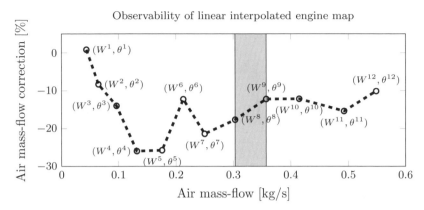

Fig. 14.2 Air mass-flow sensor error map with the grid points denoted with the pair (W^i, θ^i) corresponding to a correction factor θ^i at a mass-flow of W^i.

be compensated for when it is used in the engine control unit (ECU). The relation between the true air mass-flow W and measured air mass-flow W_{meas} can be stated as

$$W_{\text{meas}} = (1 + q_{\text{fcn}}(W, \theta)) \, W$$

where the function q_{fcn} is presented graphically in Figure 14.2. W^i are grid points, and θ^i are the corresponding correction parameters. In this case the sensor model is represented by a 1-D linear interpolation map, where W lies between grid points $W^i \leq W \leq W^{i+1}$ and the interpolation is defined by

$$q_{\text{fcn}}(W, \theta) = \theta^i + \frac{(\theta^{i+1} - \theta^i)}{W^{i+1} - W^i} \cdot (W - W^i). \tag{14.4}$$

Due to aging it is necessary to adapt this map, see [2] for a longer discussion.

$\diamond$

Note that with the construction (14.3), using parameters to track aging, the observer stores information about the operating point dependent errors, i.e. the parameters act θ as a memory.

The development of a model like (14.3) for identification and its usage for estimation entails that some new issues have to be addressed. The main concern is how to update the parameters, θ, in a controlled manner, which is the topic of Section 14.3, with respect to observability and observer tuning. Here the parameterization of q_{fcn} is given and the interested reader is referred to for example [7] for a discussion on how to find a structure and suitable regressors.

The system (14.3) is in standard state space form which means that any suitable observer design can be applied. The choice here is to use a stochastic filter which entails introduction of noise in (14.3) to describe model and measurement uncertainties. An approach for estimating states while at the same time handling unknown parameters is to apply a joint parameter and state estimating extended Kalman filter (EKF) [6]. An advantage of stochastic filters compared to deterministic observers is that, not only the state estimate, but also an estimate of the estimation error statistics is computed. The estimation error statistics is used in the computation of the filter feedback gain, which gives the stochastic filters natural tuning parameters that allow filter tailoring to handle system aging, unknown state initialization, time dependent model and measurement quality, outlier rejection etc., see Section 14.3.2 for the discussion. For these reasons the joint parameter and state estimating EKF is used.

14.3 Observability

In estimation, observability or detectability of the system at hand is central in order to ensure correct and consistent estimates. This section is devoted to the observability of (14.3), where q represents linear interpolation. The observability discussion is conducted on the continuous time system,

$$
\begin{aligned}
\dot{\theta} &= 0 \\
\dot{x} &= f(x, u, q(x, \theta)) \\
y &= h(x),
\end{aligned}
\tag{14.5}
$$

corresponding to (14.3). The results are valid also for the discrete time system as long as the sample time is chosen small enough [5].

Intuitively, a system containing linear, or higher order, interpolation with the interpolation grid points as parameter states is not locally observable. This can be seen by considering the air mass-flow adaptation map in Example 1. Considering the total system, assuming more than one interpolation interval, this can be understood by the intrinsic nature of the linear interpolation (14.4). Since not all parameters are involved in the interpolation computation in each operating point, some parameters are always locally unobservable. However, even if only one interpolation interval is studied and the non-interacting parameters are disregarded, e.g. consider only the shaded region of Figure 14.2, the ordinary rank condition, stating sufficient conditions for observability, is insufficient for assessing local observability of the reduced system. To illustrate consider (14.5) in a single interval,

$$
\begin{aligned}
\dot{\theta}^i &= 0 \\
\dot{\theta}^{i+1} &= 0 \\
\dot{x} &= f(x, u, q(x_j, \theta^i, \theta^{i+1})) \\
y &= h(x), \quad y_k = x_j.
\end{aligned}
\tag{14.6}
$$

For Example 1 this is realized by studying a stationary operating point where the interpolation variable, x_j, is in $I_i =]x_j^i, x_j^{i+1}[$, corresponding to the shaded region enclosed by W^8 and W^9 in Figure 14.2. Here the corresponding parameters θ^8 and θ^9 are unobservable according to the ordinary rank condition, i.e., the Jacobian of

$$\begin{pmatrix} h(x) \\ L_f h(x) \\ \vdots \end{pmatrix},$$

where L_f denotes the Lie derivative along the vector field f, with respect to x, θ^i, θ^{i+1} will be rank deficient in all operating points. Hence direct use of the rank condition for assessing, both local and global, observability for this system is not possible. To ensure global observability of (14.5), a trajectory has to be considered and analytical results for establishing observability are derived in [4].

14.3.1   *Unobservable Modes and Covariance Growth*

According to the previous paragraph, at any given time there are generally some parameters θ^i that are locally unobservable in (14.3). A property of EKF observers based on systems with locally unobservable states is that the estimation error covariance matrix grows linearly for the unobservable states if system noise for these modes is present, see [4]. This effect is illustrated in Figure 14.3, where the variance of three parameter states, θ^5, θ^8, and θ^{10} from Figure 14.2 are plotted versus time. Parameter θ^5 corresponds to a parameter that is not observable at all for the studied trajectory, while the parameter θ^8 is observable during the first half of the trajectory and unobservable for the second half. For the parameter θ^{10} the case is reversed. The effects of this covariance growth is two-sided: 1) it offers a way to achieve fast update of old parameters while protecting often updated parameters from spurious measurements, 2) it may cause numerical problems affecting the filter stability when considering the life-time of the system, which has to be handled.

In engine map adaptation, experience indicates that adaptation algorithms, not using the EKF and joint parameter and state estimation, have problems concerning system aging and occasional spurious measurements. For example an engine, whose trajectories do not span the entire parameter space during normal operation and only occasionally enters some areas, may suffer from undesired system behavior caused by old parameters corresponding to seldom visited operating points. In these cases, a linearly growing uncertainty for seldom updated parameters enables a fast parameter update rate of old parameters without risking large errors in the state estimates. This can in some sense be thought of as a dynamic forgetting factor similar to recursive least square (RLS) techniques [9] and is a highly desirable property in

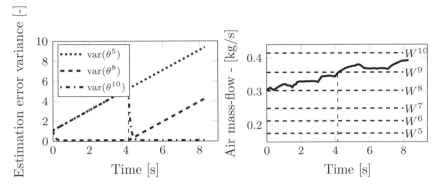

Fig. 14.3 The figure shows the development of the estimation error variance for three parameter states. One that is unobservable during the entire trajectory – θ^5, one that is observable for the first half of the trajectory – θ^8, and one that is observable for the second half of the trajectory – θ^{10}.

engine adaptation algorithms not only to handle aging parameters but also to protect updated parameters.

After discussing the positive effects induced by the locally unobservable parameter states it remains how to ensure stable observer operation even though some parameter states never may become observable, causing the corresponding covariance matrix elements to grow without bound. A direct and intuitive way of handling this linear growth of estimation error covariance of the locally unobservable parameter states is to introduce an upper limit for the corresponding estimation error covariance matrix elements. A possible such upper limit is the initial error covariance matrix, P_0, and with this choice the introduction of yet another tuning parameter is avoided [4].

14.3.2 Method for Bias Compensation

In the sense of map adaptation as a way of reducing bias, it is reasonable to compare the proposed map adaption method to other common ways of reducing operating point dependent estimation errors. One such method is presented in [3], where the state vector is augmented with a bias state that is used directly to describe the model error,

$$\begin{aligned} x_{t+1} &= f(x_t, u_t, q_t) \\ q_{t+1} &= q_t \\ y_t &= h(x_t). \end{aligned} \tag{14.7}$$

A main difference is that an observer based on (14.3) facilitates separate tracking of fast changes in operating point and slower changes due to aging. For an observer based on (14.7) this is not possible and it is necessary for the bias state to change approximately as fast as the system dynamics, otherwise it will not be able to track a change in system operating point. However, a rapidly changing bias state will also capture high frequency disturbances, and will thereby not be able to withstand spurious measurements to the same extent as the slower parameter states θ in (14.3).

14.4 Method Evaluation

To evaluate the method two studies are performed, a simulation study and a study utilizing experimental data. The two evaluations address the problem of estimating and adapting the air mass-flow into an engine from two different perspectives that are mutually exclusive. That is either the mass-flow measurement, or the volumetric efficiency is considered to be most accurate. A third alternative is to install an accurate reference sensor, which is only possible in a test cell or development vehicle and not an option for a consumer product.

The simulation study demonstrates the capability to adapt a 2-D map for the volumetric efficiency while the experimental evaluation addresses the air mass-flow sensor adaptation problem presented in Example 1. The aim of the simulations are to show the convergence of the approach and the effect of a trajectory that only partly spans the map grid. As a part of the convergence analysis an investigation of the effect of incorrect noise models when using the EKF as a parameter estimator [8] is included.

The experimental part shows the result of the method applied to experimental data with the aim to analyze robustness and performance in presence of model errors. Since the approach supplies a solution to the problem of biased default models solved in [3], the performance of that approach is presented as a comparison.

In both studies the non-linear model of a heavy duty diesel engine, developed in [11], is used together with measurements from an engine in an engine test cell. An overview of the model schematics is presented in Figure 14.4. The model has three states, intake and exhaust manifold pressures, and turbine speed which all are present in the model output together with the air mass-flow through the compressor. The data is collected during a European Transient Cycle (ETC).

In the simulation study, feedback from the intake manifold pressure and turbine speed sensors are used whilst in the experimental evaluation, feedback from all sensors except the exhaust manifold pressure sensor are used. The performance is evaluated with respect to all states and outputs, i.e. intake and exhaust manifold pressures, turbine speed, and air mass-flow through the compressor.

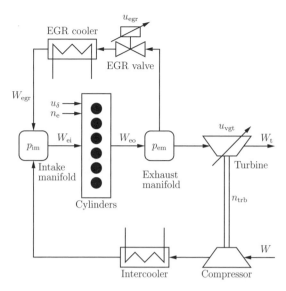

Fig. 14.4 Schematic overview of the diesel engine model with EGR and VGT [11], showing model states (p_{im}, p_{em}, and n_{trb}), inputs (u_δ, u_{egr}, u_{vgt}, and n_e), and massflows between the different components (W, W_{ei}, W_{eo}, W_{egr}, and W_t). Rectangles with rounded corners denote control volumes.

14.4.1 Study 1: Simulation

The simulation study serves three purposes; illustrating convergence of the approach, show that the method applies also to higher map dimensions, and highlighting the effect of trajectories that does not span the entire map grid.

14.4.1.1   Observer for the Simulation Evaluation

In the simulation evaluation the observer is constructed using the model developed in [11] and replacing the volumetric efficiency sub model with a six-by-six parameterized cubical spline interpolation map, $q(n_{eng}, p_{im}, \theta)$, presented in Figure 14.5, which gives a model in the form (14.3). An alternative to using a spline for the interpolation would be to use bilinear interpolation. Here the spline is chosen to favor the smoothness of the volumetric efficiency.

14.4.1.2   Simulation Evaluation

Figure 14.5 shows the true and adapted volumetric efficiency maps together with the ETC trajectory. The transparent upper surface corresponds to the true volumetric efficiency to be estimated and the lower, with a rough grid

Estimated volumetric efficiency

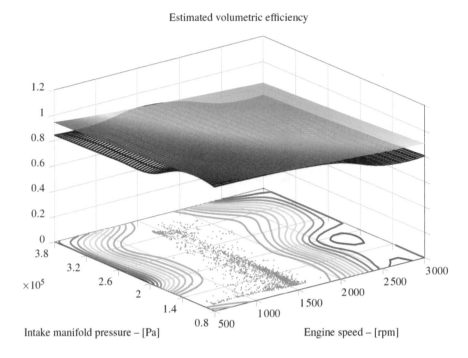

Fig. 14.5 True and estimated volumetric efficiency together with level curves of the difference between the two. The level curves are also imposed by the ETC trajectory, used as adaptation cycle, represented by the gray dots.

of 6×6, the estimated map at the end of the ETC, which was initialized 10 percent units below the true map. At the bottom of Figure 14.5 level curves of the absolute estimation error are plotted together with the ETC trajectory to better illustrate the correlation between well adapted map areas and how the grid is spanned by the trajectory. As expected, map areas where the trajectory spends much time are well adapted while grid points in areas not spanned by the trajectory is unadapted and remains at their initial values.

14.4.1.3   Noise Model Sensitivity

A theoretical analysis of the EKF as a parameter estimator is presented in [8], which concludes that the parameter estimates will be biased if an incorrect noise description is used.

Inspired by this theoretical analysis and the fact that true nature of the noise seldom is completely known a simulation study is performed with the aim to analyze the influence of incorrect noise assumptions. The analysis is executed by applying the above designed observer to several sets of simulated data where the system and measurement noise variance and color had been adjusted. Neither of these simulations show of any problem of consistent convergence of the sensor error parameters.

Even though the analysis in the studied case does not indicate this to be an issue the potential problem can not be discarded in general. Even if the main purpose of the parameters is, as in the engine applications addressed here, their function as a memory there exist applications where this is an issue. In such cases it is straightforward to use the in [8] modified EKF algorithm for improved convergence properties.

14.4.2 Study 2: Experimental Data

To assess the performance and robustness of the method an evaluation using measurements from an engine in an engine test cell is conducted. Since in this case there are other significant model errors present, besides the air mass-flow sensor error, an extra bias state is introduced to improve the state estimates. The additional bias state is introduced according to (14.8) to compensate for a known model deficiency in the compressor mass-flow. The adaptation map, $q(W, \theta)$, is in this case defined by Example 1. Note that, in this case the adaptation map is situated in the measurement equation which is formally treated in [4].

14.4.2.1   Observers for the Experimental Evaluation

The observer designs for the experimental evaluation are, starting with those without an extra bias state: An EKF based on the default model developed in [11] directly,

$$x_{t+1} = f(x_t, u_t)$$
$$y_t = \begin{pmatrix} x_{p_{\mathrm{im}},t} & x_{n_{\mathrm{trb}},t} & h_W(x_t) \end{pmatrix}^T,$$

referred to as Def. An EKF with an extra bias state introduced in the measurement equation to describe the air mass-flow measurement error from the method developed in [3],

$$x_{t+1} = f(x_t, u_t)$$
$$q_{t+1} = q_t$$
$$y_t = \begin{pmatrix} x_{p_{\mathrm{im}},t} & x_{n_{\mathrm{trb}},t} & h_W(x_t) + q_t \end{pmatrix}^T,$$

referred to as Aug. A joint state and parameter estimating EKF based on the default model and a parameterized bias,

Table 14.1 Mean estimation error using experimental data for Def., Aug. and Map.

Meas.	With $\Delta_{W_{cmp,t}}$			Without $\Delta_{W_{cmp,t}}$		
	Def.	Aug.	Map.	Def.	Aug.	Map.
p_{im} [Pa]	4208	169	-210	13875	13176	13019
p_{em} [Pa]	-10987	-13746	-14420	-2339	-3454	-3450
n_{trb} [rpm]	69	16	29	-57	-19	5
W [kg/s]	0.008	-0.020	-0.020	0.031	0.035	0.035

$$x_{t+1} = f(x_t, u_t)$$
$$\theta_{t+1} = \theta_t$$

$$y_t = \left(x_{p_{im},t} \quad x_{n_{trb},t} \quad (1 + q_{fcn}(h_W(x_t), \theta_t))h_W(x_t)\right)^T,$$

referred to as Map. where the parameterization of q_{fcn} is presented in Figure 14.2. The primary difference between the three observers are presented schematically in Figure 14.6, where the feedback part of each observer is highlighted.

The additional bias state, complementing those for handling the measurement error, is introduced according to the ideas in [3]. The purpose is to reduce the estimation errors due to a known error in the compressor model causing incorrect prediction of the compressor mass-flow, i.e. for Def.

$$x_{t+1} = f(x_t, \Delta_{W,t}, u_t)$$
$$\Delta_{W,t+1} = \Delta_{W,t} \qquad\qquad (14.8)$$
$$y_t = \left(x_{p_{im},t} \quad x_{n_{trb},t} \quad h_W(x_t, \Delta_{W,t})\right)^T.$$

14.4.2.2 Experimental Evaluation

With the introduction of an extra state, compensating for the compressor mass-flow, the estimate of the intake manifold pressure becomes significantly better at the expense of the exhaust manifold pressure estimate, while the estimates of turbine speed, and air mass-flow is almost unaffected, see Figure 14.7 and Table 14.1. While Table 14.1 shows the mean estimation error, Figure 14.7 shows the estimation error probability density functions (PDF).

In Section 14.3.2 the tuning of Aug. and Map. were discussed, especially the different philosophies of the bias describing states, i.e. the relatively fast bias state in Aug. and the slow map states in Map. In Figures 14.7 and 14.8, and Table 14.1 the similarity in estimation performance between Aug. and Map. is striking, which is an expected result. The mean estimation errors for p_{im} and n_{trb} are reduced while the mean estimation errors for p_{em} and W_{cmp} are slightly increased. That is, in absence of outliers Aug. and Map. are comparable with respect to output estimation performance. The benefit

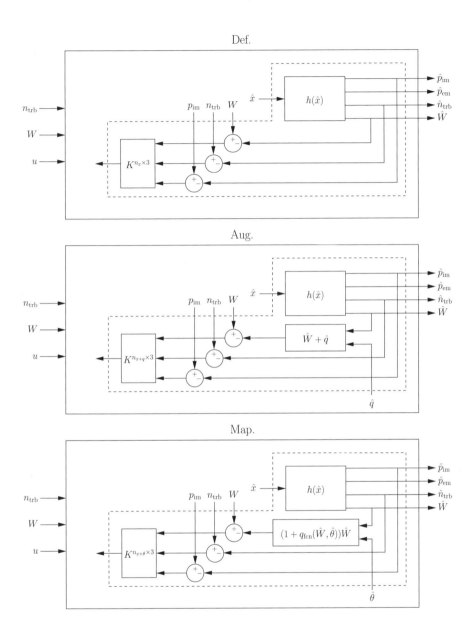

Fig. 14.6 Schematics that highlight the feedback part of the three observers, Def., Aug., and Map, studied in the experimental evaluation. The main difference between the three configurations are shown in the block on the $\hat{W}$ feedback signal, below the $h(\hat{x})$ block. To the left the inputs and feedback signals are shown, and to the right the four evaluation outputs are shown. Note that the dimension of the output signal of the Kalman gain block have different dimensions for the different observers.

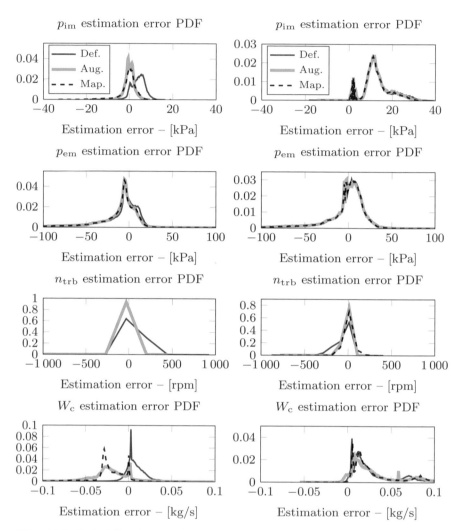

Fig. 14.7 Probability density functions of the estimation errors showing bias and estimation error statistics.

with Map. is that it produces an adaptation map, and enables the possibility to track aging and protection against corrupted measurements.

From Figure 14.8 it is seen that, even though there are unknown model errors present, besides the compressor mass-flow, the method manages to estimate a map that describes the difference between modeled and measured air mass-flow through the compressor well. Notice that there are large number of samples at air mass-flows slightly below 0.2 kg/s and that only few of these are from stationary operation.

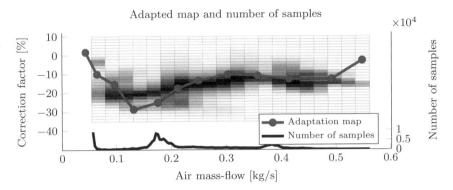

Fig. 14.8 By Aug. and Map estimated mass-flow corrections showing the agreement in bias estimation of the two methods. The solid line with dots representing the final adaptation map of Map. and a color map of the distribution of the air mass-flow correction estimated by Aug.

Finally, all this shows that an adaptation map can be estimated even though the data used is from the highly transient ETC, not specially designed for air mass-flow sensor adaptation.

14.5 Conclusions

In modern control and diagnosis systems, model based estimation has achieved increased attention and especially interesting are techniques to handle the inevitable model errors. A method for on-line map identification, based on standard observer methodology, that handles the specific issues of aging parameters and occasional corrupted measurements is discussed. A common technique in engine applications is adaptation maps that describe the model errors. A method for storing bias information from different operating points, based on the EKF as parameter estimator, is investigated. This method achieves simultaneous estimation of original model states and parameters, and applies to adaptation of engine maps.

An inherent property of the adaptation of maps is that the system is locally unobservable. Stochastic observers together with a parameterized bias that has locally unobservable states is an asset that handles seldom updated parameters and gives robustness against occasional spurious measurements in ordinary map adaptation algorithms. The linear growth of estimation error covariance, that comes as a result of local unobservability of the parameters, also forms a potential numerical problem for the observer and a way to limit this growth without introducing extra tuning parameters is provided.

The method is evaluated in a simulation study, where it is demonstrated that it estimates both the engine states and a parameterized 2-D adaptation map for the volumetric efficiency. An experimental evaluation shows that the method achieves the same estimation quality with respect to mean and maximum absolute error, as the method developed in [3], but adds value in that an engine adaptation map can be simultaneously estimated. Furthermore, the successful demonstration on experimental data, that inevitably introduces unknown model errors and noise, shows the robustness of the method. The method achieves simultaneous state estimation and map adaption, without using test cycles especially designed for adaptation, and it is therefore suitable for both on- and off-line calibration of maps.

References

[1] Guzzella, L., Amstutz, A.: Control of diesel engines. IEEE Control Systems Magazine 18(5), 53–71 (1998)
[2] Höckerdal, E., Eriksson, L., Frisk, E.: Air mass-flow measurement and estimation in diesel engines equipped with EGR and VGT. SAE Technical Paper 2008-01-0992
[3] Höckerdal, E., Frisk, E., Eriksson, L.: Observer design and model augmentation for bias compensation with a truck engine application. Control Engineering Practice 17(3), 408–417 (2009)
[4] Höckerdal, E., Frisk, E., Eriksson, L.: EKF-based adaptation of look-up tables with an air mass-flow sensor application. Control Engineering Practice (in press, 2011)
[5] Kalman, R., Ho, B.L., Narendra, K.: Controllability of linear dynamical systems. Contributions to Differential Equations 1 (1963)
[6] Kopp, R.E., Orford, R.J.: Linear regression applied to system identification for adaptive control systems. AIAA 1(10), 2300–2306 (1963)
[7] Lind, I., Ljung, L.: Regressor and structure selection in NARX models using a structured ANOVA approach. Automatica 44(2), 383–395 (2008)
[8] Ljung, L.: Asymptotic behavior of the extended kalman filter as a parameter estimator for linear systems. IEEE Transactions on Automatic Control 24, 36–50 (1979)
[9] Ljung, L.: System Identification - Theory for the user, 2nd edn. Prentice-Hall, Inc., Upper Saddle River (1999)
[10] Peyton Jones, J.C., Muske, K.R.: Identification and adaptation of linear look-up table parameters using an efficient recursive least-squares technique. ISA Transactions 48(4), 476–483 (2009)
[11] Wahlström, J., Eriksson, L.: Modeling VGT EGR diesel engines with optimization of model parameters for capturing nonlinear system dynamics. Proceedings of the Institution of Mechanical Engineers, Part D, Journal of Automobile Engineering (accepted for publication)
[12] Wu, G.: A table update method for adaptive knock control. SAE Technical Paper 2006-01-0607

Chapter 15
Identification Techniques for Control Oriented Models of Internal Combustion Engines

Daniel Pachner, David Germann, and Greg Stewart

Abstract. The use of mathematical models is widespread in the design process of modern internal combustion engines. A driving factor for this trend is the rising complexity of engine systems due to tightening emission limits and increasing demands on fuel efficiency. This chapter focuses on control oriented models for turbocharged diesel engines which were specifically developed for use in an advanced model based control design process. A number of challenges had to be overcome to guarantee stability of these models both in simulation and during identification. The solutions for these issues are presented in this chapter together with an innovative identification approach which performs the calibration of the overall engine model on top of the previously executed identification of individual engine components. Example modeling results are included for a six cylinder commercial diesel engine with EGR and dual stage turbocharging.

15.1 Introduction

Internal combustion engine systems for automotive applications are becoming increasingly complex. One of the main drivers for this trend is the ever-tightening emissions regulations. In order to meet these stringent emission limits the addition of various components to the engine design are required. Examples include turbocharging [4, 9, 17], catalytic converters [23, 15], particulate filters [11] and various topologies of EGR systems [22]. At the same time high demands are made towards cost of ownership, fuel efficiency, and

Daniel Pachner
Honeywell Prague Laboratory, V Parku 2326/18,
148 00 Prague 4, Czech Republic

David Germann · Greg Stewart
Honeywell Automation and Control Solutions, 500 Brooksbank Avenue,
North Vancouver, BC, Canada, V7J 3S4

D. Alberer et al. (Eds.): Identification for Automotive Systems, LNCIS 418, pp. 257–282.
springerlink.com © Springer-Verlag London Limited 2012

drivability of automotive and heavy duty applications. As a result the state of the art engine systems have become complex plants comprised of numerous highly nonlinear subsystems originating from various fields of technology that interact with one another in challenging ways.

Addressing all these issues in a systematic fashion during the overall design process of internal combustion engines is beyond the scope of ad hoc techniques. As consequence mathematical models of the engine have been increasingly employed over recent years in order to assist the design process.

The application area of such models is quite diverse. At one end of the modeling spectrum "proactive" models, often used before an engine or component is built, are employed to predict the behavior of potential engine configurations in order to avoid or reduce costly build-integrate-test iterations with engine components. Another important use of modeling techniques is in their key role in an efficient control design process. Accompanying the rising complexity of engine systems systematic techniques are required in making the design of high performance controllers industrially practical. A control oriented model provides crucial information in this process in order to properly characterize and quantify the multivariable dynamic interactions.

This chapter will concentrate on issues and techniques that were developed for the task of creating control oriented models of turbocharged diesel engines. In particular, these models were developed primarily for use in a proprietary control design process which allows for a systematic and efficient synthesis of explicit model predictive controllers for turbocharged diesel engines [18]. A brief overview of this controller design process is given in Section 15.2 before discussing the nature of the control oriented models required for this process in Section 15.3. In order to guarantee robust behavior of the models during simulation and identification a number of challenges had to be overcome. These issues are described in Section 15.4. The main aspect of the paper is focused on the identification of these models and is covered in Section 15.5 which treats steady state identification both on a component level as well as for the overall model. The importance for this distinction and the justification to use both techniques is discussed in Section 15.6. This chapter concludes with the obtained predictions accuracies in Section 15.7.

15.2   Background

This chapter is focused on identification techniques applied to control oriented models (COM) of turbocharged diesel engines developed for the use in a proprietary controller design process. In this section the overall control design process is briefly summarized in order to provide the context for the requirements to be satisfied by the engine models. Additional details on the applied control technology may be found in [18].

Engine systems are becoming increasingly complex in order to meet stringent emission limits. This circumstance imposes demanding requirements on the control functionalities as various multivariable interactions and nonlinearities have to to be addressed adequately while also taking into account engine variability due to production tolerance and aging. The application of traditional controller design techniques can result in a very resource intensive procedure and in some cases has even been reported to become the bottleneck of the entire engine design process. Furthermore, achieving acceptable controller performance requires a high skill level of the developer designing and tuning these conventional control functionalities.

On this background a design process and set of software tools have been developed which address these issues in the control design in a systematic and partly automated way making the process more efficient and less dependent on the skill level of the developer. A simplified work flow of this process is depicted in Figure 15.1. In general the depicted steps are followed in the order illustrated with iterations back to earlier steps being common in practice - for example to perform controller tuning iterations.

The controller design process begins with the definition of the engine layout in form of a plumbing diagram. An example of such a layout is depicted in Figure 15.2 which illustrates the engine referred to throughout this chapter. Based on this information the user configures the control oriented model of the engine. As can be seen in Figure 15.2 turbocharged diesel engine are comprised

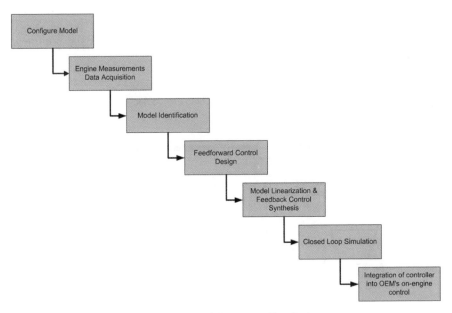

Fig. 15.1 Simplified work flow of H-ACT controller design process.

of various components including turbines, compressors, valves, coolers, the engine etc. each of which have their own mathematical submodels.

In order to identify the parameters of the nonlinear engine model a set of measurement data is required. This data will include compressor and turbine maps supplied by the turbocharger vendor and on-engine measurement data. The engine measurements can either come from preexisting data out of the standard engine mapping process or can be generated specifically for the model identification by carrying out a tailored design of experiment which typically takes two to three days to collect. Using this engine data the model identification process will fit the free parameters of the model components such that the nonlinear control oriented model quantitatively matches the behavior of the true engine. This process will be described in detail in Section 15.5.

In the next step of the design process in Figure 15.1 the nonlinear model will be used to derive the feedforward component of the engine controller. The feedforward control is a function that maps the engine operating points (characterized by engine speed, fuel injection quantity, and possibly other parameters) to a set of actuator positions such that the engine outputs will approximately match their desired setpoints at steady-state. The feedback control is based on model predictive control (MPC) techniques to generate an optimal controller. Due to the computational restrictions of current ECUs and the fast time constants of the engine the feedback controller is designed with a set of linear models which were generated from the nonlinear COM at various operating points in the speed/fuel range of the engine.

The control design process described here is inherently model based and results in a true multivariable controller which addresses the nonlinear

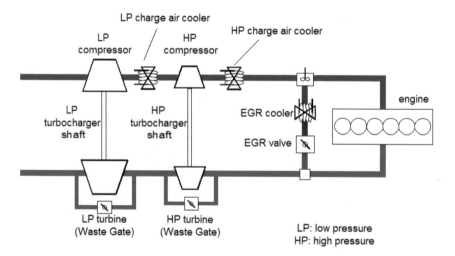

Fig. 15.2 Example diagram of engine layout used throughout this chapter.

interactions present in the internal combustion engine. As in standard MPC (see for example [10]), the feedback controller tuning allows the user to adjust weights on tracked setpoints to enforce tighter tracking of certain controlled variables while relaxing the tracking of other less important variables. Additional weights are also provided for each actuator which allows the designer to define the preference of certain actuators to carry out most of the control activity while others (e.g. throttles) are used only if absolutely necessary to achieve acceptable setpoint tracking. A technique similar to [1] is used to ensure the model predictive control delivers the closed-loop performance desired by the user.

The described process is flexible and is independent of the engine layout. At the same time different combinations of tracked variables can be selected while enforcing constraints on actuators as well as system outputs such as NOx emissions or turbocharger speed. The software solution presented in this section was developed with the goal of integration into industrial practice. Careful consideration was given to existing development processes for industrial engine control and how the control function interfaces in the hierarchical software structure.

While the models presented in this chapter can be very useful for an initial and efficient analysis of engine systems this section has endeavored to make it clear that their main purpose is to be used in the controller design process. The requirements to be satisfied by such control oriented models will be discussed in the next section.

15.3   Model Type and Requirements

In the control design process described in the previous section the control oriented model is employed for two purposes, the derivation of the feedforward controller and the synthesis of the feedback controller. The quality of the feedforward control is directly connected to the accuracy of the model, but is not required to be perfect since the feedforward control is combined with a multivariable feedback control to reduce the system sensitivity to model uncertainty and eliminate steady-state errors through integrating feedback action.

From this perspective the accuracy requirements of control oriented models can be described as medium while the model complexity is normally desired to be low for the sake of efficient development times and straightforward extraction of linear models. In addition, the prediction capabilities of a COM outside the range of the data used for identification should be fairly good. This can reduce the required amount of measurement data and thus reduce the time and cost of a test cell experiment campaign.

Besides providing an approximation to the dynamic response within the bandwidth of the closed-loop control system an important contribution of the COM is to provide an accurate prediction of the derivatives of the tracked variables with respect to the engine actuators, or in other words the system gains. Inaccurate prediction of the gains can have a detrimental effect on the control performance and even lead to system instability for example when the sign of the gain is modeled incorrectly causing the feedback controller to take action in the wrong direction (see for example the linear analysis in [16]). A correct prediction of the gain sign is a basic prerequisite of any modeling activity but it is not always straightforward due to the significant nonlinearities present in turbocharged engine systems [17, 9, 13].

In Figure 15.3 for example the influence of the variable geometry turbine (VGT) actuator is shown towards both the fresh air flow and the boost pressure for two different engines at fixed engine speed and injection quantity while all other actuators remain constant. For the engine on the right in Figure 15.3 closing the VGT increases the boost pressure initially but after a certain point the effect reverses and closing the VGT further causes the pressure to drop. A similar behavior is shown for the engine on the left of Figure 15.3 with respect to the air flow. In other words, the sign of the gain from an actuator towards a tracked variable can reverse and this behavior has to be adequately captured in the COM in order to guarantee acceptable controller performance.

There are various types of models used in practice. Deriving a categorization system in order to classify these model types is not always straightforward. A categorization which is sometimes applied groups models into white, gray and black box models [21]. This is a rather coarse classification and is used here merely for illustration purposes to distinguish the models described in this chapter from other model types. A simplified comparison of the three model categories regarding their attributes is shown in Table 15.1.

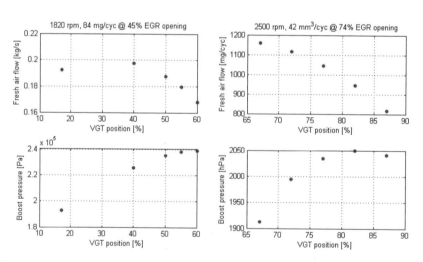

Fig. 15.3 *Left*: Diesel engine in 8 liter displacement range with high pressure EGR and VGT turbocharger. *Right*: 2.2 liter diesel engine with high pressure EGR and VGT turbocharger. For both engines VGT position is given in %-closed.

Table 15.1 Model types and attributes

	White Box	Gray Box	Black Box
Model detail	high	medium	low
Computational time	high	medium	low
Identification data	low	medium	high
Extrapolation capabilities	high	medium	low

White box models usually have a very extensive model detail allowing them to have a very limited number or even no tuning parameters. As a consequence the amount of data required for the identification of these parameters is rather small or not at all required. A disadvantage of this model category is the development time required to derive the system detail as well as the high computational burden it entails to simulate these complex models. Many types of computational fluid dynamic (CFD) models can be assigned to the category of white box models and to a certain extent GT-Power models which are widespread in the automotive sector [19]. These use large amounts of a priori information on the engine geometry and are largely based on physical principles requiring little or no data for identification.

Black box models on the other hand are usually of very simple structure and efficient in simulation but require a large amount of identification data. Essentially these models are based on experimental data alone and do not incorporate a priori known physical principles. Experimental data for this model category is usually obtained by sweeping the relevant actuators and recording the system outputs of interest. A low order dynamic model structure is then selected which adequately can reproduce the system behavior [8]. Typical examples of black box systems are Hammerstein-Wiener models [3], nonlinear polynomial structures such as NARMAX models [2] and neural network approaches [14]. The main issue with black box models is that for nonlinear systems they can only be guaranteed to be accurate in the vicinity of the region where the data was acquired. Outside of this region the extrapolation capability is rather poor. In addition minor changes to the system configuration can require to rerun the entire experiments which is a time and cost intensive process.

Gray box models can be viewed as a compromise between white and black box models. They are usually based on physical knowledge in the form of first principles combined with a small number of tuning parameters which only require a limited amount of experimental data for identification [20]. At the same time the extrapolation capabilities of gray box models are reasonably good due to the incorporation of physical principles. Because of the moderate system complexity the development effort for these models as well as their computational requirements are significantly reduced compared to white box models.

Recalling the requirements of control oriented models it is evident that out of the three categories described above gray box models are best suited for the use in the control design process. They combine reasonable computational demands with sufficient accuracy and extrapolation capabilities while requiring a limited amount of identification data. The models presented in this chapter can be considered as belonging to the category of gray box models.

As depicted in Figure 15.2 models of turbocharged engine systems are comprised of various components each of them represented by a specific mathematical submodel. To a large part these submodels can be represented by static nonlinear functions like the compressor model shown in Figure 15.4. However, a few dynamic processes also have to be considered like the inertia of the turbocharger and the mass exchange in the engine manifolds in order to capture the dynamic response of the engine.

An important aspect of the control oriented engine models presented in this chapter is that they neglect the discrete cycles of the engine assuming the main processes are spread out over an entire engine cycle. These types of engine models are often referred to as mean value models (MVM) [6]. Very fast phenomena like the combustion are treated as a static process. In addition, no spatial variation in the variables is considered allowing these so-called *lumped parameter* models [5] to be represented by ordinary differential equations (ODE).

15.4   Internal Combustion Engine Modeling Challenges

Once the gray box model structure is selected and configured based on the engine architecture as explained in Section 15.2, its parameters must be estimated from the data. Typically, the model parameters will be fit via numerical minimization of the error norm, often the weighted sum of squared errors however very often *a priori* information will be available on the parameter values. In the course of solving this optimization problem a number of difficulties were encountered and two of the main issues will be described in this section.

The model structure can be represented by a block diagram similar to Figure 15.2. To model the dynamic system equivalent to a system of first order ordinary differential state space equations, block of two types suffice: nonlinear static functions and integrators. Generally the nonlinear static functions correspond to physical *components* of the engine(compressor, heat exchanger, etc.) and host the majority of the unknown parameters of the model. The few remaining unknown parameters are associated with the integrators and typically correspond to the dynamic response of the engine model (turbocharger inertia, manifold volumes, etc.). Though the various realizations of engine models can be different, for the following it is important to keep in mind that any model in this work can always be decomposed into integrators and nonlinearities.

15.4.1 Constrained Domains

A problem comes forward immediately when dealing with such model structures: the input values to the components may be restricted to certain domains.

The first principles equations implemented by the components model the physical reality under certain conditions only. For example, the mean value model equations typically have their correct physical meaning only if the mass flow has defined direction, the turbocharger speed is positive etc. Quite often the component nonlinearities are discontinuous on the input domain boundaries, may have infinite absolute values and vertical asymptotes there. Both represent convergence problems for convex optimization techniques.

Also polynomials and rational polynomials are often being used for the empirical modeling [5, 7] of the engine components and are usually valid approximations for physically realistic input values. Thus if the model accuracy for the initial parameter estimates is not sufficient to provide physically realistic input values to all such model components, the identification may fail.

Though this problem is usually not encountered with linear or mildly nonlinear systems, an internal combustion engine model usually contains a number of nonlinear functions with restricted input domains. The compressor model defined in Section 15.5.1 can serve as an example here. Therefore, the problem has to be solved systematically: As a first step, one must determine feasible initial estimates of the model parameters. Only then, it is possible to evaluate valid model predictions and a valid gradient vector to update the model parameters. Not only should the initial estimate be feasible, but it should remain so during all the iterations of the model identification. Hence, the engine model identification is in fact a constrained optimization problem with complex nonlinear constraints. Two types of constraints imposed on the optimized parameters exist. The first type is a simple physical limit on a parameter. As an example, a manifold volume must clearly be non-negative. The second type is a physical limit on model signals. As an example, the direction of the turbocharger rotation is fixed and should not be reversed. If the direction reversal happens simulating the model with valid inputs and initial conditions, the parameter values are incorrect (they violate the constraint). However, it is not clear which parameters should be changed in what direction to satisfy the constraint. Therefore, the second constraint type is more complicated as it involves the complicated relationship between the parameter values and the model solution.

15.4.2 Ensuring Stability of Model

The turbocharged engine model contains several feedback paths. The principal feedback is closed by each turbocharger shaft. The turbine clearly affects its own inputs through the air path. This feedback is interesting because it may easily become positive, and hence unstable mathematically. However on the engine the turbocharger speed cannot become unbounded due to the finite total energy available in the fuel. There must always be a limit at which the positive feedback becomes negative and the turbocharger acceleration approaches zero.

Unfortunately, the fact that the real engine turbocharger speed cannot become unstable is not sufficient to prevent a computational engine model from doing so. The positive feedback may accelerate or decelerate the turbocharger speed easily especially when the model parameters are still inaccurate in the course of model identification. If this happens, it may be impossible or difficult to solve the model equations numerically. In practice, this will lead to slowed down evaluation of the model predictions and may not converge. Even a successful convergence, can be accompanied by signals that are quite far from the true physical solution. In which case they will likely violate the physical domain constraints and aggravate the constrained domains problem described in Section 15.4.1. Thus, the convergence of the optimization algorithm may be lost.

Another feedback path is typically created by the exhaust gas recirculation (EGR) valve. Though the feedback is not positive (as is the feedback of the turbocharger path), it complicates the causality relationships and makes the model stability analysis more challenging.

15.5 Identification of Internal Combustion Engine Models

From the preceding section, it should be clear that the engine model identification is not an easy task. Suppose the model is represented in the form of state-space equations:

$$\frac{dx(t)}{dt} = f(x(t),\ u(t),\ \theta)$$
$$y(t) = g(x(t),\ u(t),\ \theta) \tag{15.1}$$

Let N model input/output measurements be available across the engine operating range. We will denote the input output samples as follows:

$$u(t_k), \ y(t_k) \ \text{ for } \ k = 1, \dots, N \tag{15.2}$$

a reasonable cost function for model identification could be:

$$\min_{\theta, \ x(t)} \sum_k \|y(t_k) - g(x(t_k), u(t_k), \ \theta)\| \ \text{ subject to } \ \frac{dx(t)}{dt} = f(x(t), \ u(t), \ \theta)$$

$$\tag{15.3}$$

However, in practice it is extremely difficult to solve such an idealized cost function directly. As mentioned above the model (15.1) is highly nonlinear, the dimension n_θ of the parameter vector $\theta \in \mathcal{R}^{n_\theta}$ may be large (the example in Section 15.7 involves $n_\theta = 88$ parameters), and the state vector $x(t)$ may not be accurately known exactly at all measurements points $k = 1, \dots, N$. (Hence the appearance of $x(t)$ as an argument in the optimization (15.3).)

In this section we will present a stepwise approach of modifying the ideal cost function (15.3) to make the identification of the control relevant properties of the model in (15.1) more practical. For this purpose the first step of the procedure is the identification of individual model components. These results are then used to initialize the identification of the overall steady-state response of the model (i.e. with all component models interconnected with one another). In a last step the identification of the parameters responsible for the dynamic response is carried out.

As discussed above, the control oriented model at hand is composed of static nonlinear components and integrators. From an intuitive point of view, these static nonlinear components correspond to physical engine components (compressor, turbines, valves, etc.) whose inputs and outputs are typically measured in practical engine test cell work. Thus a natural decomposition of the identification problem (15.3) is suggested here. Subsection 15.5.1 presents the challenges when identifying individual static nonlinear engine components from measured engine data. Subsection 15.5.2 describes the issues involved in identifying the collective steady-state response of all components in their contribution to the overall input output behavior of the model. Subsection 15.5.3 presents the approach applied to identify those elements of the model such that (15.1) matches the dynamic response in the measured engine data.

15.5.1 Identification of Static Nonlinear Engine Components

In Subsection 15.4.1 we noted challenges posed by requiring the parameters and signals associated with the model components to satisfy constraints. In this subsection we consider this issue in the course of fitting the individual engine components to their measured input output data. With components such as valves, throttles, heat exchangers, this data is typically obtained by measurements on the engine in the test cell. Data used for fitting compressor and turbine models are provided by the turbocharger vendor [12]. In terms of the dynamic model (15.1), the functions f and g are in fact composite functions built from the individual component models according to the model structure. Hence, both the vector of the state derivatives dx/dt and the vector of measurements $y(t)$ can be equated with outputs of certain components. The components inputs are either the model external inputs $u(t)$ or the outputs of other components.

The models of the compressor presented in [7] provide a suitable example of two important considerations when designing component models. The model of compressor *flow* illustrates the importance of considering constraints in each component. The model of compressor *efficiency* shows that the selection of the functional form of the component is a key decision in the success of the modeling approach.

According to [7], the compressor flow model is based on the algebraic relation between the dimensionless head parameter Ψ and dimensionless flow rate Φ both of which are constructed from measurements of pressure, temperature, speed and flow recorded during the turbocharger mapping process. (For simplicity, the weak dependence on Mach number will be neglected here.) We will use the following equation:

$$\Phi = k_3 - \frac{k_1}{k_2 + \Psi} \tag{15.4}$$

To identify the compressor flow model, parameters k_i have to be fitted to the data $\{\Phi_j,\ \Psi_j\}$. The function (15.4) is discontinuous with vertical asymptotes. In addition, prior physical information on the function Φ includes: the function's domain is to the left of the singularity in (15.4) and it should be decreasing function of Ψ. Meaning that the following inequalities hold:

$$k_2 + \Psi < 0$$
$$k_1 < 0 \tag{15.5}$$

It can be verified the least squares identification algorithm of the k_i parameters converges to the true parameter values from any initial estimate where the constraints (15.5) are satisfied. We will prove it for the noiseless data.

Proof. We suppose the constraints (15.5) are satisfied by the true parameter values k_i^*. Then, the convex nonlinear least squares optimization method converges to the true values if the constraints are satisfied by the initial values of the parameters. The claim can be proven analyzing the sum of error squares which will be minimized by the parameter estimates:

$$\frac{1}{2}\sum_j e_j^2 = \frac{1}{2}\sum_j (\Phi(\Psi_j, k_1, k_2, k_3) - \Phi_j)^2 \tag{15.6}$$

The derivatives of just one error square with respect to the unknown parameters are products of the model prediction error and derivatives of the predictions with respect to the parameters k_i:

$$\frac{1}{2}\frac{\partial e_j^2}{\partial k_i} = e_j \frac{\partial \Phi(\Psi_j, k_1, k_2, k_3)}{\partial k_i} \tag{15.7}$$

Using (15.4, 15.5), the Φ derivatives signs are:

$$\frac{\partial \Phi}{\partial k_1} = \frac{-1}{\Psi_j + k_2} > 0, \quad \frac{\partial \Phi}{\partial k_2} = \frac{k_1}{(\Psi_j + k_2)^2} < 0, \quad \frac{\partial \Phi}{\partial k_3} = 1 > 0 \tag{15.8}$$

Hence, in (15.8), all derivative signs are known *a priori* if the constraints (15.5) hold. Then, because the derivatives (15.8) are non-zeros, the prediction error square derivative is zero only if the prediction error is zero. For noise free data, all prediction errors are zeroed by the true parameter values k_i^*. It follows that the true parameter values represent the unique stationary point in the *convex* feasible set defined by the constraints (15.5). This ends the proof.

The flow model illustrated by the solid line on the left in Figure 15.4 answers the equation (15.4) when the numerical optimization of the k_i parameters was started from a feasible point satisfying (15.5). The dashed line, in contrast, was started from an infeasible point.

Another way of addressing the constrained domains of component model inputs is by replacing the static nonlinear model components with the linearly extrapolated versions. The extrapolation is based on the linearization around the closest admissible input values. Receiving the vector of input values u_k, the component finds the closest admissible point u_k^a optimizing over the admissible set U_k with respect to its vector element v_k.

$$y_k = f_k(u_k^a, \theta_k) + \left.\frac{\partial f_k}{\partial u_k}\right|_{u_k^a} (u_k^a - u_k)$$

$$u_k^a = \arg\min_{v_k \in U_k} \|u_k - v_k\| \tag{15.9}$$

Example 1. Consider fitting data to the model with a constrained domain:

$$y = \frac{k}{x - a}, \qquad x > x_0 > a \tag{15.10}$$

The model can be linearly extrapolated outside the domain as follows:

$$x_1 = \max\{x, \ x_0\}, \qquad y = \frac{k}{x_1 - a}\left(1 - \frac{x - x_1}{x_1 - a}\right) \tag{15.11}$$

The gradient $(\mathrm{d}y/\mathrm{d}a, \mathrm{d}y/\mathrm{d}k)$ *is approximated by the gradient at* x_0 *in case the* x *value is not admissible.*

Besides the flow the other important submodel of the compressor component is its efficiency and here we use it as an example to illustrate the importance of the selection of a suitable functional form in maintaining constraints on model signals. The influential paper [7] proposed to model the compressor efficiency as a quadratic function of the dimensionless flow rate with coefficients depending on the Mach number. That model is valid only locally, where the quadratic function values modeling the compressor efficiency are positive. Such a model can be used only if it can be guaranteed that this constraint will be respected during both simulation and identification.

Instead of a quadratic functional form a two-dimensional skewed Gaussian is proposed here and is shown as a solid line on the right of Figure 15.4. The Gaussian function decreases monotonically from the maximum point along all directions but never crosses zero. The function is positive and unimodal, and provides an acceptable fit locally and asymptotically. On the other hand, the cubic multivariate polynomial $\epsilon = P(\Phi, \ \Psi)$ from [7] is plotted as the dashed line on the right of Figure 15.4. Note that the multivariate polynomial based efficiency will become invalid if the component's input values will be driven too far from the region of the usual efficiency values $\in (0.6, \ 1)$. It becomes clear that choosing an unsuitable functional form can lead to large relative errors which can lead to instability of the overall dynamic model (15.1) (see discussion in Section 15.6 below).

From the compressor flow and efficiency examples, the nature of constraints on both signal and parameters should be clear. It appears that some constraints can be satisfied implicitly selecting the functional forms carefully (as was the case with compressor efficiency). Suitable functional forms should exhibit following characteristics:

- have a justified asymptotic behavior for inputs approaching infinity along all directions,
- have minimum complexity, minimum number of extremes,
- be defined for all input values, without singularities and with bounded derivatives,
- satisfy as many of the prior assumptions implicitly as possible for any parameter value from a defined admissible domain,
- be sufficiently representative of the physics of the component.

Using such functional forms and respecting all constraints on the parameters during the optimization greatly improves the identification algorithm robustness and convergence properties.

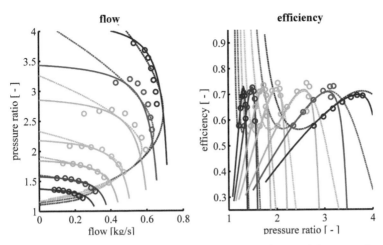

Fig. 15.4 *Left:* Curves representing fit of compressor flow model to map data. The dashed curves represent a local optimum obtained if the numerical optimization is started from an inadmissible initial condition, and the solid curves were started from an admissible initial condition. *Right:* Curves representing fit of compressor efficiency model to map data. The dashed line illustrates the cubic multivariate polynomial in [7]. The solid line indicates the proposed monotonic and unimodal skewed-Gaussian structure.

15.5.2   *Identification of Overall Model Steady State Input-Output Response*

Following the successful identification of each of the static nonlinear components of the control oriented model the next step is to ensure that the overall model (15.1) accurately represents the data measured on the engine. As discussed previously, to simplify the task we will split the overall engine model identification into two separate steps: fitting the steady-state response followed by fitting the dynamic model response. For internal combustion engine models it turns out that the majority of the parameters $\theta \in \mathcal{R}^{n_\theta}$ in (15.1) may be identified from steady-state data. For the the example in Section 15.7, 86 of the $n_\theta = 88$ parameters may be identified from the steady-state data. The next subsection will discuss the identification of the remaining parameters to determine the dynamic response.

Recalling the state-space model representation in (15.1) the static model identification may be realized by modifying the ideal cost function in (15.3) as follows:

$$\min_{\theta,\ x_k} \sum_k \|y_k - g(x_k, u_k,\ \theta)\| \text{ subject to } f(x_k, u_k,\ \theta) = 0 \qquad (15.12)$$

From (15.12) it should be clear that steady state data define all parameters of the output function g at steady state and also defining the roots of f. On the other hand the behavior of f in the neighborhood of its roots is undetermined by the steady state data. The model transient behavior will be fit using the techniques described in subsection 15.5.3 using parameters which change the f slope in the neighborhood of its roots $f = 0$.

Note also that the unknown state vector x_k is included as an argument of the optimization. Given an initial parameter vector estimate θ^i and feasible pairs x_k^i, u_k, i.e. steady state values for given states and inputs, the equality constraint $f = 0$ can be approximated linearly around the feasible points:

$$x_k - x_k^i \approx - \left(\frac{\partial f}{\partial x}\right)^{-1} \frac{\partial f}{\partial \theta}\bigg|_{x_k^i, \theta^i} \left(\theta - \theta^i\right) \tag{15.13}$$

The expression for the state differential (15.13) can be substituted to the linear approximation of the output equation. Then, we finally get the least squares *unconstrained* optimization problem with respect to θ (and no more x_k) which approximates (15.12) locally.

$$\min_\theta \sum_k \left\| g(x_k, u_k, \theta^i) - y_k + (M_k)_{x_k^i, \theta^i} \left(\theta - \theta^i\right) \right\|$$

$$M_k = \frac{\partial g}{\partial \theta} - \frac{\partial g}{\partial x} \left(\frac{\partial f}{\partial x}\right)^{-1} \frac{\partial f}{\partial \theta} \tag{15.14}$$

The approximation is based on the linear approximation of the norm $\| \cdot \|$ argument with respect to both states and parameters; essentially the Gauss-Newton method. The matrix inversion in (15.14) should exist provided the model steady state is uniquely determined by the values of inputs and parameters. We assume this condition is satisfied by the internal combustion engines under normal operating conditions. The steady state pairs x_k^i, u_k can be found simulating the engine model for a sufficiently long interval of time with input values u_k and taking the terminal state as x_k. Theoretically, it could be found minimizing the $f(x_k, u_k, \theta^i)$ norm with respect to x_k. This optimization typically terminates at a local extremum with dx/dt having a nonzero norm value, i.e. not being at steady state. That is why the simulation method is usually better. However, this method requires a stable model to be defined by the initial estimates otherwise the method will fail because the equality constraint $dx/dt = 0$ will be linearized at a point where it does not hold.

The equation (15.14) defines the cost function local approximation. It approximates the gradient and Hessian matrix. These values can be used to solve the original constrained problem (15.12) iteratively using a suitable numerical optimization method. The identification algorithm minimizing the weighted sum of steady state prediction errors can be given as follows:

1. Initialize the estimates θ^i.
2. k iterates over the engine steady state data available:
   a. Find model steady state x_k and the prediction error δy_k for the given input value u_k. This step involves model simulation.
   b. Find the numerical derivatives of the model derivatives $\mathrm{d}x_k/\mathrm{d}t$ and model outputs y_k with respect to both states and parameters and construct the matrix M_k in (15.14). This step is model linearization with respect to both states and parameters.
3. Update estimates θ^i using a numerical optimization method, cost function value J^i, gradient h^i, and Hessian H^i. The diagonal matrix W defines the weights.

$$J^i = \sum_k \frac{1}{2}\delta y'_k W \delta y_k, \quad h^i = \sum_k \delta y'_k W M_k, \quad H^i = \sum_k M'_k W M_k.$$
$$\tag{15.15}$$

4. Go to 2 or stop if $\|h^i\| \approx 0$

In practice, if the parameters θ are initialized reasonably (for example by identifying the individual components as discussed in Subsection 15.5.1), then the above algorithm typically results in obtaining parameter estimates which are close to optimal for the chosen model structure.

15.5.3   Identification of Overall Model Dynamic Input-Output Response

The remaining elements of θ in (15.1) that were not determined during the steady-state identification described in Subsection 15.5.2 typically include those parameters defining the turbocharger inertia, manifold volumes, thermal capacities, sensor and actuator dynamics. Some of these physical properties may be known with reasonable accuracy prior to the dynamic identification from engine geometry and other specifications. However, due to some of the assumptions and simplifications of the mean value modeling approach it is required to identify these parameters within reasonable bounds using transient data. It has been found that inside the typical bandwith of the feedback controller the turbocharger inertia out of all the dynamic parameters has the most significant contribution to the overall time constant of the model.

The dynamic parameters of θ define the derivatives of f around its roots. It can be shown that these *dynamic* parameters of θ change neither the function g nor roots of f in (15.1), which were matched to the steady state data according to the previous Subsection 15.5.2. Hence, the identification of dynamic parameters of θ can be done separately and the model steady state will stay unchanged.

The approach to identify these parameters based on the dynamic data is the following:

$$\min_{\theta,\ x(t_k)} \sum_k \left(\|y(t_k) - g(x(t_k),\ u(t_k),\ \theta)\| + \gamma\,\|\delta x(t_k)\| \right)$$

$$\text{subject to} \int_{t_{k-1}}^{t_k} f(x(t_{k-1}), u(t_{k-1}),\ \theta) = x(t_k) + \delta x(t_k) \tag{15.16}$$

In our approach, the cost function value, gradient and Hessian are calculated recursively using an algorithm similar to the Extended Kalman Filter. The continuity constraint on the state vector trajectory is treated as soft constraint on the error term $\delta x(t_k)$ as indicated by (15.16). This error term is treated as process noise by the Kalman filter. Knowing the gradient and Hessian, a convex optimization method can be employed. The problem structure of the dynamic data fitting is more complex compared to the steady state data fitting because it optimizes for both parameters and the whole state trajectory. However, the problem is alleviated because the number of dynamic parameters is smaller compared to the number of parameters which define the steady states. The example in Section 15.7 involved only two such parameters. Moreover, the problem is nearly linear because the dynamic parameters are typically multipliers of the state derivatives. The convergence properties are good as the model transient responses are usually sufficiently close to the measured data after the steady state identification step. From this point of view, splitting the identification into static and dynamic parts is advantageous.

15.6 A Comment on Component versus Overall Identification

The engine model is built from components as displayed on Figure 15.2 some of which contain relatively complicated nonlinearities, such as thecompressor introduced in Section 15.5.1. The input and output signals of these elements are often measured. For this reason it seems natural to use the component input and output signals to fit the component parameters. In this section we will explain that this can compromise with the model accuracy.

The reason is that the mean value modeling approach neglects some phenomena like small pressure drops, negligible heat transfers etc. Then, the overall model behavior can be significantly affected even if the key components are modeled most accurately. The second reason is that the modeling errors can get amplified mainly by the model feedback structure as discussed in Subsection 15.4.2. It is therefore necessary to study how the component errors agglomerate in the overall model behavior.

It is well known that the negative feedback can attenuate the system sensitivity to the parameters. This fact is used, for example, in feedback control. However, the feedback can easily lead to the opposite effect of increasing the system sensitivity to parameter values. As there exist feedbacks in the turbocharged engine model, the errors in component parameters can be arbitrarily amplified in the overall model behavior. When used carefully this feature can be exploited to advantage in model identification as it is possible to improve the overall model accuracy using small changes in the model components, thus preserving the intuitive accuracy of the components.

Practically, this means that certain combinations of the component model parameter values can appear almost equally good when the components are fitted individually. When the engine model is built from such independently fitted components, the overall model accuracy may be much worse compared to the component level accuracy. The reason is that the vector of parameter errors can have an unfavorable direction in the parameter space. The situation is most complicated when fitting models with both positive and negative feedback loops. In such systems, some component qualities may in fact have only marginal effects on the whole system, whereas others may be extremely important because they get amplified. Hence, good component fit does not necessarily mean a good prediction capability of the whole model and vice versa. A typical turbocharged engine model behaves in this way. One should try to achieve a good prediction capability of the whole model as well as a good component level fit but one should not rely on the hypothesis, which does not hold, that a good component fit (see Subsection 15.5.1) automatically leads to a good overall model fit.

This problem is solved naturally if fitting the whole system model to all engine data. Then, the sensitivity of the model prediction accuracy to various functions of model parameters is fully exploited. The parameters can be optimized to achieve the best overall model accuracy. Let us call such an identification method the *overall identification* in contrast to the *component identification*, where the components are fitted independently using their local input output data. We will look at these two methods using two examples, which will show that both approaches offer some advantages but also have some inherent issues.

First, we will present a simple example showing the component identification does not guarantee reasonable prediction accuracy of the overall model.

Example 2. To compare component versus overall identification consider a simple first order linear system with a feedback structure with input $u(t)$ and output $y(t)$:

$$\frac{dy(t)}{dt} = u(t) - z(t) \tag{15.17}$$

$$z(t) = \theta y(t) \tag{15.18}$$

The parameter θ can either be estimated locally from (15.18) or via its overall effect on the system (15.17), (15.18).

If we first consider the local problem, then we can logically write the "component" identification of y to z in (15.18) as an optimization at times t_j:

$$\min_{\theta} \sum_{j} \|z_{\mathrm{meas}}(t_j) - \theta\, y_{\mathrm{meas}}(t_j)\| \tag{15.19}$$

and if we further consider the case $z_{\mathrm{meas}} \approx 0$, then the optimal value for θ is close to 0 and we can see from this expression that the cost function from y to z in (15.19) is changed by an arbitrarily small amount for any θ satisfying $|\theta| < \epsilon$ for $\epsilon \to 0$.

On the other hand if we consider the overall behavior of the system (15.17), (15.18) and instead examine the transfer function from u to y, then we find that $\theta > 0, = 0, < 0$ respectively correspond to stable, marginally stable, and unstable systems - each of which represent very different overall system behaviors.

From the above example, we can see that the component-wise model identification does not guarantee small or even bounded prediction errors of the whole model if feedback exists in the system structure. Now we will construct a different example which shows that the overall model identification may require accurate initial conditions for the parameter values to converge to the global optimum.

Example 3. Consider a cascade of n nonlinearities with input $u_1(t)$ and output $y_n(t)$:

$$y_n = f_n(f_{n-1}(\dots(f_1(u_1, \ \theta_1)\dots, \ \theta_{n-1}), \ \theta_n) \tag{15.20}$$

where the kth nonlinearity is parameterized by θ_k. The nonlinearities can be understood as n model components with local outputs y_k, local inputs $u_k = y_{k-1}$, and local vectors of parameters. Suppose the components' input output data measurements are available. The component identification is therefore possible. This component method will use the gradients which define the y_k sensitivity to θ_k parameters evaluated at the current estimate θ_k^i:

$$\left. \frac{\partial y_k}{\partial \theta_k} \right|_{\theta_k^i} \tag{15.21}$$

On the other hand, the overall identification will in addition also consider the effect of the parameter θ_k along the causal chain defined by the cascade. These effects are neglected by the component gradients in (15.21). The impact of parameter θ_k on the next outputs can be evaluated using the chain rule:

$$\left. \frac{\partial y_{k+m}}{\partial \theta_k} \right|_{\theta_k^i} = \left(\prod_{j=m}^{1} \left. \frac{\partial y_{k+j}}{\partial y_{k+j-1}} \right|_{y_{k+j-1}^i} \right) \left. \frac{\partial y_k}{\partial \theta_k} \right|_{\theta_k^i}, \qquad 1 \le m \le n-k \tag{15.22}$$

The component identification will converge if the initial θ_k^i estimate is close enough to the global optimum, i.e. the negative cost function gradient vector points to the global optimum. The sufficient condition for the overall model identification is that all the gradients (15.22) have to be evaluated at points close enough (in a set which depends on the problem solved) to the optimum.

The gradients in (15.22) include the component output gradient (15.21) along with additional gradients. Thus the set of convergent initial conditions of θ_k^i for the overall model identification is a subset of the set of convergent initial conditions for the component identification.

From this example, we can see that the global identification requires more stringent sufficient conditions for convergence than does the component identification. In other words, it reinforces the intuitive notion that one must take greater care in setting the initial conditions of the global identification than the component identification. This fact is a key to the practical implementation of the proposed stepwise model identification approach outlined in this work.

The component and overall identifications represent two extreme cases of cost function formulations for identification of the model parameters θ in (15.1). In this section we have presented a two-step model identification approach that combines the good convergence of component identification along with the accuracy potential of the overall identification approach. In this way a practical nonlinear identification approach is achieved for highly nonlinear engine models.

15.7   Modeling Results

This section shows the identification results achieved with the concepts presented above. The engine model used herein corresponds to a six cylinder in-line engine with the layout shown in Figure 15.2. Some basic specifications on the engine are given in Table 15.2.

Table 15.2 Engine specifications

Manufacturer	MAN	
Model	D2676LOH27	
Displacement volume	12.42 liter	
Rated power output	353 kW	@ 1900 rpm
Max torque	2300 Nm	@ 1000-1400 rpm

The data used for the identification was obtained through a design of experiment running the engine in open loop with no feedback control active. Various combinations of actuator positions were recorded at six operating points in the fuel/speed range resulting in 798 data points. The overall time required to record this data was roughly 12 hours of test cell time.

Problem dimensions were as follows:

Parameters identified	$n_\theta = 88$
Inputs	$n_u = 10$
Measured outputs	$n_y = 20$
Steady state points	$n_k = 798$

The functional forms of model nonlinearities were selected according to recommendations given in Section 15.5.1. Specifically, the compressor and turbine efficiency functions are unimodal positive functions. All nonlinearities are extrapolated linearly outside admissible input domains according to (15.9). The identification was split into component, overall steady state and dynamic as described in Section 15.5. The overall steady state identification was based on the local approximation values given by (15.15) and optimized by the trust region algorithm. The algorithm required approximately 10 – 20 iterations to reach the desired accuracy. The fit improvement gained by doing the overall steady-state identification on top of the component identification was significant as can be seen in Figure 15.5. The model *coefficient of determination* defined as the square of the correlation coefficient between data and prediction improved from 89% to 98% on average over all

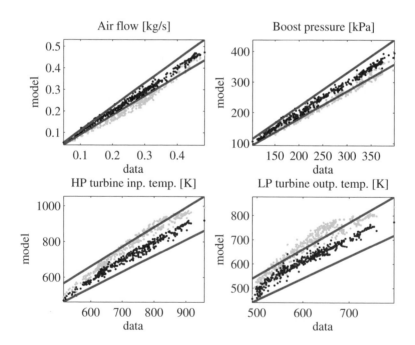

Fig. 15.5 Steady state data fit in 798 steady state points. Error bounds ±10% are shown with red solid lines. Overall steady state fit shown in black, component fit represented in gray.

$n_y = 20$ model outputs. Following the steady state identification the transient behavior of the model was calibrated by tuning the dynamic parameters, i.e. turbocharger inertia, manifold volumes etc. The resulting dynamic behavior of the model in comparison to the recorded data at one operating point is shown in Figure 15.6. Similar results are achieved at all operating points also for steps in other actuator position so that the overall dynamic behavior of the engine is captured well by the model.

Regarding simulation performance the overall computational time required to calculate the 798 data points took 3.5 seconds while the simulation of the FTP (1200 seconds in real-time) as an example of a full dynamic test cycle was executed in 2.4 seconds on a standard personal computer.

Although only the data used for the identification are depicted in Figures 15.5 and 15.6 the model has also been verified for other validation data as well and achieved good prediction accuracies. The model presented here was used in the synthesis of a model based multivariable controller (as described in Section 15.2) which performed successfully across the entire operating range of the engine.

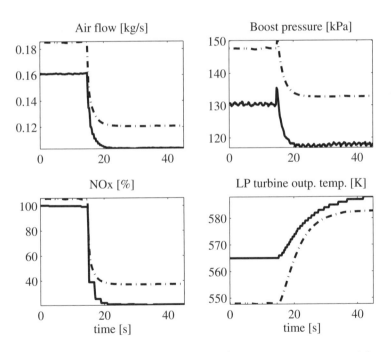

Fig. 15.6 Dynamic data fit at 1700 rpm 55mg/stroke. Responses to partial opening step in EGR valve position. Model predictions shown with dashed lines. The steady state offsets correspond to accuracies shown in Figure 15.5.

15.8   Conclusion

This chapter has discussed techniques for configuring and identifying models for internal combustion engines in particular for their use in a model based control design process. These control oriented models (COM) are based on physical first principles parameterized with several tuning parameters and as such can be assigned to the category of gray box models which typically exhibit medium model complexity while still allowing for good prediction accuracy even outside the range of data used for identification.

The model based control design process utilizing the COM synthesizes a multivariable model predictive controller (MPC) of explicit architecture which has low computational requirements in order to fit on state of the art engine control units [18]. At the same time the controller implements practical features of MPC such as constraints on system inputs and outputs as well as simultaneous tracking of multiple system variables. The controller behavior can also be influenced using specific weights on tracked variables as well as on actuator commands in order to emphasize the tracking of certain controlled variables while choosing specific actuators to carry out the majority of the control activity.

During the development of the control oriented models and the accompanying identification techniques the issue of constrained domains was raised. Some of the engine component models that are familiar from the literature were discovered to employ functional forms containing singularities and other invalid solutions for input variables outside of the normal range or an inadequate choice of parameters. This can be problematic during simulations and especially for developing robust automated model identification techniques.

In certain cases this problem was solved by modifying the functional form to eliminate the undesired behavior. Where this approach was not possible constraints on input signals and parameters were taken into account specifically during the identification.

Another challenge specific to engine models is the presence of feedback loops through turbocharger shafts and EGR systems. As with any feedback system certain choices of parameter values can cause the system to become unstable. Due to this circumstance special caution has to be given when performing the steady state identification of each individual component of the engine model separately. Although the component accuracy might appear acceptable in isolation, the errors in the parameter estimates can get amplified in the overall system behavior. To address this issue a steady state identification approach was developed which fits the entire engine model as an entity to ensure that the overall system behavior remains close the measurements. The standard challenge with the overall identification of complex nonlinear models is that it requires a good initial guess of the parameter values to guarantee the convergence to the global optimum. It was shown that this initial guess, however, can be found directly from the identification of the individual components.

The steady state identification defines the majority of the model parameters. The few remaining dynamic parameters can be identified in a separate step leading the proposed identification methodology to be a three stage process:

1. Steady state component identification
2. Steady state overall model identification
3. Dynamic overall model identification

With this identification procedure accuracy results have been presented on the basis of a six cylinder commercial diesel engine with EGR and dual staged turbocharging. Both the steady state as well as the dynamic model response match the measured engine behavior accurately and exceed the usual requirements for models developed specifically for controller design.

Acknowledgments

The authors express their gratitude to Gilles Houllé, Reinhard Lämmermann and Ralf Meyer at MAN Nutzfahrzeug AG who contributed with their expertise in engine technology and to a large part made the results presented herein possible.

References

[1] Fan, J., Stewart, G.E.: Automatic tuning method for multivariable model predictive controllers. United States Patent Number: 7,577,483 (May 2006)

[2] Landau, I.D., Zito, G.: Narmax model identification of a variable geometry turbocharged diesel engine. In: Proceedings of 2005 IEEE American Control Conference, vol. 2, pp. 1021–1026 (2005)

[3] Gomez, J.C., Baeyens, E.: Identification of multivariable hammerstein systems using rational orthonormal bases. In: Proceedings of 39th IEEE Conference on Decision and Control, Sydney, Australia (December 2000)

[4] Guzzella, L., Amstutz, A.: Control of diesel engines. IEEE Control Systems Magazine 18(2), 53–71 (1998)

[5] Guzzella, L., Onder, C.H.: Introduction to Modeling and Control of Internal Combustion Engines. Springer, Heidelberg (2004)

[6] Hendricks, E.: A compact, comprehensive model of large turbocharged, two-stroke diesel engines. SAE Technical Paper Series, 861190 (1986)

[7] Jensen, J.P., Kirstensen, A.F., Sorenson, S.C., Houba, N., Hendricks, E.: Mean value modeling of a small turbocharged diesel engine. SAE Technical Paper Series, 910070 (1991)

[8] Juditsky, A., Hjalmarsson, H., Benveniste, A., Delyon, B., Ljung, L., Sjöberg, J., Zhang, Q.: Nonlinear black-box models in system identification: Mathematical foundations (1995)

[9] Kolmanovsky, I.V., Stefanopoulou, A.G., Moraal, P.E., van Nieuwstadt, M.: Issues in modeling and control of intake flow in variable geometry turbocharged engines. In: 18th IFIP Conference on System Modelling and Optimization (1997)

[10] Maciejowski, J.M.: Predictive Control with Constraints. Pearson Education Ltd., Essex (2002)

[11] Masoudi, M., Konstandopoulos, A., Nikitidis, M.S., Skaperdas, E., Zarvalis, D., Kladopoulou, E., Altiparmakis, C.: Validation of a model and development of a simulator for predicting the pressure drop of diesel particulate filters. SAE Technical Paper Series, 2001-01-0911 (2001)

[12] Moraal, P., Kolmanovsky, I.: Turbocharger modeling for automotive control applications. SAE 1999-01-0908 (1999)

[13] Ortner, P., del Re, L.: Predictive control of a diesel engine air path. IEEE Transactions on Control Systems Technology 15(3), 449–456 (2007)

[14] Rezazadeh, S., Vossoughi, G.: Modeling and optimization of an internal combustion engine mapping using neural networks and particle swarm optimization. In: HIS, pp. 1094–1103 (2003)

[15] Schär, C.M.: Control of a Selective Catalytic Reduction Process. PhD thesis, Diss. ETH Nr. 15221, Measurement and Control Laboratory, ETH Zurich, Switzerland (2003)

[16] Skogestad, S., Postlethwaite, I.: Multivariable Feedback Control: Analysis and Design. Wiley, New York (1996)

[17] Stefanopoulou, A.G., Kolmanovsky, I., Freudenberg, J.S.: Control of variable geometry turbocharged diesel engines for reduced emissions. IEEE Trans. Contr. Syst. Technol. 8(4), 733–745 (2000)

[18] Stewart, G.E., Borrelli, F., Pekar, J., Germann, D., Pachner, D., Kihas, D.: Toward a systematic design for turbocharged engine control. In: Del Re, L., Allgower, F., Glielmo, L., Guardiola, C., Kolmanovsky, I. (eds.) Automotive Model Predictive Control. Lecture Notes in Control and Information Sciences, vol. 402, pp. 211–230. Springer, Heidelberg (2010)

[19] Gamma Technologies / GT Suite 2010, http://www.gtisoft.com/

[20] Herbert, J., Tulleken, A.F.: Grey-box modelling and identification using physical knowledge and bayesian techniques. Automatica 29(2), 285–308 (1993)

[21] Tulleken, H.J.A.F.: Grey-box modelling and identification topics. PhD thesis, TR diss 2173, Delft University of Technology (1992)

[22] van Aken, M., Willems, F.: D-J de Jong. Appliance of high EGR rates with a short and long route EGR system on a heavy duty diesel engine. SAE Technical Paper Series, 2007-01-0906 (2007)

[23] van Helden, R., Verbeek, R., Willems, F., van der Welle, R.: Optimization of urea SCR deNOx systems for HD diesel applications. SAE Technical Paper Series, 2004-01-0154 (2004)

Chapter 16
Modeling for HCCI Control

Anders Widd, Per Tunestål, and Rolf Johansson

Abstract. Due to the possibility of increased efficiency and reduced emissions, Homogeneous Charge Compression Ignition (HCCI) is a promising alternative to conventional internal combustion engines. Ignition timing in HCCI is highly sensitive to operating conditions and lacks direct actuation, making it a challenging subject for closed-loop control. This paper presents physics-based, control-oriented modeling of HCCI including cylinder wall temperature dynamics. The model was calibrated using experimental data from an optical engine allowing measurements of the cylinder wall temperature to be made. To further validate the model, it was calibrated against a conventional engine and linearizations of the model were used to design model predictive controllers for control of the ignition timing using the inlet valve closing and the intake temperature as control signals. The resulting control performance was experimentally evaluated in terms of response time and steady-state output variance.

Keywords: Homogeneous Charge Compression Ignition (HCCI), modeling, parameter identification, model predictive control.

16.1 Introduction

Homogeneous Charge Compression Ignition (HCCI), also referred to as Controlled Auto-Ignition (CAI), holds promise for reduced emissions and increased efficiency compared to conventional internal combustion engines. As HCCI lacks direct actuation of the combustion phasing, much work has been devoted to designing controllers capable of set-point tracking and disturbance rejection. Ignition timing in HCCI engines is determined by several factors [1]

Anders Widd · Per Tunestål · Rolf Johansson
Lund University, P.O. Box 118, 221 00 Lund, Sweden
e-mail: {anders.widd,per.tunestal,rolf.johansson}@control.lth.se

D. Alberer et al. (Eds.): Identification for Automotive Systems, LNCIS 418, pp. 283–302.
springerlink.com

including the auto-ignition properties of the air-fuel mixture, the intake temperature, the amount of residual gases in the cylinder, etc. As a consequence, there are many possible choices of control signals that can be used for controlling combustion timing, such as variable valve timing, intake temperature, and the amount of residuals trapped in the cylinder [1].

To facilitate model-based control design, both statistical and physical models have been considered [1]. Most control-oriented physical models of HCCI are either formulated in continuous time or on a cycle-to-cycle basis [1]. The models in the former category allow for a natural formulation, *e.g.*, of flow phenomena, while those in the latter take a form suitable for cycle-to-cycle control design. This paper presents physics-based, cycle-resolved modeling of HCCI including cylinder wall temperature dynamics. The model was initially calibrated using data from an optical engine, allowing measurements of the cylinder wall temperature to be made. To further validate the model, it was calibrated against a conventional engine and used as basis for model-based control design.

Heat transfer effects between the in-cylinder charge and the cylinder walls are important for explaining HCCI cycle-to-cycle behavior [2]. As noted recently, however, many modeling approaches for control only consider heat transfer from the gases to the walls and assume a fixed cylinder wall temperature [1]. Continuous-time models including cylinder wall temperature were presented in [2], [3]. The former also presented results on model-based control. The control results presented in this paper were obtained using linearizations of the physical model. Model Predictive Control (MPC) [4] was used with the crank angle of inlet valve closing, θ_{IVC}, and the intake temperature, T_{in}, as control signals. The controlled output was the crank angle of 50% mass fraction burned, θ_{50}. The results are evaluated in terms of response time, robustness towards disturbances, and combustion phasing variance.

Recent examples of cycle-resolved models of HCCI for control design include [5], [6], where [6] also presented experimental results of closed-loop control. Continuous-time models including cylinder wall temperature models were presented in [3], [2]. The latter synthesized an LQ controller using a cycle-resolved statistical version of the model and presented experimental results. Fast thermal management and variable compression ratio using several control methods was employed in [7]. Model predictive control of HCCI engines based on statistical models was used in [8].

16.2 Modeling

This section describes a cycle-resolved, nonlinear model of HCCI incorporating cylinder wall temperature dynamics. It uses the same basic formulation as the model in [6] with the inclusion of cylinder wall temperature dynamics and a few other modifications. The considered outputs are the indicated mean effective pressure and the crank angle where 50% mass fraction has burned. For more details, see [9]. The model structure is fairly modular in the choice of control signals as it captures the dependence on the amount of injected fuel, the amount of recycled exhausts, and the valve timings of the inlet and the exhaust.

16.2.1 Cylinder Wall Temperature Dynamics

The cylinder wall was modeled as a single mass with a slowly varying outer surface temperature T_c, a convective flow $\dot{q}_a$ between the in-cylinder gases and the cylinder wall, and a conductive flow $\dot{q}_b$ through the wall, see Fig. 16.1. The first law of thermodynamics applied to the gas when no work is performed yields the expression $\dot{T} = -\dot{q}_a/mC_v$ where T, m, and C_v are the temperature, mass, and specific heat of the gas. The Newton law $\dot{q}_a = h_c A_c (T - T_w)$ was applied, where h_c is the convection coefficient, A_c is the wall surface area and T_w is the wall surface temperature. The convection coefficient is often modeled using the following expression [1],

$$h_c = 3.26 B^{-0.2} P^{0.8} T^{-0.55} (2.28 S_p)^{0.8} \quad [\text{W}/(\text{m}^2 \cdot \text{K})] \qquad (16.1)$$

where B is the cylinder bore, P is the in-cylinder pressure, T is the gas temperature, and S_p is the mean piston speed [10]. The time derivative of the inner wall temperature, T_{iw}, is given by $\dot{T}_{iw} = (\dot{q}_a - \dot{q}_b)/m_c C_p$ where C_p is the specific heat of the cylinder wall, and m_c is the cylinder wall mass. The conductive flow is given by $\dot{q}_b = (T_w - T_c) k_c A_c / L_c$ where k_c is the conduction coefficient and L_c is the wall thickness. Assuming that the steady-state temperature condition $T_{iw} = (T_w + T_c)/2$ holds the temperature equations may be written as

$$\dot{\mathcal{T}} = A_{ht} \mathcal{T} + B_{ht} T_c, \qquad (16.2a)$$

$$A_{ht} = \begin{bmatrix} -\dfrac{h_c A_c}{m C_v} & \dfrac{h_c A_c}{m C_v} \\ 2\dfrac{h_c A_c}{m_c C_p} & -2\dfrac{h_c A_c + k_c A_c/L_c}{m_c C_p} \end{bmatrix}, \qquad (16.2b)$$

$$B_{ht} = \begin{bmatrix} 0 \\ 2\dfrac{k_c A_c}{L_c m_c C_p} \end{bmatrix}, \quad \mathcal{T} = \begin{bmatrix} T \\ T_w \end{bmatrix} \qquad (16.2c)$$

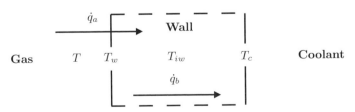

Fig. 16.1 Principle of the cylinder wall model.

Since T_c is assumed to be slowly varying, it can be assumed constant over a short time t_i. The temperature state at t_i, given an initial state $\mathcal{T}(0)$, can then be computed as

$$\mathcal{T}(t_i) = \Phi_i \mathcal{T}(0) + \Gamma_i T_c \qquad (16.3)$$

$$\Phi_i = e^{A_{ht} t_i}, \quad \Gamma_i = \int_0^{t_i} e^{A_{ht}(t_i - \tau)} B_{ht} d\tau \qquad (16.4)$$

Eq. (16.3) was used to update the gas temperature and the wall temperature after mixing, after combustion, and after expansion to capture the time constant of the cylinder wall temperature while keeping the complexity at a tractable level. Since the integration time t_i had to be adapted to each instant and some of the parameters depend on the in-cylinder state, three sets of matrices $\{\Phi_i, \Gamma_i\}$, $i = 1, 3, 5$ were used, where $i = 1$ corresponded to mixing; $i = 3$ combustion; $i = 5$ expansion. The cylinder wall temperature was assumed constant between the heat transfer events.

16.2.2   Temperature Trace

Throughout the cycle, isentropic expansion and compression [11] was assumed. The temperature T_B and pressure P_B after isentropic compression or expansion from volume V_A to a volume V_B with initial temperature and pressure T_A and P_A are given by

$$T_B = T_A \left(\frac{V_A}{V_B} \right)^{\gamma - 1}, \quad P_B = P_A \left(\frac{V_A}{V_B} \right)^{\gamma} \qquad (16.5)$$

where γ is the specific heat ratio. The temperature T_B after isentropic compression or expansion from a pressure P_A to a pressure P_B is given by

$$T_B = T_A \left(\frac{P_B}{P_A} \right)^{(\gamma - 1)/\gamma} \qquad (16.6)$$

16.2.2.1   Intake/Mixing

The gas temperature at the start of cycle k, $T_1(k)$, was modeled as the weighted average of the intake temperature and the temperature of the trapped residuals, cf. [6];

$$T_1(k) = \frac{C_{v,in} T_{in}(k) + C_{v,EGR} \chi \alpha T_{5+}(k - 1)}{C_{v,in} + \alpha C_{v,EGR}}, \qquad (16.7)$$

where $C_{v,in}$ and $C_{v,EGR}$ are the specific heats of the fresh reactants and the residual gases respectively and α is the molar ratio between trapped residuals and inducted gases. The final gas temperature of cycle $k - 1$ is denoted $T_{5+}(k - 1)$, and χ is a measure of how much the residual temperature has decreased. The initial wall temperature of cycle k was set equal to the final wall temperature of cycle $k - 1$;

$$T_{w1}(k) = T_{w5+}(k - 1) \tag{16.8}$$

Eq. (16.3) with $i = 1$ was applied to yield new temperatures $T_{1+}(k)$ and $T_{w1+}(k)$;

$$\begin{bmatrix} T_{1+}(k) \\ T_{w1+}(k) \end{bmatrix} = \Phi_1 \begin{bmatrix} T_1(k) \\ T_{w1}(k) \end{bmatrix} + \Gamma_1 T_c \tag{16.9}$$

16.2.2.2   Compression

Isentropic compression from the volume at inlet valve closing, $V_1(k)$, to the volume at auto-ignition, $V_2(k)$, was assumed, yielding temperature $T_2(k)$ and pressure $P_2(k)$.

$$T_2(k) = T_{1+}(k) \left(\frac{V_1(k)}{V_2(k)} \right)^{\gamma-1}, \quad P_2(k) = P_{in} \left(\frac{V_1(k)}{V_2(k)} \right)^{\gamma} \tag{16.10}$$

where P_{in} is the intake pressure.

16.2.2.3   Auto-ignition

The crank angle of auto-ignition, $\theta_{ign}(k)$, was modeled as in [12], so that the Arrhenius type condition

$$\int_{\theta_{IVC}(k)}^{\theta_{ign}(k)} f_k(\theta) d\theta = 1 \tag{16.11}$$

was fulfilled with

$$f_k(\theta) = A_a P_{in}^n \mathcal{V}_k(\theta)^{\gamma n} \exp \left(-\frac{E_a \mathcal{V}_k(\theta)^{1-\gamma}}{R T_{1+}(k)} \right) \tag{16.12}$$

The parameter A_a is a scaling factor corresponding to the critical concentration of radicals required for auto-ignition to occur, E_a is the activation energy for the reaction, n is the reaction sensitivity to pressure, R is the gas constant, and $\mathcal{V}_k(\theta) = V_1(k)/V(\theta)$. Note that the integrand is independent of species concentrations. To obtain an explicit expression for θ_{ign}, an approach similar to that in [6] was taken. The integrand was approximated with its maximum value, which is attained at Top Dead Center (TDC). The corresponding crank angle degree (CAD) was denoted θ_{TDC}, so that $f_k(\theta) = f_k(\theta_{TDC})$. The lower integration limit was then shifted from θ_{IVC} to θ_{TDC} and the resulting integral equation was solved for $\theta_{ign}(k)$;

$$\theta_{ign}(k) = \Delta\theta_A + \frac{1}{f_k(\theta_{TDC})} \qquad (16.13)$$

where $\Delta\theta_A$ is an offset in CAD.

16.2.2.4 Combustion

The temperature after combustion was calculated as

$$T_3(k) = T_2(k) + \frac{Q_{LHV}}{(1+\alpha)(\phi^{-1}(m_a/m_f)_s + 1)C_v} \qquad (16.14)$$

where Q_{LHV} is the lower heating value of the fuel, ϕ is the equivalence ratio, and $(m_a/m_f)_s$ is the stoichiometric air-to-fuel ratio. The denominator approximates the ratio between the total in-cylinder mass and the fuel mass [9]. Eq. (16.3) was then applied with $i = 3$ and $T_{w3}(k) = T_{w1+}(k)$ to find new temperatures $T_{3+}(k)$, $T_{w3+}(k)$. The pressure after combustion is then

$$P_3(k) = \frac{T_{3+}(k)}{T_2(k)} P_2(k) \qquad (16.15)$$

16.2.2.5 Expansion

The gas temperature and pressure after expansion, $T_4(k)$ and $P_4(k)$, were calculated assuming adiabatic expansion from $V_2(k)$ to the volume at exhaust valve opening, $V_4(k)$. At exhaust valve opening isentropic expansion from the in-cylinder pressure to atmospheric pressure was assumed, yielding temperature $T_5(k)$. Finally, Eq. (16.3) was applied with $i = 5$ and $T_{w5}(k) = T_{w3+}(k)$ to obtain the final gas temperature $T_{5+}(k)$ and the final wall temperature $T_{w5+}(k)$.

16.2.2.6 Model Summary

Table 16.1 summarizes an engine cycle. The variables in each stage can be determined from those of the previous steps.

Table 16.1 HCCI Predictive Control Model Summary. Mapping the states from the previous cycle, $T_{5+}(k-1)$ and $T_{w5+}(k-1)$, and the inputs $\theta_{\text{IVC}}(k)$ and $T_{\text{in}}(k)$ to the states at the current cycle.

---
1. Intake/Mixing with heat transfer
$$\begin{bmatrix} T_{1+}(k) \\ T_{w1+}(k) \end{bmatrix} = \Phi_1 \begin{bmatrix} \frac{C_{v,in}T_{in}(k)+C_{v,\text{EGR}}\chi\alpha T_{5+}(k-1)}{C_{v,in}+\alpha C_{v,\text{EGR}}} \\ T_{w5+}(k-1) \end{bmatrix} + \Gamma_1 T_c$$

2. Prediction of Auto-ignition and Compression
$$\theta_{\text{ign}}(k) = \Delta\theta_A + \frac{1}{f_k(\theta_{\text{TDC}})}$$
$$T_2(k) = T_{1+}(k)\left(\frac{V_1(k)}{V_2(k)}\right)^{\gamma-1}, \qquad P_2(k) = P_{in}\left(\frac{V_1(k)}{V_2(k)}\right)^{\gamma}$$

3. Combustion with heat transfer
$$\begin{bmatrix} T_{3+}(k) \\ T_{w3+}(k) \end{bmatrix} = \Phi_3 \begin{bmatrix} T_2(k) + \frac{Q_{\text{LHV}}}{(1+\alpha)(\phi^{-1}(m_a/m_f)_s+1)C_v} \\ T_{w1+}(k) \end{bmatrix} + \Gamma_3 T_c$$
$$P_3(k) = \frac{T_{3+}(k)}{T_2(k)}$$

4. Expansion
$$T_4(k) = T_{3+}(k)\left(\frac{V_2(k)}{V_4(k)}\right)^{\gamma-1}, \qquad P_4(k) = P_3(k)\left(\frac{V_2(k)}{V_4(k)}\right)^{\gamma}$$

5. Exhaust with heat transfer
$$\begin{bmatrix} T_{5+}(k) \\ T_{w5+}(k) \end{bmatrix} = \Phi_5 \begin{bmatrix} T_4(k)\left(\frac{P_{in}}{P_4(k)}\right)^{(\gamma-1)/\gamma} \\ T_{w3+}(k) \end{bmatrix} + \Gamma_5 T_c$$

16.2.3 Model Inputs and Outputs

The considered modeled outputs were the combustion phasing and the indicated mean effective pressure. The crank angle where fifty percent of the fuel energy has been released was chosen to indicate combustion phasing. The crank angle of inlet valve closing and the intake air temperature were chosen as inputs in the final closed-loop validation while the fuel amount was varied in the calibration experiments.

The combustion duration is a function of the charge temperature, composition, and θ_{ign} [12]. Around an operating point the combustion duration was assumed constant, yielding the following expression for θ_{50}, where $\Delta\theta$ is an offset in crank angle degrees.

$$\theta_{50}(k) = \theta_{ign}(k) + \Delta\theta \tag{16.16}$$

The indicated mean effective pressure (IMEP_n) was calculated from the gas temperatures [11];

$$\text{IMEP}_n(k) = \frac{mC_v}{V_d}\left(T_{1+}(k) - T_2(k) + T_{3+}(k) - T_4(k)\right) \tag{16.17}$$

where V_d is the displacement volume. The complete model is on the following form

$$x(k+1) = \mathbf{F}(x(k), u(k)) \qquad \qquad (16.18a)$$
$$y(k) = \mathbf{G}(x(k), u(k)) \qquad \qquad (16.18b)$$

where

$$x(k) = \begin{bmatrix} T_{5+}(k) \\ T_{w5+}(k) \end{bmatrix}, \quad y(k) = \begin{bmatrix} \mathrm{IMEP}_n(k) \\ \theta_{50}(k) \end{bmatrix}, \quad u(k) = \begin{bmatrix} \theta_{\mathrm{IVC}}(k) \\ T_{\mathrm{in}}(k) \end{bmatrix} \quad (16.19)$$

and the function $\mathbf{F}(y(k), u(k))$ is parametrized by the amount of injected fuel, the amount of recycled exhaust gases (EGR), the intake pressure, etc. and can be obtained by going through the stages 1-5 in Table 16.1.

16.2.4 State Selection

Using (16.3), (16.12), (16.13), (16.14), (16.16), (16.17), and the assumption of isentropic compression and expansion, the temperature state can be uniquely determined from the outputs [13]. This makes it possible to have IMEP_n and θ_{50} as states. Both quantities can be obtained from in-cylinder pressure measurements. The model then takes the following form.

$$y(k+1) = \hat{\mathbf{F}}(y(k), u(k)) \qquad \qquad (16.20)$$

This form was used in the control design descrined in Section 16.5.

16.3 Experimental Setup

Two different engines were used in this work. Data from an optical, single-cylinder, engine was used for validating the model presented in Section 16.2 while a six-cylinder conventional engine was used for the control experiments. The model was recalibrated to the conventional engine before being used for control design. Both engines were equipped with cylinder pressure sensors and operated using port fuel injection. All crank angles are given relative to combustion top dead center.

16.3.1 Optical Engine

The optical engine was a Scania heavy-duty diesel engine converted to single-cylinder operation. The engine was equipped with a quartz piston allowing measurements of the wall temperature to be made using thermographic phosphors. For further details on the measurement technique, see [14]. Table 16.2 contains geometric data and relevant valve timings for the engine. During the experiments, the engine was operated manually and only the injected fuel amount was varied in the experiments. The fuel used was iso-octane.

Table 16.2 Optical Engine Specifications

Displacement volume	1966 cm^3
Bore	127.5 mm
Stroke	154 mm
Connecting rod length	255 mm
Compression ratio	16:1
Exhaust valve open	146° ATDC
Exhaust valve close	354° ATDC
Inlet valve open	358° ATDC
Inlet valve close	569° ATDC

16.3.2 Conventional Engine

The engine was a Volvo heavy-duty diesel engine. The engine and the control system was described in detail in [15] and is based on the system used in [16, 8]. The engine specifications are presented in Table 16.3. The fuel used was ethanol.

Table 16.3 Conventional Engine Specifications

Displacement volume	2000 cm^3
Bore	131 mm
Stroke	150 mm
Connecting rod length	260 mm
Compression ratio	18.5:1
Exhaust valve open	101° ATDC
Exhaust valve close	381° ATDC
Inlet valve open	340° ATDC
Inlet valve close	variable

Controllers were designed in MATLAB/Simulink and converted to C-code using Real-Time Workshop [17]. A graphical user interface allowed enabling and disabling controllers as well as manual control of all variables. The experiments were performed at a nominal engine speed of 1200 rpm. A wide selection of possible control signals were available. The control signals used in this work were the crank angle of inlet valve closing (θ_{IVC}) and the intake temperature (T_{in}). In the robustness investigation in Section 16.6 the possibility of varying the engine speed, the amount of injected fuel, and the amount of recycled exhausts were utilized. The engine was equipped with a long-route Exhaust Gas Recirculation (EGR) system.

16.4   Calibration

The model was calibrated using measurement data from the optical engine
described in Section 16.3.1. Data from a positive and a negative step change
in the amount of fuel was available for calibration and validation. Further
validation was then done by evaluating the performance of controllers based
on linearizations of the model. Since the experiments were performed on the
conventional engine described in 16.3.2, the model was first recalibrated.

16.4.1   Calibration of Heat Transfer Parameters

The model was implemented in the Modelica language [18] and then translated into AMPL code [19] using the Optimica Compiler [20]. This allowed
the parameter calibration process to be cast as an optimization problem minimizing the error between the model output and the measured output while
respecting constraints on the parameter values. Since the model is cycle-
resolved it was implemented using difference equations describing the update
of each temperature and pressure state. The initial condition was the temperatures of the gas charge and cylinder wall at the end of an engine cycle.
Modelica does not offer support for formulation of dynamic optimization
problems. The Optimica extension was used to allow constrained optimization problems to be cast in terms of the variables of Modelica models. The
unknown parameters in the model could thus be assigned as optimization parameters and a cost function penalizing deviations from measurement data
was formulated. The Optimica extension translates the parameter identification problem from the Modelica model and Optimica code to AMPL, a
mathematical modeling language aimed at large-scale optimization [19].

Only parameters connected to the heat transfer equations were optimized
initially, and calibration of the prediction of auto-ignition was done once
the temperature trace had been determined. Due to the structure of the
matrices in (16.3) the parameters can be lumped into the following products;
$(h_c A_c)_i$, $k_c A_c / L_c$, $1/m C_v$, and $1/m_c C_p$ where $i = 1, 3, 5$. The optimization
parameters are then organized in a vector p containing these six products
and the integration times t_1, t_3, and t_5, as well as initial values, x_0, of the
states.

While some parameters can be calculated exactly, such as the wall area
at a specific crank angle, the scope of the model is to capture the main
characteristics of the temperature trace, meaning that each heat transfer
event represents the average heat transfer over a portion of the engine cycle.
Some remarks on the possibility of using pre-calculated average values are
presented in Section 16.4.3.

A stationary operating point was used for the calibration and the model
was then validated dynamically during step changes in the equivalence ratio.
The optimization problem can be written as

$$\underset{p,\,x_0}{\text{minimize}}\ (\text{IMEP}_n^m(k) - \text{IMEP}_n(k))^2 + \delta_1\,(T_w^m(k) - T_w(k))^2$$

subject to
$$x(k+1) = \mathbf{F}\,(x(k), u(k))$$
$$y(k) = \mathbf{G}\,(x(k), u(k)) \qquad\qquad (16.21)$$
$$p \in \mathcal{P}$$
$$x(k) = x_0 \quad \forall k$$

where the superscript $'m'$ denotes measured values, $\mathcal{P}$ specifies the allowed parameter values, δ_1 is a weighting parameter, and the last constraint guarantees that the model is in steady state.

16.4.2 Calibration Results

Figure 16.2 shows the model outputs and the wall temperature during a positive step in the equivalence ratio from approximately $\phi = 0.12$ to approximately $\phi = 0.27$. The model captures the qualitative behavior of all three variables. Figure 16.3 shows the model outputs and the wall temperature during a negative step in equivalence ratio from approximately $\phi = 0.27$ to approximately $\phi = 0.17$. Only parameters related to the prediction of θ_{50} were altered from the previous step response. The model somewhat overestimates the wall temperature and IMEP$_n$ at the lower load.

It should be noted that the measurement of the equivalence ratio is a potential source of errors, as a sample of ϕ was obtained approximately every 20 engine cycles, which means that some transients in the model outputs may be due to unmeasured variations in equivalence ratio.

The interaction of wall temperature and combustion phasing can be seen clearly in both experiments, where the combustion occurs earlier as the temperature increases and vice versa.

16.4.3 Calibration of Integration Times

The elements of the matrices in (16.3) all have a physical interpretation. This can be utilized to reduce the number of parameters that need to be adapted to a certain engine. In this section the possibility of using physics-based estimates of the parameters is discussed. The only remaining tuning parameters are then the integration times t_1, t_3, t_5, and the initial temperature state.

16.4.3.1 Physics-Based Parameter Estimates

The specific heat of the charge and the wall, as well as the conduction coefficient of the wall material can be found in, or calculated from, tables [11].

The gas mass can be estimated using the ideal gas law and knowledge of the thermodynamic state at inlet valve closing. The wall mass and thickness can be estimated using geometric data on the engine. The wall area, A_c is a

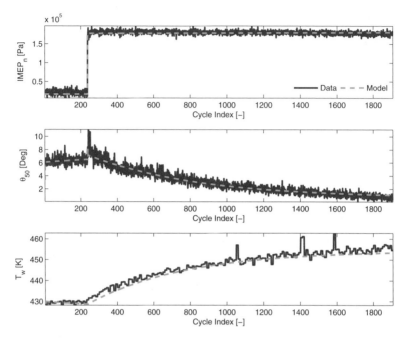

Fig. 16.2 $IMEP_n$, θ_{50} (relative to TDC), and T_w measured and model output during a positive step in equivalence ratio [21].

function of the crank angle and can be expressed as a function of the volume $V(\theta)$ as follows

$$A_c(\theta) = \frac{\pi B^2}{2} + 4\frac{V(\theta)}{B} \tag{16.22}$$

The average area during each heat transfer instant could therefore be used.

As mentioned in Section 16.2.1, the convection coefficient, h_c, is often modeled using the Woschni expression in (16.1). It is suggested in [22] that a more suitable choice for HCCI conditions is the Hohenberg expression [23]

$$h_c = \alpha_s L^{-0.2} P^{0.8} T^{-0.73} v_{\text{tuned}}^{0.8} \tag{16.23}$$

where α_s is a scaling factor for matching the expression to a specific combustion chamber geometry, L is the instantaneous chamber height, and v_{tuned} is a function of the mean piston speed, volume, temperature, and pressure. A modified version of the Woschni expression was also presented in [24]. Regardless of the expression used, average values of h_c can be determined for a specified crank angle interval.

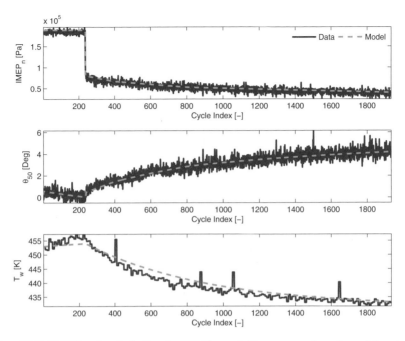

Fig. 16.3 $IMEP_n$, θ_{50} (relative to TDC), and T_w measured and model output during a negative step in equivalence ratio.

16.4.3.2   Optimization Method

The remaining parameters are the integration times, t_1, t_3, and t_5. For open-loop simulation, the initial state x_0 also needs to be selected. If a pre-determined method is used for prediction of auto-ignition, such as an Arrhenius integral with parameters adapted to the specific fuel [1, 6], the optimization criterion is

$$\underset{x_0,t_1,t_3,t_5}{\text{minimize}} \sum_{k=k_0}^{k_N} \tilde{\theta}_{50}^2(k) + \delta_1(IMEP_n^m(k) - IMEP_n(k))^2$$

subject to   (16.24)
$$x(k+1) = \mathbf{F}\left(x(k), u(k)\right)$$
$$y(k) = \mathbf{G}\left(x(k), u(k)\right)$$
$$x(k_0) = x_0$$

where k is the cycle index, $\delta_1 \geq 0$ is a weight, $\tilde{\theta}_{50}(k) = \theta_{50}^m(k) - \theta_{50}(k)$, and the last condition can be replaced with $x(k) = x_0 \, \forall k$ if only steady state measurement data is used for the calibration. To allow for some variation in the wall temperature, a regularized formulation can be used.

$$\text{minimize}_{x_0,t_1,t_3,t_5} \sum_{k=k_0}^{k_N} \tilde{\theta}_{50}^2(k) + \delta_1(\text{IMEP}_n^m(k) - \text{IMEP}_n(k))^2 + \delta_2(\Delta x(k))^2$$

$$\begin{aligned} &\text{subject to} \\ &\quad x(k+1) = \mathbf{F}\left(x(k), u(k)\right) \\ &\quad\quad\ y(k) = \mathbf{G}\left(x(k), u(k)\right) \\ &\quad\quad\ x(k_0) = x_0 \end{aligned} \tag{16.25}$$

where $\Delta x(k) = x(k) - x(k-1)$ and $\delta_2 \geq 0$ is a weight.

16.5   Control

The symbolic math toolbox in MATLAB [25] was used to obtain linearizations on the form

$$y(k+1) = Ay(k) + Bu(k) \tag{16.26}$$

where

$$A = \frac{\partial \hat{\mathbf{F}}(x(k), u(k))}{\partial x(k)}(x^0, u^0), \quad B = \frac{\partial \hat{\mathbf{F}}(x(k), u(k))}{\partial u(k)}(x^0, u^0) \tag{16.27}$$

where (x^0, u^0) is a stationary operating point. The control design was then based on the linear models.

16.5.1   *Predictive Control of Combustion Phasing*

This section reviews the fundamentals of Model Predictive Control (MPC) [4] and outlines the tuning of the controller.

16.5.1.1   Model Predictive Control

Model predictive control was shown to be a suitable control strategy for HCCI [26] due to its MIMO-capabilities and its ability to handle explicit constraints on control signals and outputs. In [8] similar closed-loop performance was demonstrated using MPC, LQ, and PID controllers in certain operating conditions. From a tuning perspective, however, the ability to enforce explicit constraints on the control signals using MPC, is advantageous. The following review is based on [4].

Consider the cost function

$$J(k) = \sum_{i=1}^{H_p} \mathcal{Y}(i|k) + \sum_{i=0}^{H_u-1} \mathcal{U}(i|k) \tag{16.28}$$

where

$$\begin{aligned} \mathcal{Y}(i|k) &= ||\hat{y}(k+i|k) - r(k+i|k)||_Q^2, \\ \mathcal{U}(i|k) &= ||\Delta \hat{u}(k+i|k)||_R^2 \end{aligned} \tag{16.29}$$

and $\hat{y}(k + i|k)$ is the predicted output at time $k + i$ given a measurement at time k, $\Delta\hat{u}(k + 1|k)$ is the predicted change in control signal, and $r(k + i|k)$ is the reference value at time $k + i$. The parameters H_p and H_u define the length of the prediction horizon and the control horizon. At each sample, the cost function in (16.33) is minimized by determining a sequence of changes to the control signal $\Delta u(k + i|k)$, $i = 0 \ldots H_u - 1$, subject to the constraints

$$y_{min} \leq y(k) \leq y_{max} \tag{16.30}$$

$$u_{min} \leq u(k) \leq u_{max} \tag{16.31}$$

$$\Delta u_{min} \leq \Delta u(k) \leq \Delta u_{max} \tag{16.32}$$

for all k. The first step of the optimal sequence is then applied to the plant and the optimization is repeated in the next step yielding a new optimal sequence [4].

16.5.1.2   Controller Tuning

The linear model in (16.26) was used to generate predictions. As only control of θ_{50} was the control objective, there was no penalty on IMEP_n, but it was used as a measurement. A small weight was added to penalize $\theta_{\mathrm{IVC}} - \theta_{\mathrm{IVC}}^r$, where θ_{IVC}^r is a set-point for the control signal, to obtain a slight mid-ranging effect as there are several combinations of control signals that achieve the same output. To avoid excitation effects related to the gas exchange, θ_{IVC} was restricted to $\theta_{IVC} \in [550, 620]$. The inlet temperature was restricted to $T_{\mathrm{in}} \in [110, 135]$. The set-point was held constant over the prediction horizon. This results in the following cost function to be minimized each engine cycle

$$J(k) = \sum_{i=1}^{H_p} ||\hat{\theta}_{50}(k + i|k) - r(k)||_{Q_1}^2 + ||\hat{\theta}_{\mathrm{IVC}}(k + i|k) - \theta_{\mathrm{IVC}}^r||_{Q_2}^2 +$$
$$+ \sum_{i=0}^{H_u - 1} ||\Delta\hat{\theta}_{\mathrm{IVC}}(k + i|k)||_{R_1}^2 + ||\Delta\hat{T}_{\mathrm{in}}(k + i|k)||_{R_2}^2 \tag{16.33}$$

subject to the constraints

$$550 \leq \theta_{\mathrm{IVC}}(k) \leq 620 \tag{16.34}$$

$$110 \leq T_{\mathrm{in}}(k) \leq 135 \tag{16.35}$$

for all k. To obtain error-free tracking, a disturbance observer was used [27]. The prediction horizon was set to 5 engine cycles and the control horizon was set to 2 engine cycles. Increasing the horizons did not have any significant impact on the results.

16.6   Experimental Results

A typical response to a sequence of steps with increasing amplitude is shown in Fig. 16.4. The response time was less than 20 cycles for all steps. The

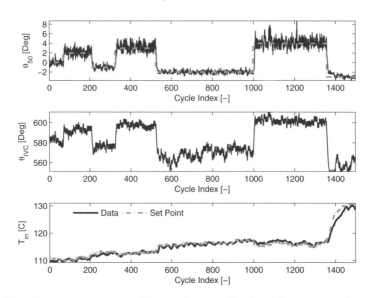

Fig. 16.4 Response to steps of increasing amplitude. All set-point changes are accomplished within 20 cycles [21].

intake temperature changed only slightly until the final step, where the inlet valve reached a constraint. The inlet air temperature was governed by a separate controller that manipulated a fast thermal management system [9] and received its set point from the model predictive controller. This set point along with the measured inlet temperature are included in Figs. 16.4-16.5.

16.6.1 Robustness Towards Disturbances

The robustness towards disturbances in the amount of fuel, the engine speed, and the amount of recycled exhaust gases was investigated experimentally. Fig. 16.5 shows the response as disturbances were added sequentially. At cycle 200 the engine speed was increased from 1200 rpm to 1400 rpm. The injected fuel energy was reduced from 1400 J to 1200 J at cycle 700. Finally, the amount of recycled exhaust gases (EGR) was increased from approximately 0% to 30% at cycle 1350. The bottom plot shows $IMEP_n$ which reflects the impact of the disturbances in engine speed and fuel energy. The disturbances were added without removing the previously added ones, so during the final phase of the experiment, all three were in effect. The combustion phasing was maintained relatively well through the whole sequence.

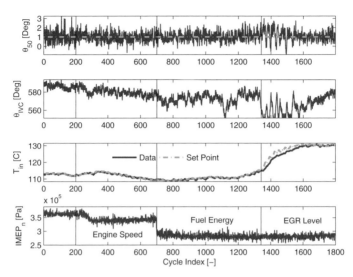

Fig. 16.5 Response to disturbances in the engine speed, the amount of injected fuel, and the amount of EGR. IMEP$_n$ is included to visualize the disturbances in engine speed and fuel energy [21].

16.7 Discussion

Heat transfer effects between the cylinder wall and the gas charge is a good candidate for explaining the slower modes of HCCI dynamics observed in experimental data. The current model is cycle-resolved and therefore captures only the main dynamics of HCCI. A continuous-time formulation, as suggested in [2, 3], allows for a more detailed heat transfer model but is less suitable for control design. The heat transfer events in the current model should be seen as averages over portions of the engine cycle. The integration time t_1, t_3, and t_5 are model parameters that determine the lengths of each event, and were included as optimization parameters in the calibration. The model captured the behavior of the studied outputs well when using data from the optical engine. Closed-loop control using model predictive control based on linearizations of the model also proved successful, further validating the model. The closed-loop system showed a fairly short response time. For step changes of a few degrees around the linearization point in θ_{50} the response time was approximately ten cycles. Only the larger steps (*e.g.*, from -2 to 4 and from 4 to -3 in Fig. 16.4) took around 15 cycles. The inlet valve closing showed to be a sufficient control signal in many operating points. An additional control signal is, however, needed to extend the range of the controller. For example, this is apparent in the end of Fig. 16.4 and during the EGR disturbance in Fig. 16.5. The closed-loop system showed good robustness to disturbances in engine speed, fuel amount, and EGR level. The

EGR seems to have the greatest impact on θ_{50} and there was a clear transient occurring in response to the increase. A possible cause for this is that the thermal properties of the charge are altered by increasing the amount of residuals. Also, as seen in Fig. 16.5, the change in EGR is not reflected in $IMEP_n$, which is used as measurement in the controller.

16.8 Conclusion

This paper has presented physics-based modeling of HCCI including cylinder wall temperature dynamics and model-based control design. The wall temperature model was based on a single zone model and accounted for convection between the charge and the wall as well as conduction through the wall. The wall temperature was updated three times per engine cycle corresponding to intake, combustion, and expansion. The model was calibrated using experimental data from an optical engine allowing measurements of the cylinder wall temperature to be made. The resulting model captured the main dynamics of combustion phasing, indicated mean effective pressure, and cylinder wall temperature. To further validate the model, it was calibrated against a conventional engine and linearizations of the model were used to design model predictive controllers for control of the combustion phasing. The crank angle of inlet valve closing and the intake temperature were used as control signals. The resulting control performance showed response times of approximately ten cycles for step changes in the reference around the linearization point. The closed-loop system also showed robustness towards disturbances in engine speed, fuel amount, and amount of recycled exhausts.

References

[1] Bengtsson, J., Strandh, P., Johansson, R., Tunestål, P., Johansson, B.: Hybrid modelling of homogeneous charge compression ignition (HCCI) engine dynamic—A survey. International Journal of Control 80(11), 1814–1848 (2007)

[2] Blom, D., Karlsson, M., Ekholm, K., Tunestål, P., Johansson, R.: HCCI engine modeling and control using conservation principles. SAE Technical Papers (2008-01-0789) (2008)

[3] Roelle, M.J., Ravi, N., Jungkunz, A.F., Christian Gerdes, J.: A dynamic model of recompression HCCI combustion including cylinder wall temperature. In: Proc. IMECE 2006, Chicago, Illinois, USA (November 2006)

[4] Maciejowski, J.M.: Predictive Control with Constraints. Prentice Hall, Essex (2002)

[5] Chiang, C.-J., Stefanopoulou, A.G., Jankovic, M.: Nonlinear observer-based control of load transitions in homogeneous charge compression ignition engines. IEEE Transactions on Control Systems Technology 15(3), 438–448 (2007)

[6] Shaver, G.M., Roelle, M., Gerdes, J.C.: A two-input two-output control model of HCCI engines. In: Proc. 2006 American Control Conference, Minneapolis, Minnesota, USA, pp. 472–477 (June 2006)

[7] Haraldsson, G.: Closed-Loop Combustion Control of a Multi Cylinder HCCI Engine using Variable Compression Ratio and Fast Thermal Management. PhD thesis, Department of Heat and Power Engineering, Lund Institute of Technology, Lund University, Sweden (January 2005)

[8] Bengtsson, J.: Closed-Loop Control of HCCI Engine Dynamics. PhD thesis, Department of Automatic Control, Lund Institute of Technology, Lund University, Sweden (November 2004)

[9] Widd, A.: Predictive control of hcci engines using physical models. Licentiate Thesis ISRN LUTFD2/TFRT--3246--SE, Department of Automatic Control, Lund University, Sweden (May 2009)

[10] Woschni, G.: A universally applicable equation for instantaneous heat transfer coefficient in the internal combustion egnine. SAE Technical Papers (670931) (1967)

[11] Heywood, J.B.: Internal Combustion Engine Fundamentals. McGraw-Hill International Editions, New York (1988)

[12] Chiang, C.J., Stefanopoulou, A.G.: Sensitivity analysis of combustion timing and duration of homogeneous charge compression ignition (HCCI) engines. In: Proc. 2006 American Control Conference, Minneapolis, Minnesota, USA, pp. 1857–1862 (June 2006)

[13] Widd, A., Tunestål, P., Johansson, R.: Physical modeling and control of homogeneous charge compression ignition (HCCI) engines. In: Proc. 47th IEEE Conference on Decision and Control, Cancun, Mexico, pp. 5615–5620 (December 2008)

[14] Wilhelmsson, C., Vressner, A., Tunestål, P., Johansson, B., Särner, G., Aldén, M.: Combustion chamber wall temperature measurement and modeling during transient HCCI operation. SAE Technical Papers (2005-01-3731) (2005)

[15] Karlsson, M.: Control structures for low-emission combustion in multi-cylinder engines. Licentiate Thesis ISRN LUTFD2/TFRT--3243--SE, Department of Automatic Control, Lund University, Sweden (March 2008)

[16] Strandh, P.: HCCI Operation - Closed Loop Combustion Control Using VVA or Dual Fuel. PhD thesis, Department of Energy Sciences, Lund University, Sweden (May 2006)

[17] Mathworks. Real-Time Workshop. User's Guide. The MathWorks Inc. (2006)

[18] The Modelica Association. The Modelica Association Home Page (2009), http://www.modelica.org

[19] AMPL: A Modeling Language for Mathematical Programming. Duxbury Press (November 2002)

[20] Åkesson, J.: Languages and Tools for Optimization of Large-Scale Systems. PhD thesis, Department of Automatic Control, Lund University, Sweden (November 2007)

[21] Widd, A., Ekholm, K., Tunestål, P., Johansson, R.: Physics-based model predictive control of HCCI combustion phasing using fast thermal management and VVA. IEEE Transactions on Control Systems Technology (2011)

[22] Soyhan, H.S., Yasar, H., Walmsley, H., Head, B., Kalghatgi, G.T., Sorusbay, C.: Evaluation of heat transfer correlations for HCCI engine modeling. Applied Thermal Engineering (29), 541–549 (2009)

[23] Hohenberg, G.F.: Advanced approaches for heat transfer calculations. SAE Technical Papers (79-08-25) (1979)

[24] Chang, J., Güralp, O., Filipi, Z., Assanis, D., Kuo, T.-W., Najt, P., Rask, R.: New heat transfer correlation for an HCCI engine derived from measurements of instantaneous surface heat flux. SAE Technical Papers (2004-01-2996) (2004)

[25] The Mathworks Inc. The Symbolic Math Toolbox User's Guide. The MathWorks Inc. (2010)

[26] Bengtsson, J., Strandh, P., Johansson, R., Tunestål, P., Johansson, B.: Model predictive control of homogeneous charge compression ignition (HCCI) engine dynamics. In: 2006 IEEE International Conference on Control Applications, Munich, Germany, pp. 1675–1680 (October 2006)

[27] Åkesson, J., Hagander, P.: Integral action – A disturbance observer approach. In: Proceedings of European Control Conference, Cambridge, UK (September 2003)

Chapter 17
Comparison of Sensor Configurations for Mass Flow Estimation of Turbocharged Diesel Engines

Tomáš Polóni, Boris Rohal'-Ilkiv, Daniel Alberer, Luigi del Re, and Tor Arne Johansen

Abstract. The increasing demands on quality of power, emissions and overall performance of combustion engines lay new goals for the hardware and the software development of control systems. High-performance embedded controllers open the possibilities for application of numerical methods to solve the problems of modeling and control of combustion engines. Algorithms for estimation of state and parameters are essential components of many advanced control, monitoring and signal processing engine applications. A widely applicable estimator is given by the Extended Kalman Filter (EKF) which defines a finite memory recursive algorithm suited for real-time implementation, where only the last measurement is used to update the state estimate, based on the past history being approximately summarized by the a priori estimate of the state and the error covariance matrix estimate. The proposed EKF uses an augmented air-path state-space model to estimate unmeasurable mass flow quantities. The EKF algorithm based on the augmented state-space model considerably reduces the modeling errors compared to the open loop estimator simulation and compared to the EKF without the augmentation which is demonstrated on a standard production Diesel engine

Tomáš Polóni · Boris Rohal'-Ilkiv
Institute of Automation, Measurement and Applied Informatics,
Faculty of Mechanical Engineering, Slovak University of Technology,
Bratislava, Slovakia
e-mail: {tomas.poloni,boris.rohal-ilkiv}@stuba.sk

Daniel Alberer · Luigi del Re
Institute for Design and Control of Mechatronical Systems,
Johannes Kepler University, Linz, Austria
e-mail: {daniel.alberer,luigi.delre}@jku.at

Tor Arne Johansen
Department of Engineering Cybernetics,
Norwegian University of Science and Technology, Trondheim, Norway
e-mail: tor.arne.johansen@itk.ntnu.no

D. Alberer et al. (Eds.): Identification for Automotive Systems, LNCIS 418, pp. 303–326.
springerlink.com © Springer-Verlag London Limited 2012

data. The experimental validation of the observed state quantities is performed against the measured pressures and turbocharger speed data. The estimated mass flow quantities are indirectly validated through the air compressor flow that is directly validated against air mass flow sensor data. Two two-sensor setups are considered in this study. In the first experiment the intake manifold pressure and the turbocharger speed is used. The second experiment uses the intake manifold pressure and the exhaust manifold pressure as the measurement information for the EKF. The second experiment gives more precise mass flow estimate in term of less bias on the estimates, but more variance due to the high frequency exhaust manifold pressure variations caused by the exhaust valves.

17.1   Introduction

The quality of control, monitoring or fault diagnoses of combustion engines relies on real-time information of the physical quantities from different places in the combustion engine. In the ideal case, to effectively control the torque and emissions, several mass flow sensors would be suitable to have in the intake, exhaust and exhaust gas recirculation (EGR) manifolds. The number of sensors placed in the combustion engines is, however, limited due to their cost and location feasibility. The combination of measured signals with the physical model is a standard approach of model-based filtration to extract unmeasurable physical quantities of the combustion engine. Mean-Value Engine Modeling (MVEM) is a state of the art physical modeling approach of the most important combustion engine phenomena for control oriented purposes [8, 5, 13]. The dynamics is based on the mass and energy balances with the filling and emptying phenomena of the intake and the exhaust manifolds [9] and turbocharger dynamics [15, 21]. Several researchers proposed different observer strategies to estimate unmeasurable mass flows of the turbocharged Diesel engine with the state-space model representation. The mass flow into the cylinders is an important quantity in order to characterize torque and emissions. The mass flow observer problem depends on the sensor configuration considered. The problem of mass flow estimation mainly focuses on in-cylinder flow where nonlinear adaptive schemes [22][23], high-gain and sliding mode schemes [24] or a set-membership algorithm [14] scheme were applied. The mass flow estimation as the replacement of mass flow sensor measurement at the beginning of the intake duct is proposed by [16]. The alternative approach is to use in-cylinder pressure information to compute the air mass flow [4], however the method is not yet fully developed for the engine transient operation.

In the references [22][23], the estimators are based on the intake manifold pressure and the intake manifold temperature measurements. The intake manifold temperature is however a slow measurement with a standard production temperature sensor. With inclusion of a proper model of the sensor dynamics this temperature information should be able to improve the observer performance, or at least not degrade it. The paper [24] gives a review of approaches which focus on different sensor setup configurations mainly with a combination of two sensors from the standard production sensor group: air mass flow (MAF), intake manifold pressure (p_i), exhaust manifold pressure and temperature sensor. The estimation approach based solely on pressure information measured in the intake and exhaust manifolds is suggested in [16]. The combination of a MAF sensor with a MAP sensor is proposed in [14].

In automotive and combustion engine applications a systematic approach for a state estimation is the Kalman Filter (KF) and its modification for nonlinear systems, the Extended Kalman Filter (EKF) [7]. The application of the EKF for the mass flow estimation of a turbocharged Diesel engine with EGR is reported in [6, 10, 18] and the air mass-flow system fault detection in [17]. One of the basic estimation problems is the mass flow bias [11] which is related to the model mismatch. A constrained derivative-free approach to mass flow estimation through unscented Kalman Filter is reported in [1], where the authors use the MAF and the p_i sensor as the measurements.

The main motivation for this work was to apply the EKF for mass flow estimation with standard production Diesel engine sensors and to explore the usability and applicability of such a state estimator. The novel result in this study is the application and comparison of the EKF with two two-sensor measurements with the design of model error bias compensation using an augmentation of the model. The first sensor setup consists of intake manifold pressure and the turbocharger speed measurement while the second is based on pressure measurements in the intake and exhaust manifolds similarly recently also studied in [11].

17.2   Turbochared Engine System Description and Mean-Value Engine Model

The simplified schematic turbocharged Diesel engine setup considered in this study is shown in Figure (17.1). According to this figure, the engine is modeled with the compressor (c), turbine (t), final volume intake manifold (i), final volume exhaust manifold (x), intercooler (ic), EGR cooler (co), EGR pipe and the engine cylinders (e). The model has five states. The first four states represent the mass dynamics and they are: p_i intake manifold pressure [kPa], m_i intake mass [kg], p_x exhaust manifold pressure [kPa] and m_x exhaust mass [kg]. The fifth state is the turbocharger speed n_{tc} [rpm]. The inputs are EGR valve position X_{egr} [%], VGT actuator position X_{vgt} [%], injected fuel mass W_f [$kg.s^{-1}$] and the engine speed n_e [rpm].

The engine model is represented by the following mass and energy balance differential equations in the intake and exhaust manifolds and by the torque balance at the turbocharger shaft [13]

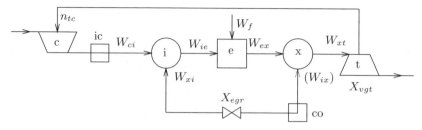

Fig. 17.1 Schematic Diesel engine model representation

$$\dot{p}_i = \frac{R\kappa}{V_i}\left(W_{ci}T_{ci} - W_{ie}T_i + W_{xi}T_{xi} - \frac{\dot{Q}_1}{c_p}\right) \tag{17.1}$$

$$\dot{m}_i = W_{ci} - W_{ie} + W_{xi} \tag{17.2}$$

$$\dot{p}_x = \frac{R\kappa}{V_x}\left((W_{ie} + W_f)T_e - (W_{xi} + W_{xt})T_x - \frac{\dot{Q}_2}{c_p}\right) \tag{17.3}$$

$$\dot{m}_x = W_{ie} + W_f - W_{xi} - W_{xt} \tag{17.4}$$

$$\dot{n}_{tc} = \left(\frac{60}{2\pi}\right)^2 \frac{1}{J}\left(\frac{P_t - P_c}{n_{tc}}\right) \tag{17.5}$$

The symbols in these equations are either input variables, state variables, constants or functions of the five states and inputs. The index associated with each variable defines the location of the variable. In the case of two indexes the first one is the upstream and the second one is the downstream location. The notation is as follows. The mass flows are denoted as W $[kg.s^{-1}]$. The EGR mass flow is modeled as W_{xi} where alternatively the backflow can be considered as W_{ix} however, under standard operating conditions $W_{ix} = 0$. The temperatures $[K]$ in the intake and exhaust manifolds are T_i and T_x. The differences of the static and dynamic pressures and temperatures are neglected because of the low mass-flows. The constant parameters for the model are the intake manifold volume V_i $[m^3]$, exhaust manifold volume V_x $[m^3]$, specific heats at constant pressure and volume $[J.kg^{-1}.K^{-1}]$ c_p, c_v, isentropic exponent of air $\kappa = \frac{c_p}{c_v}$, specific gas constant of air $R = c_p - c_v$ and the turbocharger inertia J $[kg.m^2]$. The heat loses in the Eq. (17.1), $\dot{Q}_1$ and Eq. (17.3), $\dot{Q}_2$ are assumed to be zero. For the intake manifold, this is a reasonable assumption where the temperature is usually lower than 400 K, however in the case of exhaust manifold the heat loses are significant. The exhaust gas heat loss is implicitly captured in the steady state (look-up table) out-flow engine temperature model T_e that is discussed later.

The following equations summarize the dependencies of intermediate variables in Eq. (17.1)-(17.5). Some of these dependencies are expressed as the look-up tables obtained by fitting the steady-state experimental engine data to the second order polynomial surface with parameter vector Θ. The others are obtained by physical relations. The air mass-flow from compressor to the intake manifold W_{ci}, pressure after compressor p_c and the compressor efficiency η_c are mapped as [12]

$$W_{ci} = f_{W_{ci}}(p_i, n_{tc}, \Theta_{W_{ci}}) \tag{17.6}$$

$$p_c = f_{p_c}(W_{ci}, n_{tc}, \Theta_{p_c}) \tag{17.7}$$

$$\eta_c = f_{\eta_c}(W_{ci}, p_i, \Theta_{\eta_c}) \tag{17.8}$$

with fitted polynomial surfaces shown in Figures 17.3-17.5 of the turbocharged passenger car Diesel engine shown in Figure 17.2. The temperature after the compressor is computed

$$T_c = T_a + \frac{1}{\eta_c} T_a \left[\left(\frac{p_c}{p_a}\right)^{\frac{\kappa-1}{\kappa}} - 1 \right] \tag{17.9}$$

where T_a is the ambient temperature. The following mappings for the temperature after the intercooler T_{ci} and the engine volumetric efficiency η_v are

$$T_{ci} = f_{T_{ci}}(W_{ci}, T_c, \Theta_{T_{ci}}) \tag{17.10}$$

$$\eta_v = f_{\eta_v}(n_e, p_i, \Theta_{\eta_v}) \tag{17.11}$$

with fitted polynomial surfaces shown in Figures 17.6, 17.7. The mass flow from the intake manifold to the engine cylinders is

$$W_{ie} = \frac{m_i n_e V_d}{V_i 120} \eta_v. \tag{17.12}$$

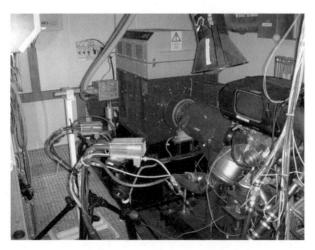

Fig. 17.2 Experimental test bench with a turbocharged passenger car Diesel engine

The temperatures in the intake and the exhaust manifolds are

$$T_i = \frac{p_i V_i}{m_i R} \tag{17.13}$$

$$T_x = \frac{p_x V_x}{m_x R} \tag{17.14}$$

The EGR flow W_{xi} is modeled by the orifice equation that is characterized by the effective flow area

$$A_{egr} = a_2 X_{egr}^2 + a_1 X_{egr} \tag{17.15}$$

where a_2 and a_1 are polynomial coefficients and by nonlinear term Ψ_{egr} that reflects the pressure conditions upstream and downstream at the orifice [9]

$$\Psi_{egr} = \begin{cases} \sqrt{\frac{2\kappa}{\kappa-1}\left[\left(\frac{p_i}{p_x}\right)^{\frac{2}{\kappa}} - \left(\frac{p_i}{p_x}\right)^{\frac{\kappa+1}{\kappa}}\right]} & \text{if } \frac{p_i}{p_x} \geq \left(\frac{2}{\kappa+1}\right)^{\frac{\kappa}{\kappa-1}} \\ \sqrt{\left(\kappa\frac{2}{\kappa+1}\right)^{\frac{\kappa+1}{\kappa-1}}} & \text{otherwise} \end{cases} \tag{17.16}$$

$$W_{xi} = \frac{A_{egr} p_x}{\sqrt{RT_x}} \Psi_{egr} \tag{17.17}$$

The temperature of the mass that flows out of the cylinders T_e, EGR mass temperature T_{xi} and the mass flow from the exhaust manifold to the turbine W_{xt} are

$$T_e = f_{T_e}(W_f, W_{ie}, \Theta_{T_e}) \tag{17.18}$$

$$T_{xi} = f_{T_{xi}}(T_x, W_{xi}, \Theta_{T_{xi}}) \tag{17.19}$$

$$W_{xt} = f_{W_{xt}}(X_{vgt}, p_x, \Theta_{W_{xt}}) \tag{17.20}$$

with fitted polynomial surfaces shown in Figures 17.8-17.10. The power of the compressor is defined as

$$P_c = W_{ci} c_p \frac{1}{\eta_c} T_a \left[\left(\frac{p_c}{p_a}\right)^{\frac{\kappa-1}{\kappa}} - 1\right] \tag{17.21}$$

The power of the turbine is calculated as

$$P_t = W_{xt} c_p \frac{1}{\eta_t} T_x \left[1 - \left(\frac{p_t}{p_x}\right)^{\frac{\kappa-1}{\kappa}}\right] \tag{17.22}$$

where the turbine efficiency

$$\eta_t = f_{\eta_t}\left(\frac{p_t}{p_x}, n_{tc}, \Theta_{\eta_t}\right) \tag{17.23}$$

is mapped with the turbine pressure

$$p_t = b_3 W_{xt}^3 + b_2 W_{xt}^2 + b_1 W_{xt} + b_0 \qquad (17.24)$$

where $b_0, \ldots, b_3$ are the polynomial coefficients.

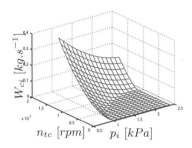

Fig. 17.3 Compressor flow

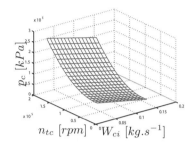

Fig. 17.4 Pressure after compressor

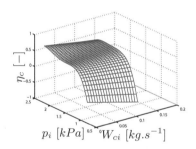

Fig. 17.5 Compressor efficiency

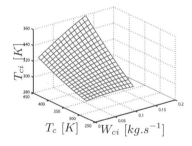

Fig. 17.6 Temperature after inter-cooler

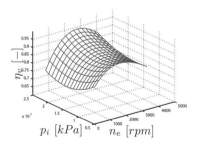

Fig. 17.7 Engine volumetric efficiency

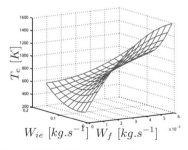

Fig. 17.8 Cylinder outflow temperature

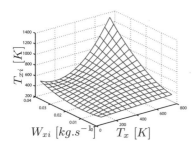

Fig. 17.9 EGR temperature

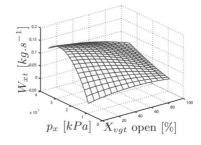

Fig. 17.10 Turbine flow

17.3   Open Loop Observer

The precision of the MVEM depends on its structure and parameters. The imprecision is given by the finite mean-value model structure that with limited accuracy gives a true picture of physical relations between different subsystems of the turbocharged Diesel engine. The identification of the maps of the MVEM is experimentally performed at the engine steady state conditions. The identification in steady state conditions introduces certain systematic errors due to not taking into account the dynamic effects of the modeled variable. This will lead to errors of modeled states compared to true engine states. To see the performance of the air-path model given by Eq. (17.1)-(17.5) the observer is compared in open loop against real measured data from the turbocharged Diesel engine shown in Figures 17.11-17.16.

The engine is operated under standard laboratory conditions. A driving cycle is performed at the engine to record the data of the intake manifold p_i, exhaust manifold p_x, turbocharger speed n_{tc} and the intake mass flow sensor data. One sampling interval is T=0.1 [s]. In the open loop mode the observer is not using any measured information to compute the estimated states except the model inputs (where n_e is a measurement). In the open loop mode, the states are purely numerically simulated through the engine model from the given initial conditions of states and the given inputs. The inputs for the engine and the model are the percentages of open positions of vanes X_{vgt} and EGR valve X_{egr}, displayed in Figure 17.11, fuel injection W_f, displayed in Figure 17.12 and the load which is directly reflected by the engine speed displayed in Figure 17.13. During the open loop performance of the observer, first three measurements p_i, p_x and n_{tc} are used to validate the model.

The open loop observer gives biased estimates as can be expected. The comparison of the estimated intake manifold pressure in open loop with the

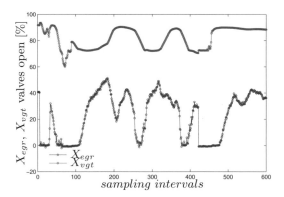

Fig. 17.11 Positions of vanes and EGR

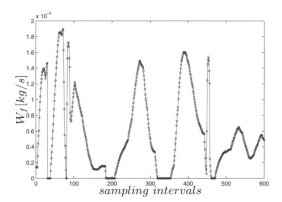

Fig. 17.12 Injected fuel mass

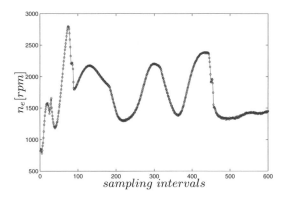

Fig. 17.13 Engine speed

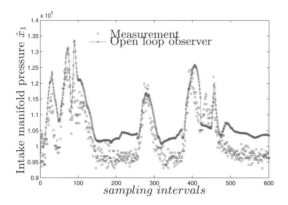

Fig. 17.14 Open loop observer of intake manifold pressure

measured intake manifold pressure is shown in Figure 17.14. The computed
state is shifted up and gives higher intake pressure than the real one. The
resulting comparison of the exhaust manifold pressure open loop and measured is shown in Figure 17.15. The modeling errors are less evident for this
state. The open loop and measured data comparison is for the turbocharger
speed shown in Figure 17.16. Similarly, as in the case of intake manifold pressure, also the turbocharger speed state estimate is obviously diverted from
its measurement. The open loop observer indicates with which states the
modeling errors are associated with. One way to remove these errors is to
consistently inspect the parameter identification of the maps, to inspect the
methods and conditions of the experiments under which the maps were obtained and identify more accurate maps. This is a considerable task and does
not account for structural model errors. Another way to compensate for these

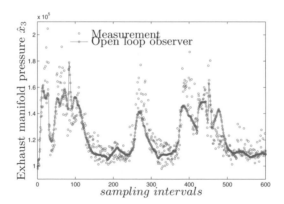

Fig. 17.15 Open loop observer of exhaust manifold pressure

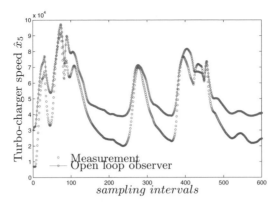

Fig. 17.16 Open loop observer of turbocharger speed

systematic modeling errors is to augment the model with bias parameters to
be on-line identified by the observer. The number of such bias parameters
is limited by the observability i. e. by the number of measurements available and by the structure of the equations [10]. Two measured variables will
be further processed by the observer which means that, at most, two biases
can be added. The system is open loop asymptotically stable and the estimates would converge if there were no model errors. Two measurements
give two additional equations that allow two unknowns (two bias variables)
to be introduced and solved. A possible augmentation of the model where
$x_s = [p_i, m_i, p_x, m_x, n_{tc}]^T = [x_1, x_2, x_3, x_4, x_5]^T$ is in the form

$$\dot{x}_1 = \frac{R\kappa}{V_i}\left(W_{ci}T_{ci} - W_{ie}T_i + W_{xi}T_{xi}\right) + w_i + z_{1,1} \tag{17.25}$$

$$\dot{x}_2 = W_{ci} - W_{ie} + W_{xi} + z_{1,2} \tag{17.26}$$

$$\dot{x}_3 = \frac{R\kappa}{V_x}\left[(W_{ie} + W_f)\,T_e - (W_{xi} + W_{xt})\,T_x\right] + z_{1,3} \tag{17.27}$$

$$\dot{x}_4 = W_{ie} + W_f - W_{xi} - W_{xt} + z_{1,4} \tag{17.28}$$

$$\dot{x}_5 = \left(\frac{60}{2\pi}\right)^2 \frac{1}{J}\left(\frac{P_t - P_c}{n_{tc}}\right) + w_{tc} + z_{1,5} \tag{17.29}$$

$$\dot{w}_i = z_{2,6} \tag{17.30}$$

$$\dot{w}_{tc} = z_{2,7} \tag{17.31}$$

where one added bias term w_i based on the open loop observer results is
associated with the intake manifold pressure state and another one w_{tc} is
associated with the turbocharger speed state. A compact form is

$$\dot{x}_s = \tilde{f}_c(x_s, w, u) + z_1 \tag{17.32}$$

$$\dot{w} = z_2 \tag{17.33}$$

where the continuous function $\tilde{f}_c$ has the mapping properties $\tilde{f}_c : \mathbb{R}^{n_s} \times \mathbb{R}^{n_w} \times \mathbb{R}^{n_u} \rightarrow \mathbb{R}^{n_s}$, n_s is the size of the state vector x_s, n_w is the size of the bias vector w and n_u is the size of the input vector u. The bias vector is $w = [w_i, w_{tc}]^T$, the state process noise vector is $z_1 = [z_{1,1}, z_{1,2}, z_{1,3}, z_{1,4}, z_{1,5}]^T$ and the augmented state (bias) process noise vector is $z_2 = [z_{2,6}, z_{2,7}]^T$. The augmented state vector $x \in \mathbb{R}^{n_s + n_w}$ is defined by

$$x = \begin{bmatrix} x_s \\ w \end{bmatrix}. \tag{17.34}$$

The state and bias equations can be combined as

$$\dot{x} = f_c(x, u) + z \tag{17.35}$$

where the continuous function f_c has the mapping properties $f_c : \mathbb{R}^{n_x} \times \mathbb{R}^{n_u} \rightarrow \mathbb{R}^{n_x}$, n_x being the dimension of the augmented state vector x and $z = [z_1, z_2]^T$. The observation equation may be written as

$$y = h_c(x, u) + v \tag{17.36}$$

where $y \in \mathbb{R}^{n_y}$ is a vector of measurements, $h_c : \mathbb{R}^{n_x} \times \mathbb{R}^{n_u} \rightarrow \mathbb{R}^{n_y}$ is a continuous measurement function and n_y is a number of measured states. The measurement errors are modeled with the noise term $v \in \mathbb{R}^{n_y}$. The most frequent situation encountered in practice is when the system is governed by continuous-time dynamics and the measurements are obtained at discrete time instances. For the problem formulation we consider the numerically discretized dynamic nonlinear system described by the equations

$$x_{t+1} = f(x_t, u_t) + z_t \tag{17.37}$$
$$y_t = h(x_t, u_t) + v_t \tag{17.38}$$

for $t = 0, 1, \ldots$, where $x_t \in \mathbb{R}^{n_x}$ is the state vector, $u_t \in \mathbb{R}^{n_u}$ is the input vector and $z_t \in \mathbb{R}^{n_z}$ is a process noise vector. The numerical discretization algorithm will be discussed in the following section. The state vector is observed through the measurement equation (17.38) where $y_t \in \mathbb{R}^{n_y}$ is the observation vector and $v_t \in \mathbb{R}^{n_y}$ is a measurement noise vector.

17.4 Extended Kalman Filter

The EKF is perhaps the most often applied algorithm for the estimation of state and parameters of nonlinear dynamic systems and the following algorithm is in the literature known as continuous-discrete or hybrid EKF [7]. The dynamic system is given by Eq. (17.37) and Eq. (17.38). The main assumption about the process and the measurement noise is that they have the white noise properties of a Gaussian distribution with zero mean

$$z_t \sim N(0, Q_t) \qquad v_t \sim N(0, R_t) \tag{17.39}$$

where Q_t is a process noise covariance matrix and R_t is a measurement noise covariance matrix. The initial condition of the state vector is $x_0 \sim N(\hat{x}_0^+, P_0^+)$. The estimate of the state vector at $t = 0$ begins with the initial state vector estimate and with the initial covariance matrix of the initial state vector estimate error

$$\hat{x}_0^+ = E[x_0] \tag{17.40}$$
$$P_0^+ = E[(x_0 - \hat{x}_0^+)(x_0 - \hat{x}_0^+)^T] \tag{17.41}$$

From time instance $t - 1$, the dynamic system is simulatively propagated one step ahead as

$$\hat{x}_t^- = f(\hat{x}_{t-1}^+, u_{t-1}) \tag{17.42}$$

where $t = 1, 2, \ldots$. This one step computation is an a priori state estimate. The time update of the covariance matrix estimate is given by

$$\dot{P} = Z(\hat{x})P + PZ^T(\hat{x}) \tag{17.43}$$

where

$$Z(\hat{x}) = \frac{\partial f_c(x)}{\partial x}\bigg|_{x=\hat{x}} . \tag{17.44}$$

The covariance matrix estimate of the state vector $\hat{x}_t^-$ estimation error is achieved by simulative propagation of Eq. (17.43)

$$P_t^- = g(P_{t-1}^+, Z(\hat{x}_{t-1}^+)). \tag{17.45}$$

The EKF gain matrix of time instant t is

$$K_t = P_t^- L_t^T [L_t P_t^- L_t^T + M_t R_t M_t^T]^{-1} \tag{17.46}$$

and the obtained measurement y_t is used for state vector estimation (a posteriori estimate)

$$\hat{x}_t^+ = \hat{x}_t^- + K_t[y_t - h(\hat{x}_t^-)] \tag{17.47}$$

The covariance matrix is computed as

$$P_t^+ = [I - K_t L_t]P_t^-[I - K_t L_t]^T + K_t M_t R_t M_t^T K_t^T + Q_t \tag{17.48}$$

where

$$L_t(\hat{x}_t^-) = \frac{\partial h(x_t)}{\partial x_t}\bigg|_{x_t=\hat{x}_t^-} \qquad M_t(\hat{x}_t^-) = \frac{\partial h(x_t)}{\partial v_t}\bigg|_{x_t=\hat{x}_t^-} \tag{17.49}$$

The numerical differential computations are the bases of dynamic state propagation Eq. (17.42), covariance matrix time update Eq. (17.45) and Jacobian matrix evaluation Eq. (17.44). In this study, the standard Matlab algorithm ode23 is used to propagate Eq. (17.42) and Eq. (17.45) which is a variable step explicit Runge-Kutta method [3]. The Jacobian is computed via an adaptive-step routine called numjac [19, 20] to evaluate Eq. (17.44).

Implementation of the EKF on a real ECU (Electronic Control Unit) poses some additional challenges and a tradeoff between computational complexity and numerical accuracy. Computations of Jacobian should either exploit the polynomial structure of the maps, or structural zeros in the Jacobian if finite differences are used. Robust numerical implementation of the EKF like Bierman's algorithm must be considered [2].

17.5 Estimation Results

The estimation of states and observation of the unmeasurable mass flows of a turbocharged Diesel engine is performed at the transient dynamic operation data from the test driving cycle. In this study, two estimation experiments were performed with different sensor setups to find out which sensor combination gives the most accurate estimates. The inputs for the EKF are shown in Figures 17.11-17.13. The computation of mass flow estimations is based on MVEM equations presented in earlier section. The mass flow computations are based on the state vector estimate computed by the EKF. After obtaining the estimated state vector from the EKF $\hat{x}_s = [\hat{p}_i, \hat{m}_i, \hat{p}_x, \hat{m}_x, \hat{n}_{tc}]^T$, the mass flows are computed as follows

$$\hat{W}_{ci} = f_{W_{ci}}(\hat{p}_i, \hat{n}_{tc}, \Theta_{W_{ci}}) \tag{17.50}$$

$$\hat{W}_{ie} = \frac{\hat{m}_i n_e V_d}{V_i 120} \eta_v \tag{17.51}$$

$$\hat{W}_{xi} = \frac{A_{egr}\hat{p}_x}{\sqrt{R\hat{T}_x}} \hat{\Psi}_{egr} \tag{17.52}$$

$$\hat{W}_{xt} = f_{W_{xt}}(X_{vgt}, \hat{p}_x, \Theta_{W_{xt}}) \tag{17.53}$$

The considered sensors for the EKF estimation were the intake manifold pressure p_i, exhaust manifold pressure p_x and the turbocharger speed n_{tc}. The only mass flow which can be validated using validation measurements on the engine is the air mass-flow from the compressor to the intake manifold. Even this MAF measurement is performed somewhat upstream of the compressor and not right after the compressor. Nevertheless, the distance is small, *i.e.*, the inertia in between the two positions is low, and it is still reasonable to compare $\hat{W}_{ci}$ (Eq. (17.50)) with the MAF sensor. It is assumed here that if the $\hat{W}_{ci}$ is accurate, then the other mass flows (Eq. (17.51)-(17.53)) are likely to be accurate as well. The performance of the EKF without a state-space model augmentation with two bias terms, as proposed by Eq. (17.25)-(17.31), was also performed. The results were just a little better than the open loop observer with similar plots as the open loop observer. Also the experiment, not reported in this study, was performed with one-sensor and one estimated bias setup, where only the turbocharger speed measurement was used. The resulting mass flow estimates were not accurate enough and similar results can be expected with any one-sensor configuration.

In the following subsections two two-sensor observers are evaluated. Two kind of measurements are considered, the measurement and the evaluation measurement. The measurement was directly used and processed by the EKF while the evaluation measurement was used to test and verify how good is the estimate of a particular state or variable. The matrices of the EKF were set as follows. The initial covariance matrix of the initial state vector estimate error was chosen as $P_0^+ = diag[10^2, 10^{-2}, 10^2, 10^{-2}, 10^2, 10^2, 10^2]$, the process noise covariance matrix $Q_t = diag[10^2, 0, 10^2, 0, 10^2, 10^3, 10^3]$ and the measurement noise covariance matrix $R_t = diag[10^2, 10^2]$.

17.5.1 Two-Sensor Observer: p_i, n_{tc}

The first EKF estimation procedure is based on the information measured in the intake manifold p_i and the turbocharger speed n_{tc}. These two measurements form the vector of measurement $y_t = [p_i, n_{tc}]^T$. The intake manifold pressure, shown in Figure 17.17, has an unbiased estimate because information about the intake manifold pressure is directly processed by the EKF. The intake manifold mass m_i is unmeasurable and its estimate is shown in Figure 17.18. The exhaust manifold pressure is the noisiest measurement in the air-path of a Diesel engine due to engine cycle pressure waves. The exhaust manifold pressure measurement was not processed by the EKF and so it served as the evaluation measurement. The estimated manifold pressure is slightly biased below the measured signal as can be seen in Figure 17.19. The estimate of the exhaust manifold pressure m_x is shown in Figure 17.20. The turbocharger speed measurement is a smooth signal which in this sensor setup was used by the EKF. The tight turbocharger speed estimate with almost no error is shown in Figure 17.21. The process noise that involves the model mismatch is compensated by the bias terms added to the intake manifold pressure equation w_i and by the turbocharger speed bias w_{tc} added to the turbocharger speed equation. The estimated bias terms are displayed in Figure 17.22. The estimated mass flows for which no evaluation measurements exist are shown in Figure 17.23. The evaluated flow from the compressor to the intake manifold W_{ci} is compared with the MAF measurement in Figure 17.24. The W_{ci} quantity is more precisely estimated for the engine speed above $2000 rpm$ where the down speeding and lower speed is accompanied by more error between the measurement and the estimate.

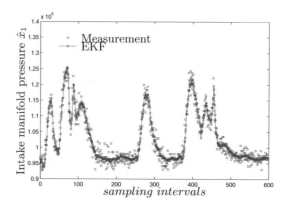

Fig. 17.17 Measured and estimated intake manifold pressure

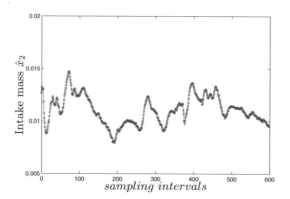

Fig. 17.18 Estimated mass of intake manifold

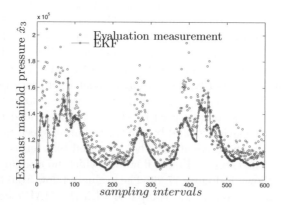

Fig. 17.19 Evaluation measurement and estimated exhaust manifold pressure

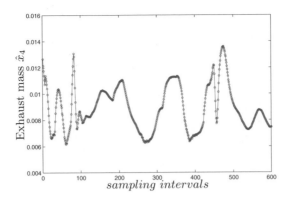

Fig. 17.20 Estimated mass of exhaust manifold

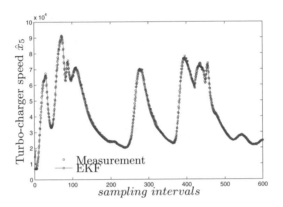

Fig. 17.21 Measured and estimated turbocharger speed

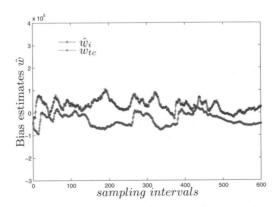

Fig. 17.22 Estimated biases

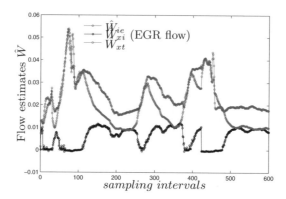

Fig. 17.23 Estimated mass flows

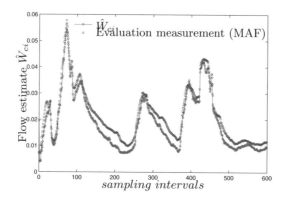

Fig. 17.24 Evaluation measurement and estimated compressor flow

17.5.2   Two-Sensor Observer: p_i, p_x

The second EKF estimation procedure is based on the pressure information measured in the intake manifold p_i and in the exhaust manifold p_x. The vector of measurement is $y_t = [p_i, p_x]^T$. Introducing the exhaust manifold pressure information to the observer causes the noise disturbance and expectingly, more noisy estimates. The filtered intake manifold pressure with the measured pressure signal used by the EKF is shown in Figure 17.25. The estimated mass in the intake manifold in shown in Figure 17.26. The exhaust manifold pressure estimate with measurement processed by the EKF and the estimated mass in the exhaust manifold are displayed in Figures 17.27 and 17.28. Both pressures were filtered satisfyingly good in comparison with their measurements, however both pressure signals were directly processed by the EKF. In this sensor setup, the EKF was not making the adjustments of the state vector based on the turbocharger speed measurement and so it is considered as the evaluation measurement. The turbocharger speed estimate is shown in Figure 17.29 where some tracking error is present during down speeding and lower speed. The estimated biases are displayed in Figure 17.30. In Figure 17.31 the estimated mass flows from different locations are displayed. The measurement noise distortion of the estimates is more evident here than in previous sensor setup. In the last Figure 17.32, the estimated compressor air mass-flow with the evaluation measurement is displayed. In this case, the estimated air mass-flow has almost no tracking error during down speeding and in lower speed regimes, however the measurement noise and the process noise introduce greater variance.

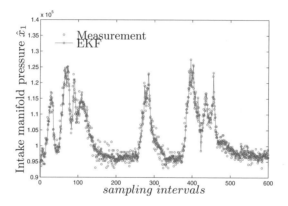

Fig. 17.25 Measured and estimated intake manifold pressure

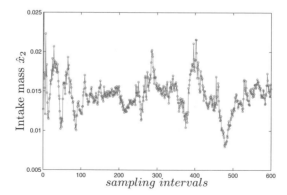

Fig. 17.26 Estimated mass of intake manifold

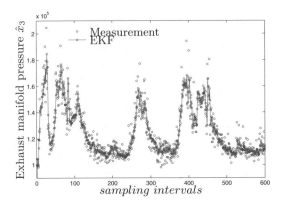

Fig. 17.27 Measured and estimated exhaust manifold pressure

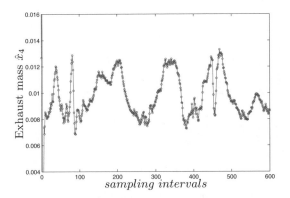

Fig. 17.28 Estimated mass of exhaust manifold

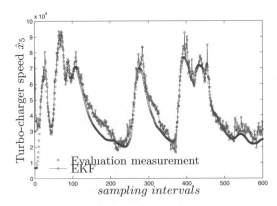

Fig. 17.29 Evaluation measurement and estimated turbocharger speed

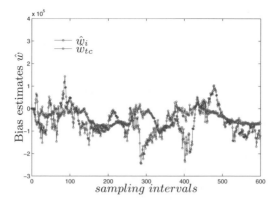

Fig. 17.30 Estimated biases

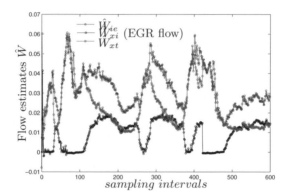

Fig. 17.31 Estimated mass flows

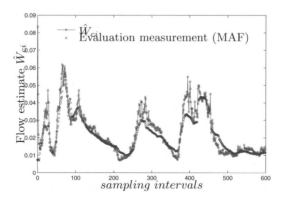

Fig. 17.32 Evaluation measurement and estimated compressor flow

17.6 Conclusion

The main idea behind the applied state estimator is to estimate mass flow quantities which can not be measured. The EKF is applied in this study to combine the model with the measured pressure information from the intake manifold, exhaust manifold and the measured turbocharger speed of a production passenger car Diesel engine. The combination of the production sensors can form the measurement vector to be processed by the EKF. Two two-sensor alternatives were considered. The first approach uses the combination of the intake manifold pressure and the turbocharger speed, the second approach uses the combination of the intake manifold pressure and the exhaust manifold pressure. The first sensor configuration lead to slightly biased estimates of the exhaust manifold pressure and the compressor air mass-flow. The second sensor configuration improved the offset of the estimates, however introducing the exhaust manifold pressure to the observer increased the variance of the estimates. The estimation of the mass flows does not rely on the intake mass flow sensor (MAF) which can be an advantage toward the MAF and thus can be seen as an alternative to using a physical MAF sensor.

Acknowledgements. This work is primarily supported by the Slovak research and development agency under project APVV-0280-06. The first author is supported by research project APVV-LPP-0118-09.

This research is also partially supported by the grant from Iceland, Liechtenstein and Norway through the EEA Financial Mechanism and the Norwegian Financial Mechanism. This project is also co-financed from the state budget of the Slovak Republic.

References

[1] Aßfalg, J., Allgöwer, F., Fritz, M.: Constrained derivative-free augmented state estimation for a diesel engine air path. In: Ninness, B., Hjalmarsson, H. (eds.) 14th IFAC Symposium on System Identification (2006)

[2] Bierman, G.J.: Factorization methods for discrete sequential estimation. Academic Press, London (1977)

[3] Butcher, J.: Numerical Methods for Ordinary Differential Equations. John Wiley & Sons, Ltd., Chichester (2003)

[4] Desantes, J., Galindo, J., Guardiola, C., Dolz, V.: Air mass flow estimation in turbocharged diesel engines from in-cylinder pressure measurement. Experimental Thermal and Fluid Science 34(1), 37–47 (2010), doi:10.1016/j.expthermflusci.2009.08.009

[5] Eriksson, L., Wahlström, J., Klein, M.: Physical Modeling of Turbocharged Engines and Parameter Identification. In: Automotive Model Predictive Control, pp. 53–71. Springer, Heidelberg (2010)

[6] Fantini, J., Peron, L., Marguerie, B.: Identification and validation of an air mass flow predictor using a nonlinear stochastic state representation. SAE paper 2000-01-0935 (2000)

[7] Gelb, A., Joseph, F., Kasper, J., Raymond, A., Nash, J., Price, C.F., Arthur, A., Sutherland, J.: Applied optimal estimation. The MIT Press, Cambridge (2001)

[8] Guzzella, L., Onder, C.H.: Introduction to Modeling and Control of Internal Combustion Engine Systems, 2nd edn. Springer, Heidelberg (2010)

[9] Heywood, J.B.: Internal Combustion Engine Fundamentals. McGraw-Hill, New York (1988)

[10] Höckerdal, E., Eriksson, L., Frisk, E.: Air mass-flow measurement and estimation in diesel engines equipped with egr and vgt. SAE paper 2008-01-0992 (2008)

[11] Höckerdal, E., Frisk, E., Eriksson, L.: Observer design and model augmentation for bias compensation with a truck engine application. Control Engineering Practice 17(3), 408–417 (2009), doi:10.1016/j.conengprac.2008.09.004

[12] Jung, M.: Mean-value modelling and robust control of the airpath of a turbocharged diesel engine. Ph.D. thesis, Sidney Sussex College, University of Cambridge (2003)

[13] Kolmanovsky, I., Moraal, P., van Nieuwstadt, M., Stefanopoulou, A.: Issues in modelling and control of intake flow in variable geometry turbocharged engines. In: 18th IFIP Conference of Systems Modelling and Optimization (1998)

[14] Kolmanovsky, I., Sivergina, I., Sun, J.: Simultaneous input and parameter estimation with input observers and set-membership parameter bounding: theory and an automotive application. International Journal of Adaptive Control and Signal Processing 20(5), 225–246 (2006)

[15] Moraal, P., Kolmanovsky, I.: Turbocharger modeling for automotive control applications. SAE paper 1999-01-0908 (1999)

[16] van Nieuwstadt, M., Kolmanovsky, I., Moraal, P., Stefanopoulou, A., Jankovic, M.: Egr-vgt control schemes: experimental comparison for a high-speed diesel engine. IEEE Control Systems Magazine 20(3), 63–79 (2000), doi:10.1109/37.845039

[17] Nyberg, M., Stutte, T.: Model based diagnosis of the air path of an automotive diesel engine. Control Engineering Practice 12(5), 513–525 (2004), doi:10.1016/S0967-0661(03)00120-5; Fuzzy System Applications in Control

[18] Pavković, D., Deur, J., Kolmanovsky, I., Hrovat, D.: Application of adaptive kalman filter for estimation of power train variables. SAE International Journal of Passenger Cars - Mechanical Systems 1(1), 480–491 (2009)

[19] Salane, D.E.: Adaptive routines for forming jacobians numerically. Tech. Rep. SAND86-1319, Sandia National Laboratories (1986)

[20] Shampine, L.F., Reichelt, M.W.: The matlab ode suite. SIAM Journal on Scientific Computing 18(1), 1–22 (1997)

[21] Sorenson, S.C., Hendricks, E., Magnússon, S., Bertelsen, A.: Compact and accurate turbocharger modelling for engine control. S A E Transactions 114(3), 1343–1353 (2005)

[22] Stefanopoulou, A.G., Storset, O.F., Smith, R.: Pressure and temperature-based adaptive observer of air charge for turbocharged diesel engines. International Journal of Robust and Nonlinear Control 14(6), 543–560 (2004)

[23] Storset, O.F., Stefanopoulou, A.G., Smith, R.: Adaptive air charge estimation for turbocharged diesel engines without exhaust gas recirculation. Journal of Dynamic Systems, Measurement, and Control 126(3), 633–643 (2004), doi:10.1115/1.1771691

[24] Stotsky, A., Kolmanovsky, I.: Application of input estimation techniques to charge estimation and control in automotive engines. Control Engineering Practice 10(12), 1371–1383 (2002), doi:10.1016/S0967-0661(02)00101-6

Chapter 18
Optimal Finite and Receding Horizon Control for Identification in Automotive Systems

Ilya Kolmanovsky and Dimitar P. Filev

Abstract. The paper illustrates the use of optimal finite horizon and receding horizon control techniques to generate input excitation to enhance parameter identification in automotive systems. Firstly, it is shown that a Design of Experiments (DoE) problem of determining transient trajectories for off-line engine parameter identification can be posed as an optimal control problem, where either the determinant or the trace of the inverse of Fisher information matrix is minimized. Then an approach to adaptation of parameters in transient feed-forward compensation algorithms is discussed. Both Fisher information matrix computation and parameter update law utilize output sensitivity with respect to parameters being identified. Finally, a receding horizon optimal control framework for on-line parameter identification is considered where through an appropriate formulation of the cost function being minimized, it is shown that the system can be controlled to maintain tracking performance, satisfy constraints, and generate persistent excitation for parameter identification. For illustration, we use throughout an example based on the identification of transient fuel model parameters. An additional example of fast engine steady-state mapping is discussed to suggest another application of the receding horizon approach to on-line parameter identification.

Ilya Kolmanovsky
Department of Aerospace Engineering,
The University of Michigan, Ann Arbor, MI
e-mail: ilya@umich.edu

Dimitar P. Filev
Ford Motor Company, Research and Advanced Engineering,
2101 Village Rd., Dearborn, MI 48121
Tel.: 313-248-1652
e-mail: dfilev@ford.com

D. Alberer et al. (Eds.): Identification for Automotive Systems, LNCIS 418, pp. 327–348.
springerlink.com   © Springer-Verlag London Limited 2012

18.1 Introduction

In this paper we discuss several techniques relevant to parameter identification and to the Design of Experiments (DoE) in automotive engines and other automotive systems. Traditionally, many of the engine model parameters are determined in the calibration phase using steady-state mapping and regression techniques. Other parameters, such as parameters in a transient fuel model of gasoline engines, are determined using collected data from transient tests in the calibration phase. Parameter identification may also be employed on-board of the vehicles to improve and adopt the compensation algorithms to specific vehicle characteristics.

The determination of input trajectories that are best suited for transient data collection and parameter identification in the calibration phase may be viewed as an optimal control problem, which can be solved numerically. Along these lines, in Section 18.2, we describe an approach to determining such optimal trajectories based on numerically minimizing the determinant or the trace of the inverse of the Fisher information matrix. In Section 18.3, we propose an approach to identifying model parameters by adapting on-line tunable parameters in the inverse model which is used for feedforward control. This approach, which is similar to the sensitivity-based identification method of [5], avoids the need to model and simultaneously identify parameters in output dynamics provided these dynamics satisfy appropriate conditions. sensitivity-based method, similar to the one in [5], for directly tuning parameters in transient feed-forward compensation algorithms during engine calibration phase. This approach is robust to uncertainties in exhaust mixing and UEGO sensor dynamics. In Section 18.4, we formulate a receding horizon approach to on-board parameter identification, wherein the control signal is determined so that to best estimate unknown states and parameters, while maintaining adequate tracking performance and *probabilistically* enforcing pointwise-in-time constraints. This approach is particularly suitable for on-board parameter identification, as it ensures fast parameter convergence and avoids unacceptable degradation in tracking performance that may negatively affect the driver.

The developments in this paper are based on an earlier work reported in the conference papers [4], [12]. Related open-loop optimal control problems were studied in the book [11] and briefly considered in [6] (p. 403).

We use transient fuel identification as an example throughout, and we discuss fast engine mapping as another potential application for the receding horizon control approach. The transient fuel dynamics are significant in port-fuel injected engines, the fuel is injected into the intake ports, typically during the exhaust stroke when the intake valves are closed. A portion of the injected fuel vaporizes and is immediately drawn into the cylinder when that cylinder intake valves open. The remainder of the injected fuel replenishes the liquid fuel puddle which is formed on the walls of the ports and on the intake valves. A fraction of that puddle evaporates and is drawn into the engine cylinder on the next engine cycle but a fraction of it remains in the liquid puddle. The Aquino's model [1] is often used to represent transient fuel dynamics. This model is combined with a first-order lag which captures the exhaust mixing and UEGO sensor dynamics. The model has the following form

$$
\begin{aligned}
\dot{m}_p &= -\frac{m_p}{\tau} + X \cdot W_{fi} \\
W_{fc} &= \frac{m_p}{\tau} + (1 - X) \cdot W_{fi}, \\
\dot{\phi}_m + a\phi_m &= a\left(\frac{W_{fc}}{W_{ac}}\right),
\end{aligned}
\tag{18.1}
$$

where m_p is the total mass of fuel puddles of all cylinders, W_{ac} is the airflow into the engine cylinders, W_{fi} is the mass flow rate of injected fuel, W_{fc} is the mass flow rate of fuel entering the engine cylinders, X, τ are the parameters in the Aquino's model (which depend on engine operating condition), ϕ_m is the measured fuel-to-air ratio and $1/a$ is the time constant of the combined exhaust mixing and UEGO sensor dynamics. During either identification or direct adaptation, trajectories of the injected fuel rate and measured fuel-to-air ratio are first shifted in time to compensate for the delay (which is a known function of engine speed and air flow rate through the engine); for this reason our formulation (18.1) does not have to explicitly include a delay. The Aquino's model can be also converted to an event-based model reflecting the behavior of the fuel puddles associated with the individual cylinders,

$$
\begin{aligned}
m_p(l + n_c) &= m_p(l) - \frac{m_p(l)}{\tau} \cdot 4 \cdot t_{event} + X \cdot W_{fi}(l) \cdot t_{event}, \\
m_{fc}(l + n_c) &= \frac{m_p(l)}{\tau} \cdot 4 \cdot t_{event} + (1 - X) \cdot W_{fi}(l) \cdot t_{event}, \\
m_p(l_k) &= \frac{X \cdot W_{fi}(l_k) \cdot \tau}{4}, \quad t_k = l_k \cdot t_{event}, \\
\phi_m(l + 1) &= e^{-at_{event}} \phi_m(l) + (1 - e^{-at_{event}}) \frac{m_{fc}(l)}{m_{ac}(l)}, \\
m_{ac}(l) &= W_{ac}(l) \cdot t_{event},
\end{aligned}
\tag{18.2}
$$

where $t_{event} = \frac{30}{N}$ is the duration of a single engine event for a four cylinder engine ($n_c = 4$), l is the event number and N is the engine speed.

18.2   Optimization of Identification Trajectory

In this section we consider a problem of optimizing the system trajectory to enhance identifiability of parameters of a given model from output measurements. The setting of this problem, including the assumptions on model representation ability and measurement noise, which justify the subsequent treatment, are similar to the ones in [3, 11]. The problem itself is related to design of experiments and input design in system identification [9].

18.2.1 General Methodology

Suppose a relevant model is given by

$$
\begin{aligned}
\dot{x} &= f(x, u, \theta), \\
y &= h(x, u, \theta), \\
x(0) &= g(\theta),
\end{aligned}
\tag{18.3}
$$

where x is a vector state, u is a vector control input, y is a scalar measured output, θ is a vector of constant parameters, and $x(0)$ is the initial state of the system. The sensitivity of the output with respect to the parameter satisfies

$$
\frac{\partial \dot{x}}{\partial \theta} = \frac{\partial f}{\partial x}\frac{\partial x}{\partial \theta} + \frac{\partial f}{\partial \theta},
$$

$$
\frac{\partial y}{\partial \theta} = \frac{\partial h}{\partial x}\frac{\partial x}{\partial \theta} + \frac{\partial h}{\partial \theta}
\tag{18.4}
$$

$$
\frac{\partial x}{\partial \theta}(0) = \frac{\partial g}{\partial \theta},
$$

where all partial derivatives are evaluated at the nominal trajectory $x(t)$, induced by the control input $u(t)$ and nominal parameter value θ.

Suppose our goal is to select the control input $u(t)$, $0 \leq t \leq T$, which provides optimal excitation to the system (18.3) to best identify the value of θ. One approach, which can be used for this purpose (see *e.g.*, [3] and references therein), is to minimize a trace or a determinant of the inverse of the Fisher information matrix. The Fisher information matrix is defined as

$$
F = \int_0^T \left(\frac{\partial y}{\partial \theta}(t)\right)^{\mathrm{T}} R^{-1}\left(\frac{\partial y}{\partial \theta}(t)\right) dt,
\tag{18.5}
$$

where $R = R^{\mathrm{T}} > 0$ is a weight matrix and the cost which is minimized has the form,

$$
I = det\left(F^{-1}\right) \text{ or } I = trace\left(F^{-1}\right).
\tag{18.6}
$$

As in the case of classical optimal DoEs, this approach aims to minimize the variance of the parameter estimates induced by measurement noise or by unmodelled additive perturbations.

Minimizing the cost of the type defined in (18.6), (18.5) is an *optimal control problem* which can be solved numerically. Subsequently, we employ a linear B-spline to approximate the control input,

$$u(t) = \sum_{j=0}^{J} \alpha_j \Phi\left(\frac{t}{T}\frac{J-1}{2} + \frac{1-j}{2}\right), \tag{18.7}$$

$\Phi(t) = 1 - 2|t|$ for $|t| \leq 0.5$ and $\Phi(t) = 0$ for $|t| > 0.5$, and we treat the values, α_j, $j = 0, \cdots, J$, of the B-spline at the equally spaced knot points as optimization variables. With this representation, (18.7) defines a piecewise linear function passing through the knot points. While linear B-splines provide a convenient and simple parametrization of the input, other choices of the parametrization (see e.g., [13]) can also be considered.

18.2.2   *Optimal Trajectories for Transient Fuel Identification*

A typical approach to identify parameters in the model (18.1) or (18.2) relies on the use of fuel step inputs and least squares optimization to match modeled and measured fuel-to-air ratio. Figures 18.1 and 18.2 illustrate the results of fitting X and τ in the model (18.2), for given $1/a$, so that the air-to-fuel ratio response matches measured data at 1500 rpm, $t_{event} = 0.02$ sec). These Figures point to the conclusion that multiple sets of X, τ and $1/a$ can provide a close match of experimental data, and that X and τ may not be correctly determined if $1/a$ is not accurately known.

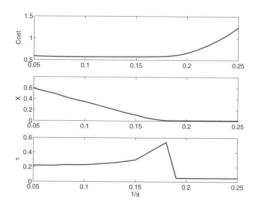

Fig. 18.1 The minimum least-squares cost (top), optimal X (middle) and optimal τ (bottom) as functions of the assumed value for $1/a$ for step input identification. Artificial bounds $0.01 \leq X \leq 0.6$, $0.01 \leq \tau \leq 1.4$ were imposed.

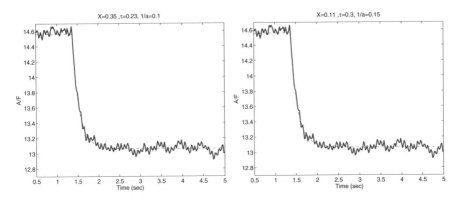

Fig. 18.2 Measured (solid) versus modeled air-to-fuel ratio (dashed) in response to the step in the fueling rate assuming $1/a = 0.1$ (left) and $1/a = 0.15$ (right) and optimal X and τ. Specifically, $X = 0.35, \tau = 0.23$ if $1/a = 0.1$ and $X = 0.11$, $\tau = 0.3$ if $1/a = 0.15$.

It is thus of interest to investigate a choice of input which improves simultaneous identifiability of X, τ and $1/a$. We used the approach based on minimizing costs in (18.6), where $y = \phi_m$ and $W_{fi} = u \cdot W_{ac} \cdot \phi_d$, and $\phi_d = \frac{1}{14.6}$ is the desired fuel-air ratio. Note that u can be viewed as a fuel injector rate multiplier. We selected $T = 5$ sec, imposed bounds $0.9 \leq u(t) \leq 1.1$, and chose $J = 50$ so that with the B-spline parameterization of $u(t)$ the fueling rate is the same for all cylinders in the same engine cycle at the engine speed of interest (1200 rpm). Further, we constrained two knots of $u(t)$ at the beginning and five knots at the end of the interval to 1, to start and end at nominal conditions. The results of the numerical optimization on the basis of the model (18.1), assuming nominal model parameters $X = 0.2$, $\tau = 0.3$, $1/a = 0.15$, are shown in Figure 18.3 and 18.4. The optimal trajectory for $u(t)$ consists of a sequence of steps and impulses when either the trace or the determinant of the inverse of the Fischer information matrix are minimized. Unfortunately, as Figure 18.5, obtained on the basis of the model (18.2) with the fueling trajectory resulting from minimizing the trace of Fischer information matrix, shows even the optimal fueling trajectories are not able to significantly improve the simultaneous identifiability of X, τ and $1/a$. It is nevertheless interesting that a rapid increase in optimal τ occurs when a starts to exceed the nominal value of 0.15, and this property may be useful in pinpointing the value of a.

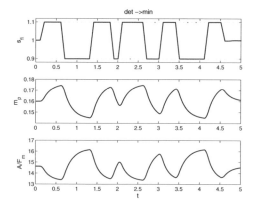

Fig. 18.3 Optimal trajectories of fuel injection rate multiplier, fuel puddle mass and air-to-fuel ratio resulting from minimizing the determinant of the inverse of the Fischer information matrix at 1500 rpm.

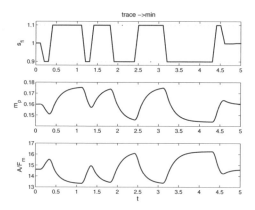

Fig. 18.4 Optimal trajectories of fuel injection rate multiplier, fuel puddle mass and air-to-fuel ratio resulting from minimizing the trace of the inverse of the Fischer information matrix at 1500 rpm.

18.3   Parameter Identification by On-Board Tuning of an Inverse Model

18.3.1   General Methodology

Consider the model (18.3) and an iteratively tuned inverse model for the control input parameterized as

$$u(t) = u(\hat{x}_k(t), \hat{\theta}_k, y_d(t)), \tag{18.8}$$

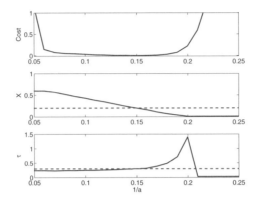

Fig. 18.5 The minimum least-squares cost (top plot), optimal X (middle plot) and optimal τ (bottom plot) as a function of the assumed value for $1/a$. Fuel injection rate determined from minimizing the trace of Fischer information matrix. Artificial bounds $0.6 \geq X \geq 0.01$ and $1.4 \geq \tau \geq 0.01$ were imposed. Dashed lines show nominal parameter values.

where $\hat{\theta}_k$ is an estimate of the parameter during the k-th parameter adjustment iteration; $\hat{x}_k$ is the estimate of the state resulting by integrating (18.3) with $\theta = \hat{\theta}_k$ and with $u(t)$ defined by (18.8) over the kth time interval $t_k \leq t \leq t_{k+1}$ over which the data are collected[1] to yield an improved parameter estimate, $\hat{\theta}_{k+1}$; and y_d is the desired value of the output. Note that different trajectories for y_d may be employed over different time intervals, $t_k \leq t \leq t_{k+1}$.

Define

$$e_k(t) = e(t; \theta, \hat{\theta}_k) = y_k(t) - y_d(t), \tag{18.9}$$

where $y_k(t)$ is the measured output from the real system excited by (18.8) over the time interval $t_k \leq t \leq t_{k+1}$. Note that (18.9) represents the deviation of the measured output from the desired output over the kth time interval.

We assume that if there is no parameter mismatch, *i.e.*, if $\hat{\theta}_k = \theta$, then the control (18.8) results in the exact output tracking so that $y_k(t) = y_d(t)$, $e(t; \theta, \hat{\theta}_k) = 0$, $t_k \leq t \leq t_{k+1}$. Using Taylor's series expansion, it then follows that

$$\begin{aligned} e_k(t) &= e(t; \theta, \hat{\theta}_k) = e(t; \hat{\theta}_k, \hat{\theta}_k) + L_k(t)(\theta - \hat{\theta}_k) + O(\|\theta - \hat{\theta}_k\|^2) \\ &= L_k(t)(\theta - \hat{\theta}_k) + O(\|\theta - \hat{\theta}_k\|^2), \\ t_k &\leq t \leq t_{k+1} \end{aligned} \tag{18.10}$$

---

[1] At the beginning of the time interval $t_k \leq t \leq t_{k+1}$ the system needs to be brought in to the same initial condition $x(t_k) = g(\theta)$ while the estimator for $\hat{x}$ must be started with the initial condition $\hat{x}(t_k) = g(\hat{\theta}_k)$.

where

$$L_k = \frac{\partial y}{\partial \theta}\big|_{\theta = \hat{\theta}_k},$$

is the sensitivity (row vector) of the output of (18.3) with respect to the parameter θ evaluated at the current parameter estimate, $\hat{\theta}_k$, and where the control input $u(t)$ is defined by (18.8). Note that the control input remains fixed, as a function of time, in computing this output sensitivity.

Using (18.10), we can try to reduce the error by finding a new guess, $\hat{\theta}_{k+1}$, for θ. For instance, we can minimize a cost function

$$I(\theta) = \int_{t_k}^{t_{k+1}} (e_k(t) - L_k(t)(\theta - \hat{\theta}_k))^2 dt + (\theta - \hat{\theta}_k)^{\mathrm{T}} R_k (\theta - \hat{\theta}_k), \quad (18.11)$$

with respect to the choice of $\theta = \hat{\theta}_{k+1}$. The matrix R_k is the weight which determines the aggressiveness of the parameter adjustment. The solution to minimizing (18.11) produces the following update law for the parameter estimate,

$$\hat{\theta}_{k+1} = \hat{\theta}_k + (\Gamma_k + R_k)^{-1} \cdot \int_{t_k}^{t_{k+1}} L_k(t)^{\mathrm{T}} e_k(t) dt,$$
$$\Gamma_k = \int_{t_k}^{t_{k+1}} L_k(t)^{\mathrm{T}} L_k(t) dt. \quad (18.12)$$

Note that (18.12) improves the parameter estimate by convolving the measured error, $e_k(t)$, with the output sensitivity with respect to the parameter, $L_k(t)$, which is evaluated at the parameter estimate, $\hat{\theta}_k$. The matrix Γ_k may be viewed as an adaptation gain, and it is related to the Fischer information matrix (18.5) for the system $\dot{x} = f(x, u(\hat{x}, \hat{\theta}, y_d), \theta), y = h(x, u(\hat{x}, \hat{\theta}, y_d), \theta)$, at $\theta = \hat{\theta}$, where $\dot{\hat{x}} = f(\hat{x}, u(\hat{x}, \hat{\theta}, y_d), \hat{\theta}), \hat{x}(0) = g(\hat{\theta})$ and where y_d is viewed as an adjustable input which can, in fact, be selected to minimize the trace or the determinant of Γ_k.

The convergence mechanism and robustness properties of (18.12) were examined in [4]. It was argued that updates (18.12) can lead to the error reduction even if e_k in (18.12) is contaminated by measurement noise or if it is replaced by $\tilde{e}_k$, where $\tilde{e}_k = H[\tilde{e}_k]$ is a filtered version of e_k through the additional and possibly uncertain dynamics, H, satisfying appropriate passivity assumptions. For instance, in the transient fuel identification case, H may account for first order uncertain exhaust mixing and UEGO sensor dynamics, which, based on the discussion in Section 18.2, can obscure the actual transient fuel parameters. With the approach of tuning parameters in the inverse model, the need to simultaneously identify parameters of H is avoided.

18.3.2 Identifying Transient Fuel Parameters

We define the adaptation procedure, based on the model (18.1) without exhaust mixing/UEGO sensor dynamics. We let $y = W_{fc}$, $u = W_{fi}$, $x = m_p$,

$\theta = [\theta_1, \theta_2]^T, \theta_1 = X, \theta_2 = \frac{1}{\tau}$. Thus in the model (18.1), $f = -\theta_2 \cdot m_p + \theta_1 u$, $h = \theta_2 \cdot m_p + (1 - \theta_1) \cdot u$, $g = \frac{\theta_1}{\theta_2} \cdot W_{fi}(t_k)$, and $y_d(t) = W_{fd}(t)$ is the desired cylinder fueling rate, defined as $W_{fd}(t) = \frac{W_{ac}(t)}{14.64}$, where $W_{ac}(t)$ is the cylinder air flow rate controlled by adjusting throttle and valve timing. The transient fuel compensation algorithm in the form (18.8) can be expressed as

$$u(t) = \frac{W_{fd}(t) - \hat{\theta}_2 \hat{m}_p(t)}{1 - \hat{\theta}_1},$$

where $\hat{\theta}_1$, $\hat{\theta}_2$ are estimates of θ_1 and θ_2, respectively.

We defined $\tilde{e}_k(t) = W_{ac}(t) \cdot \phi_m(t) - W_{fd}(t)$, $t_k \leq t \leq t_{k+1}$, where ϕ_m is the measured fuel-to-air ratio modeled as in (18.1) with $a = 1/0.3$. Consistently with our preceding discussion, we used $\tilde{e}_k$ in place of e_k in (18.12). A trajectory for the desired cylinder fuel flow rate, W_{fd}, was defined as shown in Figure 18.6-top. It corresponded to the cylinder air flow trajectory, $W_{ac} = 14.6 \cdot W_{fd}$. The weight R_k in the cost (18.11) was chosen as $R_k = 10^{-2} \cdot I$. The simulated trajectories in Figure 18.7 correspond to parameter values $\theta_1 = 0.4$, $\theta_2 = 1/1.2$, while the initial parameter estimates were set far apart to $\hat{\theta}_1(0) = 0.2$ and $\hat{\theta}_2(0) = 1/0.3$, and they demonstrate rapid convergence of the air-to-fuel ratio error and of the parameter estimates within just four iterations of our algorithm. Similar results are obtained if the values of parameter values and parameter values are interchanged as $\theta_1 = 0.2$, $\theta_2 = 1/0.3$, $\hat{\theta}_1(0) = 0.4$ and $\hat{\theta}_2(0) = 1/1.2$.

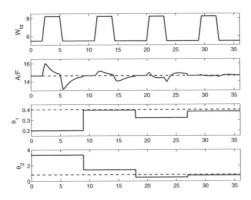

Fig. 18.6 The time histories of (a) W_{fd}; (b) the cylinder air-to-fuel ratio $\frac{W_{ac}}{W_{fc}}$ (solid) and desired air-to-fuel ratio (dashed); (c) $\hat{\theta}_1$ (solid) and θ_1 (dashed); and (d) $\hat{\theta}_2$ (solid) and θ_2 (dashed). Here $\theta_1 = 0.4$, $\theta_2 = 1/1.2$, $\hat{\theta}_1(0) = 0.2$ and $\hat{\theta}_2(0) = 1/0.3$.

An experimental assessment of our algorithm has been performed on an engine in a dynamometer. There were several differences in the algorithm we implemented experimentally. First, the error, $\tilde{e}_k$ was defined as the measured

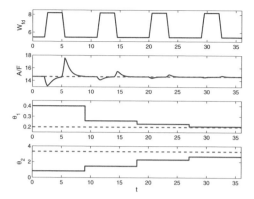

Fig. 18.7 The time histories of (a) W_{fd}, (b) the cylinder air-to-fuel ratio $\frac{W_{ac}}{W_{fc}}$ (solid) and desired air-to-fuel ratio (dashed); (c) $\hat{\theta}_1$ (solid) and θ_1 (dashed); and (d) $\hat{\theta}_2$ (solid) and θ_2 (dashed). Here $\theta_1 = 0.2$, $\theta_2 = 1/0.3$, $\hat{\theta}_1(0) = 0.4$ and $\hat{\theta}_2(0) = 1/1.2$.

fuel-to-air ratio by UEGO sensor, and assumed to be equal to the in-cylinder fueling rate divided by the airflow, compensated for the delay, t_d, and lost fuel fraction (or equivalent steady-state offset in air charge), LFF,

$$\tilde{e}_k = \frac{W_{fc}(t - t_d)(1 - LFF)}{W_{ac}(t - t_d)} - \frac{1}{14.6},$$

where 14.6 is the stoichiometric air-to-fuel ratio value. The t_d and LFF have been estimated prior to the application of the algorithm. Second, our definition of the parameter θ_2 was $\theta_2 = \tau$ as opposed to $\theta_2 = \frac{1}{\tau}$. Third we defined separate sets of parameters X and τ to capture tip-in (X_a, τ_a) and tip-out (X_d, τ_d) behavior of the engine, consistently with the existing structure in the Engine Management System (EMS). The adaptation algorithm was adjusted to accommodate these differences and logic was added to adapt tip-in parameters only from the tip-in portion of the trajectory and tip-out parameters only from the tip-out portion of the trajectory. Furthermore, to simplify the on-line implementation we used a slightly modified algorithm which uses a constant gain and avoids gain calculation via matrix inversion in (18.12).

The requested torque profile consisting of multiple repetition of tip-ins and tip-outs $(t_{k+1} - t_k = 15 \text{ sec})$ was commanded. The requested torque was translated into the cylinder air flow request and, ultimately, into commands to engine actuators to deliver this air flow request. The cylinder air flow estimate, W_{ac}, was also calculated within EMS. The desired cylinder fuel flow rate was then defined as $W_{fd} = \frac{W_{ac}}{14.6}$ to match stoichiometric air-to-fuel ratio value of 14.6. In our implementation we sampled the fuel-to-air ratio data so that to always select them for the same cylinder in each engine cycle; this helped reducing the influence of cylinder-to-cylinder differences.

The experimental results are shown in Figure 18.8. The parameters $"tau_a"$ and $"tau_d"$ plotted in Figure 18.8 were defined as $"tau_a" = \dfrac{1}{1 - e^{\frac{-t_{event}}{\tau_a}}}$ and $"tau_d" = \dfrac{1}{1 - e^{\frac{-t_{event}}{\tau_d}}}$ where t_{event} is the duration of a single engine event, consistently with the existing structure in the Engine Management System (EMS).

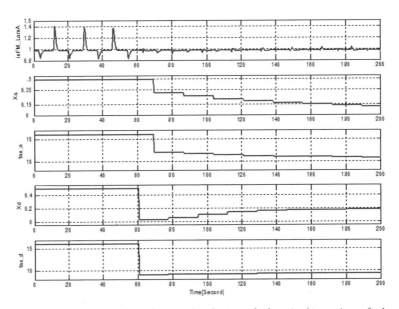

Fig. 18.8 The time histories of normalized air-to-fuel ratio (*i.e.*, air-to-fuel ratio divided by stoichiometric value), tip-in compensation parameters X_a, $"tau_a"$ and tip-out compensation parameters X_d, $"tau_d"$. The algorithm is manually enabled around 58 sec.

In lieu of using (18.12) there is an alternative approach, which is based on perturbation of the current parameter estimates vector to compute a numerical approximation of the gradient of the cost in (18.11), and then perform a gradient descent step to update the parameters. With this approach, sensitivity computations, reliant on the nominal model of transient fuel behavior and current parameter estimates, are not required. The algorithm settings are the parameter perturbations amounts for the purpose of the gradient approximation and the gain multiplying the estimate of the gradient to determine the amount of parameter estimate adjustment. Given that we have two adjustable parameters, the most basic approach to estimate the gradient is to utilize four parameter perturbations. Figure 18.9 illustrates this approach in simulations. Recognizing that the performance of our direct adaptation algorithm and this gradient search algorithm can be quite dependent upon each algorithm settings, our observation based on the simulations was nevertheless that the total time spent is less with the direct adaptation algorithm.

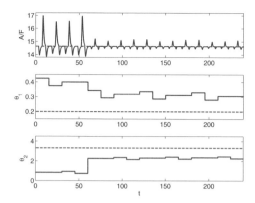

Fig. 18.9 The time histories obtained through the gradient search algorithm of (a) the cylinder air-to-fuel ratio ($\frac{W_{ac}}{W_{fc}}$ (solid) and desired air-to-fuel ratio (dashed) in the top plot; (b) $\hat{\theta}_1$ (solid) and θ_1 (dashed) in the middle plot; and (c) $\hat{\theta}_2$ (solid) and θ_2 (dashed) in the bottom plot.

18.4   Receding Horizon Control For Tracking with On-Line Parameter Identification

18.4.1   General Methodology

We consider state and parameter estimation problems for a class of discrete-time nonlinear systems in the general form,

$$x(t+1) = f(x(t), u(t), \nu(t)),$$
$$y(t) = h(x(t), \xi(t)). \tag{18.13}$$

Here $x(t)$ is a vector of dynamic states and parameters in the system which needs to be estimated, $u(t)$ is a vector of control inputs, $y(t)$ is a vector of measurements, and $\nu(t)$, $\xi(t)$ are process and measurement noise vectors, respectively. In our formulation, the unknown parameters are treated as states with constant dynamics.

Since f and h in (18.13) are nonlinear, they are linearized to enable the application of the Extended Kalman Filter (EKF) to predict the error covariance. The approximations of the state and output measurement are as follows,

$$x(t+k+1) \approx f(\hat{x}^-(t+k), u(t+k), 0) + A(t+k) \cdot (x(t+k) - \hat{x}^-(t+k)) + V(t+k) \cdot \nu(t+k),$$

$$y(t+k) \approx h(\hat{x}^-(t+k), 0) + H(t+k) \cdot (x(t+k) - \hat{x}^-(t+k)) + G(t+k) \cdot \xi(t+k),$$

where $A(t + k)$ is the linearized dynamics system matrix, $H(t + k)$ is the linearized output matrix, and $V(t + k)$, $G(t + k)$ are the linearized process and measurement noise matrices at $x = \hat{x}^-(t + k)$, $u = u(t + k)$, $\nu = 0$ and $\chi = 0$. The $\hat{x}^-(t + k)$ denotes the predicted state estimate computed according to (18.13) with $\nu = 0$.

The signals $\nu(t)$ and $\xi(t)$ are assumed to be zero mean, independent white-noise processes: $\nu(t) \sim N(0, \Sigma_\nu)$, $\xi(t) \sim N(0, \Sigma_\xi)$. The *a priori* error covariance matrix, $P^-(t+k)$, and the *a posteriori* error covariance matrix, $P^+(t+k)$, are predicted using EKF equations:

$$P^-(t + k + 1) = A(t + k)P^+(t + k)A(t + k)^{\mathrm{T}} + V(t + k)\Sigma_\nu V(t + k)^{\mathrm{T}},$$

$$K(t + k) = P^-(t + k)H^{\mathrm{T}}(t + k) \cdot$$
$$\cdot \left\{ H(t + k)P^-(t + k)H^{\mathrm{T}}(t + k) + G(t + k)\Sigma_\xi G^{\mathrm{T}}(t + k) \right\}^{-1},$$

$$P^+(t + k) = \left(I - K(t + k)H(t + k)\right)P^-(t + k).$$
$$\tag{18.14}$$

Since (18.13) is nonlinear, the linearized dynamics, output and noise matrices depend on the control input. Therefore, $P^-(t + k)$ and $P^+(t + k)$ depend on the chosen control sequence.

The receding horizon optimization [2, 8, 10] is applied to a cost function which penalizes the tracking error, the control effort, and the predicted state error subject to constraints. For instance, the cost minimized at a time instant t can have the following form,

$$J = \sum_{k=0}^{T} \left\{ (\hat{x}^-(t + k) - x_d)^{\mathrm{T}}Q(\hat{x}^-(t + k) - x_d) + (u(t + k) - u_d)^{\mathrm{T}} \right. \tag{18.15}$$
$$\left. R(u(t + k) - u_d) + trace(SP^+(t + k)) \right\},$$

where Q, R and S are weighting matrices of appropriate dimensions, and x_d, u_d are the desired values of the state and of the control input, respectively. The first term in the cost encourages tracking, the second term in the cost discourages large control effort, while the third term in the cost encourages rapid convergence of the estimation error. More general formulations of the cost (18.15) may use time-dependent weights and, if a preview is available, time-dependent set-points, $x_d(k)$, $u_d(k)$.

The cost is minimized on-line with respect to the control sequence $\{u(t), \cdots, u(t + T)\}$. The optimal value of the cost, $J^*(\hat{x}(t), P(t))$, is a function of the estimate of the state of the system at the time instant t, $\hat{x}(t)$, and the error covariance matrix estimate, $P(t)$, at the time instant t. To predict $\hat{x}^-(t + k)$, $P^+(t + k)$, we initialize $\hat{x}^-(t) = \hat{x}(t)$ and $P^-(t) = P^+(t) = P(t)$ and perform the updates according to (18.14). After the optimization is completed, the control $u(t)$ is applied to the system, the output, $y(t + 1)$, is measured, the estimate $\hat{x}(t + 1)$ is updated as

$$\hat{x}(t+1) = \hat{x}^-(t+1) + K(t+1) \cdot (y(t+1) - h(\hat{x}^-(t+1), 0)), \quad (18.16)$$

the assignment $P(t+1) = P^+(t+1)$ is made, and the same process is repeated at the time instant $(t+1)$.

During minimization of (18.15) pointwise-in-time constraints can be probabilistically enforced. For instance, suppose a scalar variable $z = Lx$ is to be maintained between the limits,

$$z_{min} \leq z = Lx \leq z_{max}. \quad (18.17)$$

The constraint (18.17) induces the following constraint,

$$z_{min} + \frac{n_\sigma}{2}\sqrt{L\bar{P}^-(t+k)L^T} \leq L\hat{x}^-(t+k) \leq z_{max} - \frac{n_\sigma}{2}\sqrt{L\bar{P}^-(t+k)L^T}, \quad (18.18)$$

where $\bar{P}^-(t+k)$ is generated according to $\bar{P}^-(t+k+1) = A(t+k)\bar{P}^-(t+k)A(t+k)^T + V(t+k)\Sigma_\nu V(t+k)^T$, $\bar{P}^-(t) = P(t)$ and larger values of n_σ tend to discourage more the occurrence of constraint violation. Note that since the estimate $\hat{x}(t)$ is obtained from (18.16), $\hat{x}^-(t) = \hat{x}(t)$ can violate the constraints. Since the control sequence $\{u(t), u(t+1), \cdots, u(t+T)\}$ being determined during the receding horizon optimization does not affect the value of $\hat{x}^-(t)$, the constraint (18.18) should, as a practical measure, only be applied for $k > 0$ to avoid infeasibility.

Remark 1: If dynamics of certain states are unaffected by process noise and these states are accurately measured (*i.e.*, their measurements are unaffected by the measurement noise), only a reduced-order EKF is necessary for the remaining states. For systems in which such a state decomposition into "noise-free" states and "noise-affected" states holds, pointwise-in-time constraints on noise-free states can be enforced deterministically. Clearly, control input constraints can also be enforced deterministically.

Remark 2: The covariance matrix, $P(t)$, may need to be reset if the error between the measured output and estimated output grows due to unmodelled changes in the system. Such a reset re-institutes aggressive excitation to the system and results in more rapid estimation of states and parameters. Abrupt change detection algorithms, such as CUSUM, may be used to determine on-line when a reset of $P(t)$ is necessary.

Remark 3: For systems linear in unmeasured states an alternative formulation of the Kalman filter for conditionally Gaussian processes (see *e.g.*, [6], p. 490) may be used to avoid approximation errors in the covariance matrix, caused by the linearization in the EKF formulation. This approach can be used in conjunction with systems that can be put in the following form (compare with (18.13): $x(t+1) = A(y(t))x(t) + V(y(t))\nu(t)$, $y(t) = H(y(t))x(t) + G(y(t))\xi(t)$. The predicted error covariance matrix can still be determined according to

(18.14), with $A(t + k) = A(y(t + k))$, $V(t + k) = V(y(t + k))$, $H(t + k) = H(y(t + k))$ and $G(t + k) = G(y(t + k))$ and where $y(t + k)$ needs to be predicted.

18.4.2 Estimating Transient Fuel Model Parameters

Adopting the Aquino's model (18.1) for the fuel puddle behavior, and accounting for engine cycle delays, exhaust transport delays as well as fuel-to-air ratio sensor dynamics one can consider the following model,

$$
\begin{aligned}
m_p(t + 1) &= \tau m_p(t) + X m_{f,inj} + \nu_1(t), \\
\tau(t + 1) &= \tau(t) + \nu_2(t), \\
X(t + 1) &= X(t) + \nu_3(t), \\
z_1(t + 1) &= \frac{(1 - \tau)m_p(t) + (1 - X)m_{f,inj}(t)}{m_{a,c}(t)}, \\
z_2(t + 1) &= z_1(t), \\
z_3(t + 1) &= \lambda_u z_3(t) + (1 - \lambda_u)z_2(t), \\
y(t) &= z_3(t) + \xi_1(t).
\end{aligned}
\tag{18.19}
$$

Here $m_{a,c}(t)$ denotes the air-flow into the engine cylinders during the tth engine cycle (which is assumed to be accurately known), and λ_u is the time-constant of the fuel-air ratio sensor. The system (18.19) is in the form (18.13) with $x = [m_p, \tau, X, z_1, z_2, z_3]^T$. The process noise ν_1 can capture modelling inaccuracies, while ν_2, ν_3 capture drift in the values of τ and X, respectively. The measurement noise is ξ_1. The noise statistics are $\Sigma_\nu = diag(0.01^2, 0.015^2, 1, 0.015^2)$ and $\Sigma_\xi = 0.01^2$.

The cost function being minimized has the form (18.15) with

$$
R = 1, Q = diag(\frac{1}{100}, 0, 0, 0, 0, 0), S = diag(0, 1000, 1000, 0, 0, 0).
$$

The horizon $T = 6$, chosen in part because it was a short horizon still exceeding the time delay of the model, appeared to be effective in simulations. The control constraints were imposed in the form $0.9 \cdot m_{f,d} \le u \le 1.1 \cdot m_{f,d}$, where $m_{f,d}$ is desired fuel flow (which is equal to the air flow multiplied by the stoichiometric fuel-air ratio). Although not done here, additional constraints on the deviations of the in-cylinder fuel-air ratio or engine torque could also have been imposed with the help of our framework.

The resulting time histories of the parameter estimates and of the control input are shown in Figure 18.10. The control input resulting from the receding horizon optimization is seen to effect aggressive excitation of the fueling trajectory which results in rapid parameter estimation. The aggressiveness of the excitation can be reduced by increasing the control weight to $R = 10$. As can be seen from Figure 18.11 fast parameter convergence is still retained even with lower excitation levels.

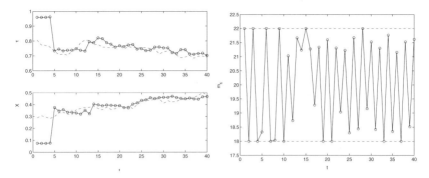

Fig. 18.10 Left: Actual (dashed) and estimated (solid, "o") τ and X for $R = 1$. Right: control input (*i.e.*, injected fuel mass), u, for $R = 1$. The control constraints are shown by the dashed lines.

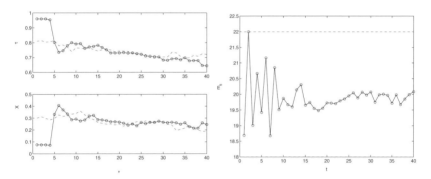

Fig. 18.11 Left: Actual (dashed) and estimated (solid, "o") τ and X for $R = 10$. Right: Control input (*i.e.*, injected fuel mass), u, for $R = 10$. The constraints are shown by the dashed lines.

18.4.3 Steady-State Engine Mapping

During mapping, a system, such as an internal combustion engine, is driven through a sequence of operating points, and the response data, such as engine air charge, torque, fuel consumption and emissions, are recorded. From the collected data, static regression models are generated for one or more of the responses. The models are typically of the form

$$y = \phi^{\mathrm{T}}(v)z + \xi, \tag{18.20}$$

where y is a scalar modelled response,

$$\phi(v) = [\phi_1(v), \cdots, \phi_m(v)]^{\mathrm{T}},$$

is a vector of known basis functions, v is the vector of system operating conditions, z is the vector of coefficients to be estimated and ξ is the measurement noise. If $v^1, \cdots, v^n$ are operating points at which data, $y^1, \cdots, y^n$, have been collected, then the least squares estimator of z has the form

$$\hat{z} = (\Phi^T \Phi)^{-1} \Phi^T Y,$$

where

$$\Phi = [\phi(v^1), \cdots, \phi(v^n)]^T, Y = [y^1, \cdots, y^n]^T.$$

Suppose that $\zeta(t) \sim N(0, \sigma^2)$ then the estimation error covariance satisfies $E[(z - \hat{z})(z - \hat{z})^T] = \sigma^2 (\Phi^T \Phi)^{-1}$. The methods of optimal Design of Experiments (DoE) often select the location of the operating points $v^1, \cdots, v^n$ within an admissible region so that the size of $(\Phi^T \Phi)^{-1}$, as measured by the determinant or by the trace of that matrix, is minimized.

The dynamics of the system, as it is driven through the selected operating points, or the associated cost (*e.g.*, control effort or time involved) are not taken into account in the traditional DoE approaches. These dynamics and cost can be included in deciding on the operating points to follow with our receding horizon approach. We also note that our approach permits to simultaneously consider and identify *multiple* scalar models of the form (18.20), where each model can have different number and types of basis functions.

To illustrate the basic ideas, we consider an example where the objective is to *simultaneously* estimate the coefficients z_1, z_2, z_3, z_4 and z_5 in two scalar regression models of the form,

$$\begin{aligned} y_3 &= 1 \cdot z_1 + v_1 \cdot z_2 + sin(v_2) \cdot z_3 + \xi_3, \\ y_4 &= v_2 \cdot z_4 + z_5 + \xi_4, \end{aligned} \tag{18.21}$$

with $\Sigma_{\xi_3, \xi_4} = diag(0.3^2, 0.3^2)$. We consider the following model for the operating point change dynamics,

$$v(t+1) = v(t) + u(t) + \Gamma(u(t))\nu(t), \tag{18.22}$$

where $u(t)$ is viewed as the control input. The function $\Gamma(u)$ can be used to model the degradation in the operating point error for larger values of $||u||$, reflecting the perspective that lower accuracy is expected if the operating points, between which the transition is taking place, within a given time window, are further apart. For instance, we hypothesize that not achieving stabilized engine temperature conditions during engine mapping may be accounted for via an appropriate model of $\Gamma(u(t))$. Augmenting the constant dynamics of the coefficients, $z(t+1) = z(t)$, and defining $x = [v^T, z^T]^T$, we obtain a system in the form (18.13).

Since the dynamics (18.22) and the outputs (18.21) are linear in *unmeasured* states, Remark 3 applies and an alternative formulation of the Kalman filter can be used instead of EKF to accurately predict error covariance matrix.

If $\Gamma(u) = 0$ and v is measured exactly, Remark 1 applies and a reduced order EKF can be implemented just for the elements of z based on the output measurements y_3 and y_4. This approach was followed and the cost function (18.15) was defined with $R = diag(0.01, 0.01)$, $S = diag(1, 1, 1, 1, 1)$ and $Q = 0$. The constraints of the form $v(t) \in [0.5, 6] \times [-\pi, \pi]$ were imposed. Since the dynamics of v are not affected by process noise and v is accurately measured, these constraints can be enforced deterministically. The resulting time histories of v, u and $\hat{z}$ for $T = 4$ are shown in Figures 18.12-18.13, respectively. As can be seen, the constraints are enforced and rapid parameter convergence is achieved.

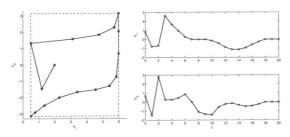

Fig. 18.12 Left: The trajectory of $v_1(t)$ and $v_2(t)$ on the v_1-v_2 plane in the case $\Gamma(u) = 0$. The rectangle is the region allowed by the constraints. Initially, $v_1(0) = 2$, $v_2(0) = 0$. Right:The time histories of the control inputs, u_1 and u_2

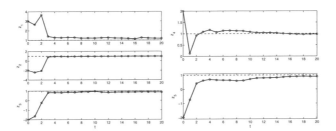

Fig. 18.13 Left: The time histories of the parameter estimates, $\hat{z}_1(t)$, $\hat{z}_2(t)$ and $\hat{z}_3(t)$ in the case $\Gamma(u) = 0$. True parameter values are shown by the dashed lines. Right: The time histories of the parameter estimates, $\hat{z}_4(t)$, and $\hat{z}_5(t)$.

We next considered the case when $\Gamma(u) = 0.2 \cdot diag(1, 1)\|u\| \neq 0$, $\Sigma_\nu = diag(1, 1)$ and for the measured outputs, $y_1 = v_1 + \xi_1$, $y_2 = v_2 + \xi_2$, y_3 and y_4 in (18.21), the measurement noise statistics were $\Sigma_\xi = diag(0.1^2, 0.1^2, 0.3^2, 0.3^2)$. The cost function (18.15) was defined with $R = diag(0.01, 0.01)$, $S = diag(1, 1, 1, 1, 1, 1, 1)$ and $Q = 0$. In this case, we used an implementation based on a full order EKF and enforced the constraints probabilistically according to (18.18) with $n_\sigma = 3$. Figures 18.14-18.15 confirm that the constraints are satisfied and fast parameter convergence is achieved. As can be seen from Figures 18.14 and 18.12, the system now explores the operating points near the constraint boundary cautiously and slowly, in order to guard against process noise causing constraint violation.

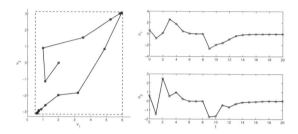

Fig. 18.14 Left: The trajectory of $v_1(t)$ and $v_2(t)$ on the v_1-v_2 plane in the case $\Gamma(u) \neq 0$. The rectangle is the region allowed by the constraints. Initially, $v_1(0) = 2$, $v_2(0) = 0$. Right:The time histories of the control inputs, u_1 and u_2.

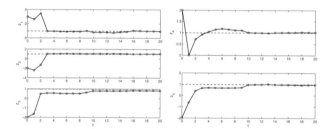

Fig. 18.15 Left: The time histories of the parameter estimates, $\hat{z}_1(t)$, $\hat{z}_2(t)$ and $\hat{z}_3(t)$ in the case $\Gamma(u) \neq 0$. True parameter values are shown by the dashed lines. Right: The time histories of the parameter estimates, $\hat{z}_4(t)$, and $\hat{z}_5(t)$.

18.5   Concluding Remarks

We reviewed finite horizon and receding horizon control problems which arise in identification of engine parameters. We firstly showed that an optimal control problem can be formulated and solved numerically to determine transient trajectories for engine parameter identification. The approach is based on minimizing the determinant or trace of the inverse of Fisher information matrix.

We next considered the identification of parameters in transient models, and considered applying the adaptation to the parameters in the inverse model. With this approach, the need to simultaneously identify un-modeled output dynamics, satisfying appropriate assumptions, may be avoided while often it is the inverse model which is of the primary interest anyway as it can be used in the transient compensation algorithms.

Finally, we demonstrated that a receding horizon optimal control problem can be formulated and solved numerically to generate control signals which enhance on-board parameter identification, maintain acceptable tracking performance and probabilistically enforce pointwise-in-time constraints. To achieve these results, the receding horizon cost functional optimized online includes a penalty on the error covariance matrix predicted using EKF update equations.

We have applied the preceding theoretical approaches to a problem of transient fuel identification based on the classical Aquino's model. We observed that exhaust mixing and sensor dynamics may obscure identifiability of parameters in the transient fuel model and the parameter identifiability can be poor even with the optimal trajectories. Fast convergence of the air-to-fuel ratio error and parameter estimates with our proposed parameter adaptive algorithm were compared with a gradient search algorithm which required larger total amount of time to reduce errors to similar levels. These approaches may be used with higher fidelity transient fuel models. In addition, we have discussed an application of receding horizon optimal control framework to determine engine mapping points.

Acknowledgement

The authors would like to thank Dr. Gary Song and Dr. Vincent Winstead for past collaboration on the developments presented in this paper.

References

[1] Aquino, C.: Transient A/F Control Characteristics of 5-liter Central Fuel Injection Engine. SAE Paper 810494 (1981)

[2] Camacho, E., Bordons, C.: Model Predictive Control, 2nd edn. Advanced Textbooks in Control and Signal Processing. Springer, Heidelberg (2004)

[3] Jauberthie, C., Denis-Vidal, L., Coton, P., Joly-Blanchard, G.: An optimal input design procedure. Automatica 42, 881–884 (2006)

[4] Kolmanovsky, I., Winstead, V.: A receding horizon optimal control approach to active state and parameter estimation in automotive systems. In: Proc. of IEEE Conference on Control Applications, Munich, Germany, pp. 2796–2801 (2006)

[5] Liu, S., Bewley, T.: Adjoint-based system identification and feedforward control optimization in automotive powertrain subsystems. In: Proc. of American Control Conference, Denver, Colorado, pp. 2566–2571 (2003)

[6] Afanasiev, V.N., Kolmanovskii, V.B., Nosov, V.R.: Mathematical Theory of Control System Design. Kluwer, Dordrecht (1996)

[7] Grewal, M.S., Andrews, A.P.: Kalman Filtering: Theory and Practice. Prentice-Hall, Englewood Cliffs (1993)

[8] Kwon, W., Han, S.: Receding Horizon control: Model Predictive Control for State Models. Springer, Heidelberg (2005)

[9] Ljung, L.: System Identification: Theory for the User. Prentice-Hall, Englewood Cliffs (1999)

[10] Maciejowski, J.: Predictive Control with Constraints. Prentice Hall, Harlow (2002)

[11] Malyshev, V.V., Krasilshikov, M.N., Karlov, V.I.: Optimization of Observation and Control Processes. AIAA Education Series (1992)

[12] Song, G., Kolmanovsky, I., Gibson, A.: Sensitivity equations based experiment design and adaptive compensation of transient fuel dynamics in Port-Fuel Injection engines. In: Proc. of IEEE Multi-Conference on Systems and Control, Suntec City Convention Centre, Singapore (2007)

[13] Subchan, S., Zbikowski, R.: Computational Optimal Control: Tools and Practice. Wiley, Chichester (2009)

Author Index

Subject Index

Lecture Notes in Control and Information Sciences

Edited by M. Thoma, F. Allgöwer, M. Morari

Further volumes of this series can be found on our homepage:
springer.com

Vol. 397: Yang, H.; Jiang, B.;
Cocquempot, V.:
Fault Tolerant Control Design for
Hybrid Systems
191 p. 2010 [978-3-642-10680-4]

Vol. 396: Kozlowski, K. (Ed.):
Robot Motion and Control 2009
475 p. 2009 [978-1-84882-984-8]

Vol. 395: Talebi, H.A.; Abdollahi, F.;
Patel, R.V.; Khorasani, K.:
Neural Network-Based State
Estimation of Nonlinear Systems
appro. 175 p. 2010 [978-1-4419-1437-8]

Vol. 394: Pipeleers, G.; Demeulenaere, B.;
Swevers, J.:
Optimal Linear Controller Design for
Periodic Inputs
177 p. 2009 [978-1-84882-974-9]

Vol. 393: Ghosh, B.K.; Martin, C.F.;
Zhou, Y.:
Emergent Problems in Nonlinear
Systems and Control
285 p. 2009 [978-3-642-03626-2]

Vol. 392: Bandyopadhyay, B.;
Deepak, F.; Kim, K.-S.:
Sliding Mode Control Using Novel Sliding
Surfaces
137 p. 2009 [978-3-642-03447-3]

Vol. 391: Khaki-Sedigh, A.; Moaveni, B.:
Control Configuration Selection for
Multivariable Plants
232 p. 2009 [978-3-642-03192-2]

Vol. 390: Chesi, G.; Garulli, A.;
Tesi, A.; Vicino, A.:
Homogeneous Polynomial Forms for
Robustness Analysis of Uncertain
Systems
197 p. 2009 [978-1-84882-780-6]

Vol. 389: Bru, R.; Romero-Vivó, S. (Eds.):
Positive Systems
398 p. 2009 [978-3-642-02893-9]

Vol. 388: Jacques Loiseau, J.; Michiels, W.;
Niculescu, S-I.; Sipahi, R. (Eds.):
Topics in Time Delay Systems
418 p. 2009 [978-3-642-02896-0]

Vol. 387: Xia, Y.;
Fu, M.; Shi, P.:
Analysis and Synthesis of
Dynamical Systems with Time-Delays
283 p. 2009 [978-3-642-02695-9]

Vol. 386: Huang, D.;
Nguang, S.K.:
Robust Control for Uncertain
Networked Control Systems with
Random Delays
159 p. 2009 [978-1-84882-677-9]

Vol. 385: Jungers, R.:
The Joint Spectral Radius
144 p. 2009 [978-3-540-95979-3]

Vol. 384: Magni, L.; Raimondo, D.M.;
Allgöwer, F. (Eds.):
Nonlinear Model Predictive Control
572 p. 2009 [978-3-642-01093-4]

Vol. 383: Sobhani-Tehrani E.;
Khorasani K.;
Fault Diagnosis of Nonlinear Systems
Using a Hybrid Approach
360 p. 2009 [978-0-387-92906-4]

Vol. 382: Bartoszewicz A.;
Nowacka-Leverton A.;
Time-Varying Sliding Modes for Second
and Third Order Systems
192 p. 2009 [978-3-540-92216-2]

Vol. 381: Hirsch M.J.; Commander C.W.;
Pardalos P.M.; Murphey R. (Eds.)
Optimization and Cooperative Control Strategies:
Proceedings of the 8th International Conference
on Cooperative Control and Optimization
459 p. 2009 [978-3-540-88062-2]

Vol. 380: Basin M.
New Trends in Optimal Filtering and Control for
Polynomial and Time-Delay Systems
206 p. 2008 [978-3-540-70802-5]

Vol. 379: Mellodge P.; Kachroo P.;
Model Abstraction in Dynamical Systems:
Application to Mobile Robot Control
116 p. 2008 [978-3-540-70792-9]

Vol. 378: Femat R.; Solis-Perales G.;
Robust Synchronization of Chaotic Systems
Via Feedback
199 p. 2008 [978-3-540-69306-2]

Vol. 377: Patan K.
Artificial Neural Networks for
the Modelling and Fault
Diagnosis of Technical Processes
206 p. 2008 [978-3-540-79871-2]

Vol. 376: Hasegawa Y.
Approximate and Noisy Realization of
Discrete-Time Dynamical Systems
245 p. 2008 [978-3-540-79433-2]